Numerical Analysis of Singular Perturbation Problems

Edited by

P.W. HEMKER
Mathematical Centre
Amsterdam
The Netherlands

J.J.H. MILLER
School of Mathematics
Trinity College
Dublin, Ireland

1979

ACADEMIC PRESS

London New York San Francisco

A Subsidiary of Harcourt Brace Jovanovich, Publishers

ACADEMIC PRESS INC. (LONDON) LTD.
24/28 Oval Road,
London NW1

United States Edition published by
ACADEMIC PRESS INC.
111 Fifth Avenue
New York, New York 10003

Library of Congress Catalog Card Number: 78-72554
ISBN: 0-12-340250-6

Printed in Great Britain by
Whitstable Litho Ltd., Whitstable, Kent

Numerical Analysis of Singular Perturbation Problems

Proceedings of the Conference held at the
University of Nijmegen, The Netherlands
May 30th to June 2nd, 1978

CONTRIBUTORS

L.R. Anderson, *Dept. of Aerospace and Mechanical Engineering, University of Arizona, Tucson, Arizona 85721, U.S.A.*

M. Aslam Noor, *Mathematics Department, Kerman University, Kerman, Iran.*

L. Auslander, *City University of New York, New York, New York, U.S.A.*

O. Axelsson, *Dept. of Computer Sciences, Chalmers University of Technology, S-402 20 Göteborg, Sweden.*

J. Baranger, *Mathématiques Appliquées, Université de Lyon 1, 69621 Villeurbanne, France*

K.E. Barrett, *Mathematics Dept., Lanchester Polytechnic, Coventry CV1 5FB, U.K.*

E. Bohl, *Institut für numerische und instrumentelle Mathematik, University of Münster, Roxelerstrasse 64, Münster, Germany.*

A. Bourgeat, *Centre de Mathématiques, 303, I.N.S.A., 20 Avenue A. Einstein, 69621 Villeurbanne, France.*

A. Brandt, *Weizmann Institute of Science, Rehovot, Israel.*

I. Christie, *Dept. of Mathematics, University of Dundee, Dundee DD1 4HN, Scotland, U.K.*

G.M. Côme, *Institut National Polytechnique de Lorraine, E.R.A. no 136 du C.N.R.S., 1 Rue Grandville, 54042 Nancy, France.*

G. Demunshi, *Mathematics Dept., Lanchester Polytechnic, Coventry CV1 5FB, U.K.*

Ph. Destuynder, *Centre de Mathématiques Appliquées, Ecole Polytechnique, 91128 Palaiseau, France.*

A.L. Dontchev, *Institute of Mathematics, Bulgarian Academy of Sciences, P.O.B. 373, Sofia, Bulgaria.*

J.E. Flaherty, *Dept. of Mathematical Sciences, Rensselaer Polytechnic Institute, Troy, New York 12181, U.S.A.*

L.S. Frank, *Mathematics Institute, University of Nijmegen, Nijmegen, The Netherlands.*

D.F. Griffiths, *Dept. of Mathematics, University of Dundee, Dundee DD1 4HN, Scotland, U.K.*

P.P.N. de Groen, *Dept. of Mathematics, University of Technology Eindhoven, Eindhoven, The Netherlands.*

CONTRIBUTORS

I. Gustafsson, *Dept. of Computer Sciences, Chalmers University of Technology, S 402 20 Göteborg, Sweden.*

J.C. Heinrich, *Dept. of Civil Engineering, University College of Swansea, Swansea SA2 8PP, Wales, U.K.*

P.W. Hemker, *Dept. of Numerical Mathematics, Mathematical Centre, Tweede Boerhaavestraat 49, Amsterdam, The Netherlands.*

G.C. Hsiao, *Dept. of Mathematics, University of Delaware, Newark, Delaware 19711, U.S.A.*

K. Inayat Noor, *Mathematics Department, Kerman University, Kerman, Iran.*

K. Ingólfsson, *Science Institute, University of Iceland, Dunhaga 3, Reykjavik, Iceland.*

S. Kesavan, *IRIA-LABORIA, Domaine de Voluceau, Rocquencourt, BP 105, 78150 Le Chesney, France* and *School of Mathematics, Tata Institute of Fundamental Research, Bombay, India.*

K.E. Jordan, *Dept. of Mathematics, University of Delaware, Newark, Delaware 19711, U.S.A.*

J. Lorenz, *Institut für numerische und instrumentelle Mathematik, University of Münster, Roxelerstrasse 64, Münster, Germany.*

R.E. O'Malley, Jr., *Dept. of Mathematics and Program in Applied Mathematics, University of Arizona, Tucson, Arizona 85721, U.S.A.*

R.M.M. Mattheij, *Mathematics Institute, University of Nijmegen, Nijmegen, The Netherlands.*

J.J.H. Miller, *School of Mathematics, Trinity College, Dublin, Ireland.*

W.L. Miranker, *IBM T.J. Watson Research Center, P.O.B., 218, Yorktown Heights, New York 10598, U.S.A.*

A.R. Mitchell, *Dept. of Mathematics, University of Dundee, Dundee DD1 4HN, Scotland, U.K.*

S.L. Paveri-Fontana, *Istituto Matematico "Ulisse Dini", Viale Morgagni 67a, 50134 Florence, Italy.*

H.-J. Reinhardt, *Fachbereich Mathematik, Goethe-Universität, Robert Mayerstrasse 10, Frankfurt am Main, Germany.*

R. Rigacci, *Istituto Matematico "Ulisse Dini", Viale Morgagni 67a, 50134 Florence, Italy.*

A.Y. Le Roux, *Laboratoire d'Analyse Numerique, I.N.S.A., BP 14a, 35301 Rennes, France.*

D.N. Shields, *Mathematics Dept., Lanchester Polytechnic, Coventry CV1 5FB, U.K.*

R. Tapiéro, *Centre de Mathématiques, 303, I.N.S.A., 20 Avenue A. Einstein, 69621 Villeurbanne, France.*

M. van Veldhuizen, *Wiskundig Seminarium, Free University, De Boelelaan 1081, Amsterdam, The Netherlands.*

O.C. Zienkiewicz, *Dept. of Civil Engineering, University College of Swansea, Swansea SA2 8PP, Wales, U.K.*

PREFACE

A conference on the Numerical Analysis of Singular Perturbation Problems was held in the Mathematics Institute, Catholic University, Nijmegen, The Netherlands, from 30 May to 2 June 1978. The total number of participants was 65, among whom were representatives from 17 countries.

The analytic theory of singular perturbation problems is a well-established area of research, which now has been developing for many years. On the other hand, the numerical analysis of such problems seems to have received relatively little attention and most of this has been in the last few years. The main purpose of this conference was to bring together the mathematicians working at present on these problems. Judging by the attendance at the conference - the first on this topic - interest now appears to be growing rapidly.

This volume contains the full text of 14 lectures by the invited speakers and 14 shorter contributions from the other speakers. The papers were received from the authors in camera-ready form and were prepared according to specific instructions to ensure as uniform an appearance as possible. We are indebted to those authors who were able to comply with these instructions, thus reducing substantially the time taken in publication. In order to keep this volume within reasonable proportions, the length of short contributed papers was restricted to eight pages in length.

We are most grateful to all participants at the conference, especially the speakers, for what they contributed to the meeting. In addition we want to thank the Mathematical Centre, Amsterdam, for its assistance and all those from the Faculty of Science, Catholic University of Nijmegen, who helped so enthusiastically to make this event a success.

March, 1979

P.W. Hemker
J.J.H. Miller

Editors

ACKNOWLEDGEMENTS

This conference was sponsored by the Catholic University Nijmegen, the Dutch Mathematical Society and the United States Army Research and Development Group (Europe), to whom we are grateful.

CONTENTS

Hemker/Miller
Numerical Analysis of Singular Perturbation Problems

ERRATUM

Due to a printer's error, the first pages of the chapter by Dontchev and that by Flaherty & O'Malley have been transposed. Page 143 should be 151 and vice versa.

CONTENTS

CONTENTS

Part I

INVITED LECTURES

ALGEBRAIC METHODS IN THE STUDY OF STIFF DIFFERENTIAL EQUATIONS

Louis Auslander

City University of New York, New York, New York

Willard L. Miranker

IBM T. J. Watson Research Center, Yorktown Heights, New York

ABSTRACT

The asymptotic form of solutions of the highly oscillatory system depends on the average, $\bar{B} = \lim_{T\to\infty} \frac{1}{T} \int_0^T \exp(-At)B\exp(At)dt$, where A and B are matrices characterizing the system. We give an algebraic study of this averaging process which leads to several alternative ways for determining $\bar{B}$.

1. INTRODUCTION

The model problem for differential equations whose solutions are highly oscillatory is

$$\varepsilon\frac{du}{dt} = (A + \varepsilon B)u, \; t \in (0,L], \tag{1.1}$$

where A is an oscillatory matrix. That is when the matrix A is similar to a diagonal matrix all of whose eigenvalues are purely imaginary. In this case the matrizant of (1.1) has the following asymptotic representation

$$\Phi(t) \equiv e^{A\frac{t}{\varepsilon}+Bt} = e^{A\frac{t}{\varepsilon}} e^{\overline{B}t}(1+O(\varepsilon)) \tag{1.2}$$

for $t \in (0, L/\varepsilon]$. Here $\overline{B}$ is the average of B given by

$$\overline{B} = \lim_{T\to\infty}\frac{1}{T}\int_0^T e^{-A\sigma}Be^{A\sigma}d\sigma\ . \tag{1.3}$$

Thus to within $O(\varepsilon)$ and on the interval $(0, L/\varepsilon]$, the solution of (1.1) consists of a carrier wave, varying on the fast time scale, t/ε, and which is modulated by a signal varying on the slow time scale, t. Thus to $O(\varepsilon)$, the information in the solution is characterized by the eigenvalues of A and the matrix $\overline{B}$.

When ε is small the solution is highly oscillatory and this seems to preclude its numerical approximation for all practical purposes by customary finite difference techniques. A possible numerical approach is suggested by the asymptotic description, (1.2) and (1.3), of the solution.

The asymptotic result is sometimes called the Bogoliubov method of averaging (cf. Volosov (1962)). Hoppensteadt and Miranker (1974) and (1976); discuss this method and by means of a different approach they extend it to classes of nonlinear differential equations in which the matrix A has eigenvalues with nonzero real parts (positive or negative). Finite difference techniques for these more general problems are likewise precluded in the case of small ε . The natural approach to replace the more general system, piecewise in time, by a system of the form of the model problem (1.1) suggests itself. Proceeding in this way and using the results in Hoppensteadt and Miranker (1974), we are led once more to the determination of the average $\bar{B}$ of a matrix B as in (1.3).

In this paper we will study the algebraic aspects of the averaging process (1.3) in the case that A is an oscillatory matrix. This will lead to a solution of the problem of computing the average (1.3) in the sense that we will provide several alternative ways for giving $\bar{B}$.

We begin in section 2 with the introduction of several algebraic constructs which lead to a geometric characterization of a space, S , spanned by the integrand in (1.3). Then we give our main theorem characterizing $\bar{B}$ in terms of a basis for S as well as an appropriate projection operator associated with S . In section 3 we give a proof of the Main Theorem. Section 4 contains an example giving the determination of $\bar{B}$ when A corresponds to the generic autonomous mechanical system. In section 5 we propose a method of preconditioning A (block diagonalizing it) so that the average $\bar{B}$ is determinable almost without effort. This latter method is particularly suitable when there are many matrices B to be averaged as in the case of the piecewise numerical solution of the nonlinear differential equations of the type of interest.

2. ALGEBRAIC CHARACTERIZATION OF AVERAGING

In this section we show that the integrand, $e^{-A\sigma}Be^{A\sigma}$ which is being averaged in (1.3) spans a subspace S, of n^2-dimensional Euclidean space. To do this we make use of three algebraic constructs or mappings called exp, ad and Ad, respectively. We describe a base for and then invariant properties of S. This leads to our Main Theorem, which gives the value of the average $\overline{B}$.

The mappings exp, ad and Ad

Let $\mathbf{E}(n)$ denote the set of all $n \times n$ matrices considered as a vector space over the field of definition which will be either $\mathbf{R}$ or $\mathbf{C}$. Let $GL(n) \subset \mathbf{E}(n)$ be the group of all nonsingular linear transformations.

Let GL denote the group of nonsingular linear transformations of $\mathbf{E}(n)$ and let $\mathbf{E}$ denote the space of linear endomorphisms of $\mathbf{E}(n)$. Clearly $\mathbf{E}$ can be identified with the space of all $n^2 \times n^2$ matrices.

For $A \in \mathbf{E}(n)$ (or for $A \in \mathbf{E}$) we define $\exp(A) \equiv e^A$ as

$$\exp(A) = I + A + \frac{A^2}{2!} + \dots \, . \tag{2.1}$$

Clearly $\exp(A) \in GL(n)$ or $\exp(A) \in GL$ as the case may be.

It will be important to note that if a linear vector space V has the property that $A(V) \subset V$, then $\exp(tA)(V) \subset V$, $t \in \mathbf{R}$.

It is well known that $\exp(tA)$, $t \in \mathbf{R}$, is a one parameter subgroup of $GL(n)$ (or of GL) and that

$$\frac{d}{dt}\exp(tA)\,|_{t=0} = A \ .$$

Further for any one parameter subgroup g(t) of GL(n), we have that if

$$\frac{d}{dt}g(t)\,|_{t=0} = A\text{ , then } g(t) = \exp(tA) \ .$$

The form of the integrand in (1.3) leads us to the concept of the adjoint representations. For $g \in GL(n)$ we define ad(g) by the formula

$$ad(g)(X) = gXg^{-1}, \ X \in \mathbf{E}(n) \ .$$

Multiplication in the right member here denotes matrix multiplication. That $ad(g) \in \mathbf{E}$ follows from the following relation which expresses a standard property of matrix multiplication.

$$g(a_1X_1 + a_2X_2)g^{-1} = a_1gXg^{-1} + a_2gX_2g^{-1} \ .$$

Here $a_1, a_2 \in \mathbf{R}$ or $\mathbf{C}$ and $X_1, X_2 \in \mathbf{E}(n)$. Note also that since $ad(g_1g_2) = ad(g_1)\ ad(g_2)$, then $ad(g^{-1})$ is the inverse of ad(g). Thus $ad(g) \in GL$.

We will make use of an additional mapping denoted by Ad(A) and defined as follows. Let $A \in \mathbf{E}(n)$. Then define

$$Ad(A)(Y) = [A,Y] = AY - YA\text{ , } \ Y \in \mathbf{E}(n) \ .$$

Clearly Ad(A) may be viewed as a linear map of $\mathbf{E}(n)$ into $\mathbf{E}$. (Most of these algebraic ideas may be found discussed in Chevelley (1946).)

Since GL $\subset$ **E** it is meaningful to consider the mapping exp: **E** $\rightarrow$ GL . Thus we may state the following well known theorem connecting the three maps, AD, ad and exp.

Theorem 2.1: Let A $\in$ **E**(n) , then exp(Ad(A)) = ad(exp(A))

The proof of this theorem is by, direct computation which uses the definitions of the three mappings, exp, ad and AD .

The subspace S:

Let B $\in$ **E**(n). The quantity ad(exp(At))B is what is being averaged in (1.3) to determine $\bar{B}$. Thus we are let to describe the span of ad(exp(At))B as t ranges over **R** . We will see that this span is a subspace S = S(B) of **E** .

Let V_0 = B and $V_k = Ad(A)V_{k-1}$, k = 1, 2, ... Let N be the first index such that $V_0, \ldots , V_{N+1}$ are linearly dependent and let

$$V_{N+1} = \sum_{i=0}^{N} a_i V_i \ .$$

Let S be the subspace of **E** spanned by $\{V_i \mid i = 0, \ldots, N\}$.

The following lemma connects S with the averages of interest.

Lemma 2.2: Each of the following two sets span S .

(*i*) $$\left\{ \frac{d^k}{dt^k} ad(\exp(tA))B \mid_{t=0} \mid k = 0,1,.. \right\}$$

(*ii*) ad(exp(tA))B, $t \in$ **R**

Proof: (i) easily follows since

$$\frac{d^k}{dt^k} ad(\exp(tA))B\,|_{t=0} = Ad^k(A)(B) \ .$$

To show (ii) we begin by noting that S is Ad(tA) invariant so that it is exp(Ad(tA)) invariant as well. By Theorem 2.1, exp(Ad(tA)) = ad(exp(tA)). Thus S is ad(exp(tA)) invariant. Thus since $B \in S$ then $\exp(Ad(tA))(B) \in S$, $\forall t$. Now in fact part (i) shows that exp(Ad(tA))(B) spans S for t close to t = 0 ; a fortiori it spans S for all t . This completes the proof of the lemma.

By construction S is a cyclic subspace for Ad(A), with cyclic vector B . Thus the restriction of Ad(A) to S has the matrix representation

$$\begin{bmatrix} 0 & 1 & 0 & . & . & . & & 0 \\ 0 & 0 & 1 & 0 & . & . & . & 0 \\ & & . & . & . & & & \\ 0 & & . & . & . & & 0 & 1 \\ a_0 & & . & . & . & & & a_N \end{bmatrix}^T$$

relative to the base $V_0, \ldots, V_N$ of S . Thus the determinant of Ad(A) is $\pm a_0$.

Main Theorem

We can now state our main result

Theorem 2.3: Let A be an oscillatory matrix. Let

$$\bar{B} = \lim_{T \to \infty} \frac{1}{T} \int_0^T ad(\exp(tA))Bdt \ ,$$

and let $V_0, \ldots, V_N$, $V_{N+1} = \sum_{i=0}^{N} a_i V_i$ and S be as defined above.

(i) If $a_0 \neq 0$ then $\overline{B} = 0$.

(ii) If $a_0 = 0$, there exists a nonzero $W \in S$ which is unique up to a scalar multiple such that $Ad(A)W=0$. If $W = \sum_{i=0}^{N} b_i V_i$ then $\overline{B} = -\frac{W}{b_0}$.

An invariant statement of the main result

Before proceeding to the next section and a proof of Theorem 2.3, we will rephrase the main result in an invariant terminology which frees us from the coordinate system $V_0, \ldots, V_N$ used in the statement of the main result.

According to Theorem 2.3, if $Ad(A)$ is nonsingular then $\overline{B} = 0$ while if $Ad(A)$ is singular then $\overline{B} = -\frac{W}{b_0}$. Since $W = \sum_{i=0}^{N} b_i V_i$ and $B = V_0$ we have

$$B = -\frac{W}{b_0} + \frac{1}{b_0} \sum_{i=1}^{N} b_i V_i \ .$$

The uniqueness of W implies that the first term in this sum is a vector spanning the one dimensional eigenvector space $S_0 \subset S$ corresponding to the eigenvalue zero of $Ad(A)$. By definition of the V_i, the second sum here lies in the subspace $S_1 \subset S$ which is the range of $Ad(A)$ (restricted to S). Since $S = S_0 \oplus S_1$, if $B = B_0 + B_1$ with $B_0 \in S_0$ and $B_1 \in S_1$, then $\overline{B} = B_0$. That is to say $\overline{B}$ is the projection along the range of $Ad(A)$ (restricted to S) onto the one dimensional null space of $Ad(A)$ (restricted to S).

This invariant statement of the main result depends on S and the construction of its base $V_0, V_1, \ldots, V_N$. In fact even this dependence can be eliminated.

To see this we begin by noting that $\mathrm{Ad}(A)$ is a completely reducible transformation acting on $\mathbf{E}(n)$. Thus

$$\mathbf{E}(n) = \mathbf{E}_0(n) \oplus \mathbf{E}_1(n) .$$

where $\mathbf{E}_0(n)$ is the null space of $\mathrm{Ad}(A)$ and $\mathbf{E}_1(n)$ is its range.

Clearly $S_0 \subset \mathbf{E}_0(n)$ and $S_1 \subset \mathbf{E}_1(n)$. Thus since $B = B_0 + B_1$ with $B_0 \in S_0$ and $B_1 \in S_1$, we have $B_0 \in \mathbf{E}_0(n)$ and $B_1 \in \mathbf{E}_1(n)$. Thus $B = B_0 + B_1$ is also a decomposition of B into a part in $\mathbf{E}_0(n)$ and a part in $\mathbf{E}_1(n)$. But any such decomposition is well defined. Thus B_0 is the projection along $\mathbf{E}_1(n)$ onto $\mathbf{E}_0(n)$. If P denotes this projection operator we have that

$$\bar{B} = P(B) = \lim_{T\to\infty} \frac{1}{T} \int_0^T e^{At} B e^{-At} dt .$$

In section 5 we will show how to carry out the computation, $P(B)$, when the matrix A has a particularly simple form.

3. PROOF OF THEOREM 2.3

In this section we prove the Main Theorem.

We will make use of the following lemma.

Lemma 3.1: If A has purely imaginary eigenvalues then so does $\mathrm{Ad}(tA)$.

Proof: By hypothesis $\exp(tA)$ is contained in the orthogonal group (which is compact). The C = closure of $\exp(tA)$ (in the orthogonal group) is compact. Then $\mathrm{ad}(C)$ is a compact group operating as linear transformations on $\mathbf{E}(n)$. But any compact group is similar to a subgroup of the orthogonal group on $\mathbf{E}(n)$. (Cf. Hochshild (1965).) Thus every matrix in $ad(C)$ is similar to a diagonal matrix whose jj-th entry is $\exp(ik_j t)$ for some $k_j \in \mathbf{R}$.

Using Theorem 2.1, we have

$$\mathrm{Ad}(A) = \frac{d}{dt}\exp(\mathrm{Ad}(tA))\,|_{t=0} = \frac{d}{dt}\mathrm{ad}(\exp(tA))\,|_{t=0} \;.$$

That is $\mathrm{Ad}(A)$ is similar to a diagonal matrix whose jj-th entry is $\frac{d}{dt}\exp(ik_j t)\,|_{t=0} = ik_j$. This completes the proof of the lemma.

We turn now to the proof of theorem 2.3. In fact we will prove it in its invariant terminology form.

Proof: Lemma 3.1 shows that there exists a coordinate system in S relative to which $\mathrm{Ad}(tA)$ has the matrix representation:

$$\mathrm{Ad}(tA) = \begin{bmatrix} i\lambda_0 t & & & 0 \\ & \cdot & & \\ & & \cdot & \\ & & & \cdot \\ 0 & & & i\lambda_N t \end{bmatrix} .$$

In this coordinate system let B have the coordinates

$$\begin{bmatrix} \beta_0 \\ \cdot \\ \cdot \\ \cdot \\ \beta_N \end{bmatrix} .$$

Then using Theorem 2.1,

$$\exp(\mathrm{Ad}(tA))B = \mathrm{ad}(\exp(tA))B = \begin{bmatrix} e^{i\lambda_0 t}\beta_0 \\ \cdot \\ \cdot \\ \cdot \\ e^{i\lambda_N t}\beta_N \end{bmatrix} .$$

This and part (i) of Lemma 2.2 imply that the following vectors,

$$\begin{bmatrix} \beta_0 \\ \cdot \\ \cdot \\ \cdot \\ \beta_N \end{bmatrix}, \quad \begin{bmatrix} \lambda_0\beta_0 \\ \cdot \\ \cdot \\ \cdot \\ \lambda_N\beta_N \end{bmatrix}, \quad \ldots \quad, \begin{bmatrix} \lambda_0{}^N\beta_0 \\ \cdot \\ \cdot \\ \cdot \\ \lambda_N{}^N\beta_N \end{bmatrix}$$

span S. Thus the matrix whose columns are these vectors is nonsingular. The latter may be written as the following product of a diagonal matrix and a van der Monde matrix, viz:

$$\begin{bmatrix} \beta_0 & & 0 \\ & \ddots & \\ 0 & & \beta_N \end{bmatrix} \begin{bmatrix} 1 & \lambda_0 & \cdots & \lambda_0{}^N \\ \vdots & \vdots & & \vdots \\ 1 & \lambda_N & \cdots & \lambda_N{}^N \end{bmatrix} .$$

The nonsingularity of this matrix product implies that no β_j vanishes and that the λ_j are distinct.

Evaluating $\overline{B}$ amounts to evaluating

$$\lim_{T\to\infty} \frac{1}{T} \int_0^T e^{i\lambda_j t}\beta_j \, dt , \quad j = 0,\ldots,N .$$

This limit is zero if $\lambda_j \neq 0$ and equals β_j if $\lambda_j = 0$, $j = 0$, ..., N .

If no $\lambda_j = 0$, then Ad(A) is nonsingular and the proof is complete. Suppose then that Ad(A) is singular. Since the λ_j are distinct, at most one λ_j can vanish. Thus the space S_0 annihilated by Ad(A) has dimension one. By relabelling we may suppose that $\lambda_0 = 0$ so that the range S_1 of Ad(A) consists of vectors of the form

$$\begin{bmatrix} 0 \\ y_1 \\ \cdot \\ \cdot \\ \cdot \\ y_N \end{bmatrix} .$$

Thus writing $B = B_0 + B_1$ with $B_0 \in S_0$ and $B_1 \in S_1$, we have

$$B_0 = \begin{bmatrix} \beta_0 \\ 0 \\ \cdot \\ \cdot \\ \cdot \\ 0 \end{bmatrix} \quad \textit{and} \quad B_1 = \begin{bmatrix} 0 \\ \beta_1 \\ \cdot \\ \cdot \\ \cdot \\ \beta_N \end{bmatrix} .$$

Thus $\overline{B} = \beta_0$, completing the proof of the theorem.

4. AN EXAMPLE

Let us apply the theory just developed to the differential equation describing a canonical mechanical system.

$$M\ddot{q} + Cq = 0 .$$

q is the position vector, M is an inertial matrix and C is the stiffness matrix. M and C are symmetric and positive definite. We introduce matrices L and R through the following relations

$$M=L^{-1}R^{-1},\ LC=R^{T} .$$

Then introducing a momentum vector, p , the differential equation becomes

$$\begin{bmatrix} \dot{q} \\ \dot{p} \end{bmatrix} = \begin{bmatrix} 0 & R \\ -R^T & 0 \end{bmatrix} \begin{bmatrix} q \\ p \end{bmatrix} .$$

Typically $M = I$ so that we may take

$$L^{-1} = C^{1/2},\ R^{-1} = C^{1/2} \text{ and } R^T = LC = C^{1/2} \quad .$$

Setting $a = C^{1/2}$, so that a is symmetric and nonsingular, the differential equation becomes

$$\begin{bmatrix} \dot{q} \\ \dot{p} \end{bmatrix} = \begin{bmatrix} 0 & a \\ -a & 0 \end{bmatrix} \begin{bmatrix} q \\ p \end{bmatrix} .$$

Then for this example

$$A = \begin{bmatrix} 0 & a \\ -a & 0 \end{bmatrix} .$$

Let B be the matrix

$$B = \begin{bmatrix} \alpha & \beta \\ \gamma & \delta \end{bmatrix} ,$$

where α , β , γ and δ are blocks corresponding to the blocks of A . Let $V_0 = B$ be taken in the following form

$$V_0 = \begin{bmatrix} \alpha \\ \delta \\ \beta \\ \gamma \end{bmatrix} .$$

Further introduce the matrices E , F and G .

$$E = \begin{bmatrix} I & I \\ I & I \end{bmatrix}, \quad F = \begin{bmatrix} I & -I \\ -I & I \end{bmatrix}, \quad G = \begin{bmatrix} I & I \\ -I & -I \end{bmatrix},$$

where the blocks here also correspond to the blocks of A . Then we find that for $n = 1, 2, \ldots,$

$$(\mathrm{Ad}(A))^n = 2^{n-1} a^n \begin{cases} (-1)^{\frac{n-1}{2}} \begin{bmatrix} 0 & G \\ -G^T & 0 \end{bmatrix}, & n \text{ odd} \\ \\ (-1)^{\frac{n}{2}} \begin{bmatrix} F & 0 \\ 0 & E \end{bmatrix}, & n \text{ even} . \end{cases}$$

Here and in what follows this notation means that the matrix a multiplies each of the blocks in the other indicated matrices. (In this equation there are 16 such blocks per matrix.)

Then

$$\text{Range Ad}(A) = \left\{ V \mid V = a \begin{bmatrix} 0 & G \\ -G^T & 0 \end{bmatrix} \begin{bmatrix} \alpha \\ \delta \\ \beta \\ \gamma \end{bmatrix} \right\}$$

$$= \left\{ V \mid V = a \begin{bmatrix} \beta + \gamma \\ -\beta - \gamma \\ -\alpha + \delta \\ -\alpha + \delta \end{bmatrix} \right\} .$$

Thus, if x and w are blocks of the size of a, we have

$$\text{Range Ad(A)} = \left\{ V \mid V = a\begin{bmatrix} w \\ -w \\ x \\ x \end{bmatrix} \right\} .$$

We similarly find that if u and v are blocks of size a, that

$$\text{Null Ad(A)} = \left\{ a\begin{bmatrix} u \\ u \\ v \\ -v \end{bmatrix} \right\} .$$

Thus writing B as a sum of vectors in the null space and range respectively, requires

$$\begin{bmatrix} \alpha & \beta \\ \gamma & \delta \end{bmatrix} = a\begin{bmatrix} u & v \\ -v & u \end{bmatrix} + a\begin{bmatrix} w & x \\ x & -w \end{bmatrix} .$$

Thus

$$\begin{aligned} u &= \frac{1}{2}a^{-1}(\alpha+\delta) \\ v &= \frac{1}{2}a^{-1}(\beta-\gamma) \\ w &= \frac{1}{2}a^{-1}(\alpha-\delta) \\ x &= \frac{1}{2}a^{-1}(\beta+\gamma) \end{aligned} .$$

Thus for the projection, P(B), of B along the range of Ad(A) and onto the null space of Ad(A), we find

$$P(B) = \frac{1}{2}\begin{bmatrix} \alpha+\delta & \beta-\gamma \\ \gamma-\beta & \alpha+\delta \end{bmatrix} .$$

Of course this is the value of the average $\overline{\text{B}}$ as well.

5. PRECONDITIONING

The discussion concluding section 2 showed that the average $\overline{B} = P(B)$, the projection along the range of $Ad(A)$ onto the null space of $Ad(A)$. We may carry out this computation when A has a particularly simple form.

To begin with, suppose that A is J where

$$J = \begin{bmatrix} 0 & 1 \\ -1 & 0 \end{bmatrix} .$$

Then we have the following lemma whose proof follows by direct computation.

Lemma 5.1 $Ad(J)$ is defined on $\mathbf{E}(2) = \text{Null}\ (Ad(J)) \oplus \text{Range}\ (Ad(J))$

where

$$\text{Null}\ (Ad(J)) = \begin{bmatrix} a & b \\ -b & a \end{bmatrix} .$$

and

$$\text{Range}\ (Ad(J)) = \begin{bmatrix} c & d \\ d & -c \end{bmatrix} .$$

Thus, given the arbitrary matrix $\begin{bmatrix} \alpha & \beta \\ \gamma & \delta \end{bmatrix}$ we compute

$$a = \frac{\alpha+\delta}{2}, b = \frac{\beta-\gamma}{2}, c = \frac{\alpha-\delta}{2}, d = \frac{\beta+\gamma}{2} .$$

Then

$$\begin{bmatrix} \alpha & \beta \\ \gamma & \delta \end{bmatrix} = \begin{bmatrix} a & b \\ -b & a \end{bmatrix} + \begin{bmatrix} c & d \\ d & -c \end{bmatrix}$$

and

$$P\begin{bmatrix}\alpha & \beta \\ \gamma & \delta\end{bmatrix} = \begin{bmatrix}\frac{\alpha+\delta}{2} & \frac{\beta-\gamma}{2} \\ \frac{\gamma-\beta}{2} & \frac{\alpha+\delta}{2}\end{bmatrix}.$$

Let J_s be the block diagonal matrix with s blocks of J along the main diagonal. Then for an arbitrary skew symmetric matrix of even order we have the following lemma. (Cf. Bellman (1960).)

Lemma 5.2 An even ordered skew symmetric matrix is similar to a block diagonal matrix with the blocks $k_i J_{s_i}$, $i = 1, \ldots, N$ along the main diagonal.

(We will suppose that the skew symmetric matrix is nonsingular, so that no k_i vanishes. Moreover since the eigenvalues of J are $\pm$ i , we may assume without loss of generality that $k_i > 0$, $i = 1, \ldots, N$.)

Remark: The similarity transformation, $R^T AR$, where R is the orthogonal matrix whose columns are the real parts and the imaginary parts of the eigenvectors of A , will produce the block diagonal form referred to in the lemma. (Cf. Bellman (1960), p 64.) Determining R may be viewed as a preconditioning of the averaging problem when there are many matrices B to be averaged (as is the case of solving nonlinear differential equations piecewise). Indeed with this preconditioning having been performed, the averaging process becomes very simple indeed, as we will presently see.

Suppose that A has the block diagonal form described in Lemma 5.2. Given any matrix B of order $2\sum_{i=1}^{N} s_i$, (i.e. the order of A , the skew symmetric matrix under consideration), let it be viewed as a matrix of blocks B_{ij}, $i,j = 1, \ldots, N$, corresponding to the blocks of A . Thus B_{ij} is a submatrix of order $2s_i \times 2s_j$, $i, j = 1, \ldots, N$. Each such block corresponds to a subspace of $\mathbf{E}(n)$ which is invariant under $\mathrm{Ad}(A)$. To see this let $\hat{B}_{ij}$ be the matrix obtained from B by setting every element in B equal to zero except for those in the one block B_{ij}, $i, j = 1, \ldots, N$. Then in particular

$$B = \sum_{i,j=1}^{N} \hat{B}_{ij} .$$

Now a computation shows that $\mathrm{Ad}(A)\, \hat{B}_{ij}$ is zero everywhere except in the $ij-$ th block where we have

$$k_i J_{s_i} B_{ij} - k_j B_{ij} J_{s_j} .$$

Now let each block B_{ij} be composed of 2×2 subblocks C^{ij}_{lm}, $l = 1, \ldots, s_i$, $m = 1, \ldots, s_j$, viz :

$$B_{ij} = \begin{bmatrix} C^{ij}_{11} & \cdots & C^{ij}_{1s_j} \\ \vdots & & \vdots \\ C^{ij}_{s_i 1} & \cdots & C^{ij}_{s_i s_j} \end{bmatrix} .$$

Then the $lm-$th 2×2 block of $k_i J_{s_j} B_{ij} - k_j B_{ij} J_{s_j}$ is

$$k_i J C^{ij}_{lm} - k_j C^{ij}_{lm} J, \ \forall i,j,l,m . \tag{5.1}$$

In particular if $\hat{\hat{B}}_{ij;lm}$ is the matrix obtained from $\hat{B}_{ij}$ by setting every element equal to zero except the $lm-$th 2×2 subblock of B_{ij} itself, then

$$B = \sum_{i,j=1}^{N} \sum_{l=1}^{s_i} \sum_{m=1}^{s_j} \hat{\hat{B}}_{ij;lm} .$$

Moreover, a computation shows that $\mathrm{Ad}(A)\hat{\hat{B}}_{ij;lm}$ is every where zero except in lm–th 2×2 subblock of the subblock B_{ij} where it has the value (5.1). That is every 2×2 subblock C^{ij}_{lm} of B corresponds to an invariant subspace of Ad(A).

Thus to find Ad(A)B it is only necessary to compute $\mathrm{Ad}(A)\hat{\hat{B}}_{ij;lm}$ (cf. (5.1)) and add.

To find P(B) the projection of B along the range of Ad(A) onto the null space of Ad(A) we may use the invariant subspaces, as just noted, and find this projection for each one in turn. In fact, they are only two simple possibilities.

If $i = j$ so that B_{ij} is a diagonal block, $P(\hat{\hat{B}}_{ii;lm})$ is everywhere zero except in the nonvanishing 2×2 subblock to which $\hat{\hat{B}}_{ii;lm}$ corresponds where Lemma 5.1 tells us that $P(\hat{\hat{B}}_{ii;lm})$ has the value

$$\begin{bmatrix} \dfrac{C^{ii}_{lm;11}+C^{ii}_{lm;22}}{2} & \dfrac{C^{ii}_{lm;12}-C^{ii}_{lm;21}}{2} \\ \dfrac{C^{ii}_{lm;21}-C^{ii}_{lm;12}}{2} & \dfrac{C^{ii}_{lm;11}+C^{ii}_{lm;22}}{2} \end{bmatrix} .$$

Here $C^{ij}_{lm;pq}$, $p, q = 1, 2$ are the four components of the 2×2 submatrix C^{ij}_{lm}.

If $i \neq j$ we seek the null space of the transformation

$$\mathrm{Ad}(A)\hat{\hat{B}}_{ij;lm} = k_i J C^{ij}_{lm} - k_j C^{ij}_{lm} J, \quad i \neq j .$$

Setting this to zero gives the following matrix equation

$$k_i \begin{bmatrix} C_{21} & C_{22} \\ -C_{11} & -C_{12} \end{bmatrix} = k_j \begin{bmatrix} -C_{12} & C_{11} \\ -C_{22} & C_{21} \end{bmatrix} , \quad i \neq j ,$$

where we have suppressed the index pairs ij and lm for convenience.

These equations yield $k_i = \mp k_j$. Since $k_i > 0$, $i = 1, \ldots,$ N the null space we seek is empty. Thus

$$P\left(\hat{\hat{B}}_{ij;lm}\right) = 0 \ .$$

Summary: Suppose A has the simple form specified in Lemma 5.2. Then B may be determined by considering the block structure given by A . In the off-diagonal blocks, $\overline{B}$ vanishes. In the diagonal blocks, simply apply Lemma 5.1 to each 2×2 subblock.

REFERENCES

Bellman, R. (1960). *Introduction to Matrix Analysis.* McGraw Hill, New York.

Chevelley, C. (1946). *Theory of Lie Groups.* Princeton Press.

Hochshild, G. (1965). *The Structure of Lie Groups.* Holden-Day.

Miranker, W.L. and Hoppensteadt, F.C. (1974). Numerical methods for stiff systems of differential equations related with transistors, tunnel diodes, etc., in "Computing Methods in Applied Science and Engineering, International Symp., Versailles," Vol I, (R. Glowinski and J. L. Lions, Eds.), pp 416-432.

Hoppensteadt, F.C. and Miranker, W.L. (1976). Differential equations having rapidly changing solutions: Analytic Methods for Weakly Nonlinear Systems. *Journal Differential Equations*, 22, 237-249.

Volosov, V.M. (1962). Averaging in systems of ordinary differential equations, Russ. Math. Surveys, Vol 17, pp 1-126.

INVERSE MONOTONICITY IN THE STUDY OF CONTINUOUS AND DISCRETE SINGULAR PERTURBATION PROBLEMS

E. Bohl
University of Münster, Fachbereich Mathematatik
44 Münster, W. Germany

ABSTRACT

This paper deals with nonlinear boundary layer problems for ordinary differential equations with a boundary layer on the right end of the underlying interval. We summarize new and known results on the continuous problem and on the upwind discrete analogue. Finally, we introduce three improved versions of the upwind scheme, discuss them theoretically and test them on numerical examples.

0. INTRODUCTION

In this paper we consider the boundary value problem

$$-x''=\lambda(p(t)x'+f(t,x)) \text{ on } [0,1],\ x(0)=x(1)=0 \qquad (1)$$

under the basic assumptions

$$p\in C[0,1],\ f(t,v),\ D_2f(t,v)\in C([0,1]\times\mathbb{R}) \qquad (1a)$$

$$p(t) \leq -1, \ D_2 f(t,v) = \partial/\partial v f(t,v) \leq q_0 \text{ on } [0,1] \ \mathbb{R} \quad (1b)$$

for some real $q_0 \geq 0$. The parameter λ in (1) is assumed to be >> 1.

In this situation it is well known [6] that "for λ sufficiently large" the problem (1) has a unique solution $\bar{x}_\lambda \in C^2[0,1]$ and for $\lambda \to \infty$ the function $\bar{x}_\lambda(t)$ tends uniformly in any interval $[0,\bar{t}]$ with $0 < \bar{t} < 1$ to a solution $\bar{z}$ of the so called reduced problem

$$p(t)x' + f(t,x) = 0, \ x(0) = 0 \quad (2)$$

if $\bar{z} \in C^2[0,1]$ exists. This last property describes a boundary layer phenomenon at the right endpoint 1 of the underlying interval [0,1].

For actual applications it seems desirable to replace the vague term "for λ sufficiently large" by a quantitative assumption in the above results. This can be done: what we need is simply

$$0 \leq 4q_0 < \lambda \quad (1c)$$

where q_0 comes from (1b). In section 1 we are going to study the operator

$$T_\lambda x = -x'' - \lambda(p(t)x' + f(t,x)) \quad (3)$$

from the subset

$$V = \{x \in C^2[0,1] : x(0) = x(1) = 0\} \quad (4)$$

into C=C[0,1]. Assuming (1a), (1b), (1c) we shall list seven properties of T_λ, among them the two above and the stability inequality

$$|x(t)-y(t)| \leq \int_0^1 G_\lambda(t,s)\,|T_\lambda x(s)-T_\lambda y(s)|\,ds \qquad (t\in[0,1]) \tag{5a}$$

which holds for all $x,y\in V$. The kernel $G_\lambda(t,s)$ is a Green's function satisfying

$$0\leq G_\lambda(t,s)\leq(\lambda^2-4q_0\lambda)^{-1/2}\begin{cases}\exp(-\frac{\lambda}{2}(s-t)) & t\leq s\\ \exp(2q_0) & s<t.\end{cases} \tag{5b}$$

Since (5a) is a pointwise inequality rather than a stability inequality in a norm (cf. [8] for that) it yields a very simple proof of the convergence of $\bar{x}_\lambda$ to $\bar{z}$ stated above (see section 5).

(5a) and (5b) describe an extreme stability of T_λ as $\lambda \to \infty$. Any numerical scheme approximating (1) must enjoy stability of this type since this avoids oscillations of the numerical results in the boundary layer area if λ is large (cf. [5,9,14]). The so called upwind scheme (cf. [5,9,10,14] and section 2) is probably the simplest finite difference scheme for an approximation of $\bar{x}_\lambda$ on the underlying grid. In section 2 we obtain six properties of the discrete version T_λ^h of T_λ given by the upwind scheme where the step width $h > 0$ is fixed. The properties are analogous to six of the seven properties stated in section 1 for T_λ. In particular the upwind scheme is stable as $\lambda\to\infty$ for any $h > 0$

fixed. The numerical results show a clearly marked boundary layer and no oscillations. However, the accuracy is normally poor since the upwind scheme uses only first and second order formulae to substitute x'' and x', respectively. To improve the accuracy we use standard finite difference formulae of order 4 for x'' and x' in the boundary layer region where the mesh width satisfies $\lambda h < 1$. Outside the boundary layer we use a much larger mesh width so that the total number of grid points is limited. For our examples this number is between 37 and 55 and is hence much smaller than in [15] and about the same as in [11] (cf. the numerical examples in sections 3 and 4).

In this way we set up three schemes (see section 4) based upon the upwind scheme. Our examples show a tolerable error for $|\bar{x}_\lambda(t)-\bar{x}_\lambda^h(t)|$ ($\bar{x}_\lambda^h$ = solution of the discrete problem) much smaller than for the upwind scheme. An analysis of the schemes for $h \to 0$ is not interesting since they coincide outside the boundary layer region with the upwind scheme so that an improvement in terms of order of convergence cannot be expected. More important is for fixed $h > 0$ the stability as $\lambda \to \infty$. Since eventually λh is no longer < 1 the only thing we can hope for is the stability inequality

$$\|x-y\|_{V^h} \leq \lambda^{-1} \|T_\lambda^h x - T_\lambda^h y\|_{V^h} \qquad (6)$$

for fixed λ and h such that $0 < \lambda h < 1$. We indeed can prove this for our schemes if (1) is linear. Here, $\| \ \|_{V^h}$ denotes the maximum norm with respect to the grid points in (0,1). Experiments with the improved upwind schemes show no oscillations as long as λ and h are adjusted the way that (6) holds. If (6) is violated the oscillations show up again.

We feel that it is important for the numerical treatment of boundary layer phenomena as described by (1) to set up schemes which produce a tolerable accuracy inside and outside the boundary layer region using as less grid points as possible. If we want information in the boundary layer region we necessarily have to work with an irregular grid (cf. also [1]). It seems that a stability inequality (6) for λ and h fixed plays a central role where h is the smallest step width of the underlying grid. For schemes which are even uniformly stable in $\lambda > 0$ and $h \in (0,h_0)$ see[13].

1. RESULTS ON T_λ

In this section we consider the operator T_λ as given by (3) from V (see (4)) into $C = C\ [0,1]$ assuming (1a), (1b) and (1c). Then the following properties hold:

P1: T_λ^{-1} exists from C into V.
In particular (1) has a unique solution $\bar{x}_\lambda = T_\lambda^{-1}\theta$.

P2: T_λ^{-1} is monotone with respect to the natural partial ordering on C.

P3: Let $G_\lambda(t,s)$ be the Green's function of the operator $-x''+\lambda x'-\lambda q_o x$ and the boundary constraints $x(0)=x'(1)=0$ then the inequalities (5b) and the stability inequality (5a) hold. In particular we have the apriori bound

$$|\bar{x}_\lambda(t)| \leq \int_0^1 G_\lambda(t,s)|T_\lambda(\theta)(s)|ds \tag{7}$$

$$\leq (1-4q_o\lambda^{-1})^{-1/2}(1+\exp(2q_o))\int_0^1 |f(s,0)|ds$$

for $0 \leq t \leq 1$. Let $R_\lambda \geq 0$ be an upper bound for the right hand side of (7) then we put

$$q_\lambda = \mathrm{Min}\{D_2 f(t,v):\ 0 \leq t \leq 1,\ |v| \leq R_\lambda\} \tag{8}$$

$$g(t,v)= \begin{cases} f(t,-R_\lambda) & \text{if } v \leq -R_\lambda \\ f(t,v) & \text{if } |v| \leq R_\lambda \\ f(t,R_\lambda) & \text{if } R_\lambda \leq v \end{cases} \tag{9}$$

for $t\in[0,1]$, $v\in\mathbb{R}$.

P4: The parallel chord method

$$-x''_{n+1}-\lambda p(t)x'_{n+1}-\frac{\lambda}{2}(q_\lambda+q_o)x_{n+1}$$

$$=\lambda(g(t,x_n)-\frac{1}{2}(q_\lambda+q_o)x_n)$$

$$x_{n+1}(0)=x_{n+1}(1)=0$$

converges for any $x_o\in V$ in the maximum norm of C to $\bar{x}_\lambda$.

Recall that $\bar{z} \in C^2$ denotes a solution of the reduced problem (2) then we have

P5: For any $\bar{t} \in (0,1)$ there exists a constant $c(\bar{t})$ independent of λ such that

$$|\bar{x}_\lambda(t) - \bar{z}(t)| \le c(\bar{t})\lambda^{-1} \quad \text{for} \quad 0 \le t \le \bar{t}.$$

P6: $f(t,0) \ge 0$ and $\bar{z}''(t) \le 0$ on $[0,1]$ imply $0 \le \bar{x}_\lambda(t) \le \bar{z}(t)$ on $[0,1]$.

P7: $f(t,v) > 0$ for $0 < t < 1$ and $v \ge 0$ implies $0 \le \bar{x}_\lambda(t)$ on $[0,1]$ and there exists a unique $t_\lambda \in (0,1)$ such that $\bar{x}'_\lambda(t_\lambda)=0$. Furthermore, t_λ tends to 1 as $\lambda \to \infty$.

Hence, if we assume (1a), (1b) and (1c) as well as

$$f(t,v) > 0 \quad \text{for} \quad 0 < t < 1,\ v \ge 0 \tag{10}$$

then P1, P5 and P7 tell us that figure 1 qualitatively shows the graph of $\bar{x}_\lambda(t)$. Here, t_λ travels to 1 as $\lambda \to \infty$.

It seems reasonable to define the boundary layer region by the interval $(t_\lambda, 1]$. This will be convenient for the application of the schemes to be discussed in sections 3 and 4. If (1) is a linear problem, i.e.

$$f(t,v)=q(t)v+r(t) \ , \ q,r \in C, \tag{11}$$

then the "canonical assumptions" yielding the qualitative picture of figure 1 for $\bar{x}_\lambda$ read

$$p(t) \leq -1, \; 0 \leq 4q(t) < \lambda \text{ on } [0,1], \qquad (12)$$
$$0 < r(t) \text{ on } (0,1).$$

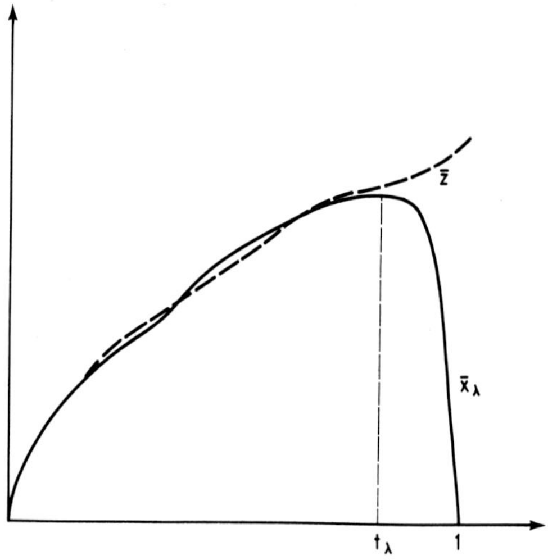

Fig. 1

Just apply (1a), (1b), (1c) and (10) to f(t,v) as given by (11).

We shall discuss the main ideas of the proofs for P1 - P7 in section 5.

2. THE UPWIND SCHEME ON AN IRREGULAR MESH

For $M \in \mathbb{N}$, $k_j \in \mathbb{N}$ $(j=1,\ldots,M)$ we define the step width h via

$$h^{-1} = \sum_{j=1}^{M} k_j. \qquad (13a)$$

The corresponding grid on [0,1] is

$$\Omega_h=\{t_o=0,\ t_j=t_{j-1}+k_jh:\ j=1,..,M\} \tag{13}$$

Since we use symmetric formulae at any $t_j\in\Omega_h$ with the step width k_jh we need at least $t_j+k_jh\in\Omega_h$. To ensure this we assume that to any $j\in\{1,..,M-1\}$ there exists $\sigma_j\in\mathbb{N}$ such that

$$k_j=\sum_{i=j+1}^{j+\sigma j} k_i. \tag{13b}$$

For $k_j=1$, $j=1,..,M$ we have $\sigma_j=1$, $j=1,..,M-1$. This leads to the regular grid. A typical choice for an irregular grid as we will need it is

$$k_i=k_1\ (i=1,..,N_1),\ 2k_i=k_{i-1}\ \ (i=N_1+1,..,N_2),$$
$$k_i=1\ \ (i=N_2+1,..,M)$$

where we assume $1\leq N_1 < N_2 < M$ and $k_{N_2}=2$. Note that (13b) implies that $\sigma_{M-1}=1$ and that the grid points accumulate at the endpoint 1 of the interval [0,1].

Consider the problem (1). The upwind scheme [5,9,10,14] substitutes x'' by the second order central difference formula and x' by a first order one-sided formula yielding the system

$$\begin{aligned}
&x(0) &= 0\\
&(k_jh)^{-2}(-x(t_{j-1})+2x(t_j)-x(t_{j+\sigma_j}))\\
&-\lambda\{p(t_j)(k_jh)^{-1}(-x(t_{j-1})+x(t_j))\\
&\qquad\qquad +f(t_j,x(t_j))\} &= 0 \qquad (14a)\\
&\qquad\qquad x(1) &= 0
\end{aligned}$$

Let B_h=diag $(0,1,..,1,0)$ $\in L^h = L[\mathbb{R}^{\Omega_h}]$ = {linear mappings of $\mathbb{R}^{\Omega_h}$ into itself}, P_h=diag $(p(t_0),..,p(t_M))$ $\in L^h$ and let F_h be the nonlinear mapping on $\mathbb{R}^{\Omega_h}$ defined via

$$(F_h x)(t)=f(t,x(t)) \qquad (t\in\Omega_h) \tag{15}$$

for all $x\in\mathbb{R}^{\Omega_h}$. Then the braces in (14a) are described as $B_h(P_hD_h+F_h)x$ taking care of the fact that the first and the last equation are free of the braces. Here, D_h describes the substitution of x'. The discretisation of x'' as well as of the boundary operators form the matrix $A_h\in L^h$ so that (14a) is equivalent to

$$T_\lambda^h x:=(A_h-\lambda B_h(P_hD_h+F_h))x=\Theta \text{ on } \mathbb{R}^{\Omega_h}. \tag{14b}$$

This equation also defines the discrete version T_λ^h of T_λ as given by the upwind scheme. Analogous to V of the continuous case we define
$V^h=\{x\in \mathbb{R}^{\Omega_h}: x(0)=x(1)=0\}$.

As in section 1 we are going to list properties P_hi for T_λ^h corresponding to Pi for T_λ. But more restrictive than in section 1 we assume $q_0=0$. Hence, what we need is

$$p\in C[0,1],\ f(t,v),\ D_2f(t,v)\in C([0,1]\times\mathbb{R}) \tag{16a}$$

$$p(t)\le -1,\ D_2f(t,v)\le 0 \text{ on } [0,1]\times\mathbb{R},\ 0<\lambda. \tag{16b}$$

Then the following holds true:

P_h1: $(T^h_\lambda)^{-1}$ exists from V^h into V^h.

In particular (14) has a unique solution $\bar{x}^h_\lambda=(T^h_\lambda)^{-1}\theta\in V^h$.

P_h2: $(T^h_\lambda)^{-1}$ is monotone with respect to the natural partial ordering on V^h.

P_h3: $G^h_\lambda=(A_h-\lambda B_h P_h D_h)^{-1}$ exists, is nonnegative and $|x-y| \leq G^h_\lambda|T^h_\lambda x-T^h_\lambda y|$ for all $x,y\in\mathbb{R}^{\Omega h}$, where $|x|$ is defined componentwise, i.e. $|x|(t)=|x(t)|$ on Ω_h. Furthermore, if $\| \; \|_{V^h}$ denotes the maximum norm on V_h and $\|G^h_\lambda\|_{V^h}$ the corresponding operator norm, we have

$$\|G^h_\lambda\|_{V^h} \leq \lambda^{-1}.$$

In particular $\bar{x}^h_\lambda$ satisfies

$$\|\bar{x}^h_\lambda\|_{V^h} \leq \|G^h_\lambda|T^h_\lambda\theta|\|_{V^h} \leq \lambda^{-1}\|T^h_\lambda\theta\|_{V^h}=\|F_h\theta\|_{V^h}.$$

Let $R_\lambda \geq \|F_h\theta\|_{V^h}$ and define q_λ and g as in (8) and (9), respectively. Finally, the mapping G_h on $\mathbb{R}^{\Omega h}$ is given via (15) with f replaced by g.

P_h4: The parallel chord method

$$(A_h-\lambda B_h P_h D_h-\tfrac{\lambda}{2}q_\lambda B_h)x^{n+1}=\lambda B_h G_h x^n-\tfrac{\lambda}{2}q_\lambda B_h x^n$$

converges for any $x^0\in V^h$ to $\bar{x}^h_\lambda$.

P_h5: There exists a unique solution $\bar{z}^h\in V^h$ of the "reduced problem"

$$B_h(P_hD_h+F_h)x=\Theta$$

and there is a constant c^h independent of λ such that

$$\|\bar{x}^h_\lambda-\bar{z}^h\|_{V^h} \leq c^h\lambda^{-1}.$$

P_h6: $F_h\Theta \geq \Theta$ and $A_h\bar{z}^h \geq \Theta$ implies $\Theta \leq \bar{x}^h_\lambda \leq \bar{z}^h$.

As in the continuous case we leave the proofs to section 5.

3. NUMERICAL EXAMPLES

Our model problem has the simple form

$$-x''-\lambda(r(t)-x')=0 \text{ on } [0,1],\ x(0)=x(1)=0 \tag{17}$$

where we assume $r\epsilon C^1[0,1]$, $r(t) > 0$ on $(0,1)$. This meets the "canonical assumptions" of section 1 and the solution behaves qualitatively as shown in figure 1 with

$$\bar{z}(t)=\int_0^t r(s)ds.$$

The value $t_\lambda\epsilon(0,1)$ where the solution $\bar{x}_\lambda(t)$ assumes its only maximum travels to 1 as $\lambda\rightarrow\infty$. Finally, $\bar{x}_\lambda$ exists for any $\lambda > 0$ by P1.

We use the upwind scheme on a grid Ω_h based on natural numbers $k_1,k_2,..,k_N \geq 2$ followed by a number of ones $k_i=1$ $(i=N+1,..,M)$ where $1 \leq N < M$

(see section 2). Then t_λ^h is the grid point where the discrete solution $\bar{x}_\lambda^h$ assumes its maximum (if this is unique!). We consider the system based upon $k_1,..,k_N,1,...,1$ adjusted if

$$t_N=\sum_{j=1}^{N} k_j \leq t_\lambda^h$$

that means if the discrete boundary layer region $[t_\lambda^h,1]$ is covered by the regular grid of the smallest step width $h > 0$ from (13a). To achieve this situation we possibly have to add ones at the end by keeping $k_1,..,k_N$ fixed. This brings the number M of grid points up but does not substantially change the value of $h > 0$ from (13a).

We have used the following three distributions of grid points given by

$k_i=256$ $(i=1,..,7)$, $2k_i=k_{i-1}(i=8,..,12)$, $k_{13}=k_{14}=4$,
$k_{15}=k_{16}=2$, $k_i=1$ $(i=17,..,36)$

$k_1=512$, $k_{2i}=k_{2i-1}(i=1,..,9)$, $2k_{2i+1}=k_{2i-1}(i=1,..,8)$,
$k_i=1$ $(i=19,..,38)$

$k_i=128$ $(i=1,..,15)$, $2k_i=k_{i-1}$ $(i=16,..,21)$,
$k_i=1$ $(i=22,..,41)$.

We lable them by the number M=36, M=38, M=41, respectively. Note that we have M-1 grid points in (0,1) and that M is the total number of k's. We take $\lambda=1000$ then $\lambda h \sim 0.48$ in all three cases of k's. In the following table t_λ denotes the grid point of the respective grid where $\bar{x}_\lambda$ assumes its maximum on that grid.

r(t)	M		t_λ^h	t_λ	Error % in $[0,t_\lambda)$	$[t_\lambda,1]$
1	38	14	0.9917	0.9932	0.30	15.03
cos	41	18	0.9918	0.9927	1.68	16.39
exp	36	16	0.9922	0.9937	6.31	10.41
exp	38	16	0.9922	0.9937	12.93	8.99

The respective systems are adjusted in the sense defined above in all four cases. The figures in the third column give the number of grid points in the boundary layer region $[t_\lambda,1]$. In the three examples the true solutions are known:

$$r=1\quad: \bar{x}_\lambda(t)=t-\psi_\lambda(t)$$

$$r=\cos: \bar{x}_\lambda(t)=\lambda(1+\lambda^2)^{-1}(\varphi_\lambda(t)-\varphi_\lambda(1)\psi_\lambda(t)+e^\lambda\psi_\lambda(t-1))$$

$$r=\exp: \bar{x}_\lambda(t)=\lambda(\lambda-1)^{-1}(e^t-e\psi_\lambda(t)+e^\lambda\psi_\lambda(t-1))$$

$$\psi_\lambda(t)=(e^{\lambda t}-1)(e^\lambda-1)^{-1},\qquad \lambda\sin t+\cos t=\varphi_\lambda(t)$$

Hence, we can obtain the error $\bar{x}_\lambda(t)-\bar{x}_\lambda^h(t)$ on the respective grid. The maximal percentage of the error for the grid points in $[0,t_\lambda)$, $[t_\lambda,1]$ is given in the sixth, seventh column, respectively (see the table above).

The results show three characteristics:

i) $t_\lambda^h < t_\lambda$: the discrete solution shows a boundary layer phenomenon much too early.

ii) The accuracy is poor.

iii) The figures show no oscillations due to the stability described by P_h3.
The property iii) was confirmed by other examples and values of λ up to 10^{10}. However, the same is true for the properties i) and ii) if we add more ones at the end of the sequences of k's. ii) and iii) have also been observed in [5] where the model problem for $r(t)\equiv 1$ has already been studied with the upwind scheme.

By the stability inequality in P_h3 we have

$$|\bar{x}_{\lambda h}-\bar{x}_\lambda^h| \leq G_\lambda^h |T_\lambda^h \bar{x}_{\lambda h}| \tag{18}$$

where $\bar{x}_{\lambda h}$ denotes the restriction of the solution $\bar{x}_\lambda$ to the grid. Since G_λ^h behaves like λ^{-1} (cf. P_h3) we only have to worry about the consistency error $T_\lambda^h \bar{x}_{\lambda h}$ which is of the form

$$\begin{aligned} &T_\lambda^h \bar{x}_{\lambda h}(t_j) \sim (k_j h)^m D^{m+2} \bar{x}_\lambda(s_j) \\ &\text{where } |t_j - s_j| \leq k_j h \end{aligned} \tag{19}$$

(j=1,..,M-1). Here, m=2 for the upwind scheme. Outside the boundary layer region the derivatives of $\bar{x}_\lambda$ are of moderate absolut value so that we can choose $k_j > 1$ there. In the boundary layer the derivatives of $\bar{x}_\lambda$ may have extremely large values and there we have $k_j=1$, $h \sim \lambda^{-1}$, hence

$$T_\lambda^h \bar{x}_{\lambda h}(t_j) \sim \lambda^{-m} D^{m+2} \bar{x}_\lambda(s_j) \tag{20}$$

by (19). The application of higher order formulae

in this part of the interval will bring up the number $m > 2$ in (20) so that λ^{-m} decreases and (hopefully) $T_\lambda^h \bar{x}_{\lambda h}(t_j)$ decreases as well. Then (18) shows that $\bar{x}_\lambda^h$ and $\bar{x}_{\lambda h}$ will be cl osed together if the stability inequality of $P_h 3$ is still true with $\|G_\lambda^h\|_{V^h} \leq \lambda^{-1}$. We stress the point that we do not need this for all $\lambda > 0$, it is enough to have it for the fixed $\lambda > 0$ under consideration.

4. IMPROVED UPWIND SCHEMES

We are now going to describe three possibilities to improve the upwind approximation in the boundary layer region on the lines of the discussion at the end of section 3.

Scheme 1: For any $t \in \Omega_h$ such that $t \pm h \in \Omega_h$ substitute $-x''(t), x'(t)$ by the second order formulae [7]

$$-h^2 x''(t) \sim -x(t-h)+2x(t)-x(t+h)$$

$$2hx'(t) \sim -x(t-h)+x(t+h),$$

for all other grid points use the upwind substitution

Scheme 2: For any $t \in \Omega_h$ such that $t \pm ih \in \Omega_h (i=1,2)$ substitute $-x''(t)$, $x'(t)$ by the fourth order formula [7]

$$-12h^2 x''(t) \sim x(t-2h)-16x(t-h)+30(t)-16x(t+h) +x(t+2h)$$

$$12hx'(t) \sim x(t-2h)-8x(t-h)+8x(t+h)-x(t+2h),$$

for all other grid points use the substitution as

in scheme 1.

Scheme 3: For any $t\in\Omega_h$ such that $t-h$, $t+ih\in\Omega_h$ $(i=1,2,3,4)$, $t-2h\notin\Omega_h$ substitute $-x''(t)$, $x'(t)$ by the fourth order formula [7,12]

$$-12h^2x''(t)\sim -10x(t-h)+15x(t)+4x(t+h)-14x(t+2h) +6x(t+3h)-x(t+4h)$$

$$12hx'(t)\sim -3x(t-h)-10x(t)+18x(t+h)-6x(t+2h) +x(t+3h).$$

For any $t\in\Omega_h$ such that $t-ih$, $t+h\in\Omega_h$ $(i=1,2,3,4)$, $t+2h\notin\Omega_h$ substitute $-x''(t)$, $x'(t)$ by the fourth order formula

$$-12h^2x''(t)\sim -x(t-4h)+6x(t-3h)-14x(t-2h)+4x(t-h) +15x(t)-10x(t+h)$$

$$12hx'(t)\sim -x(t-3h)+6x(t-2h)-18x(t-h)+10x(t) +3x(t+h).$$

For all other grid points use the substitution as in scheme 2.

We postpone theoretical aspects of these schemes to the end of this section and take first a closer look at their numerical performance. To this end we consider the model problem (17) of section 3 with $\lambda=1000$, $r(t)\equiv 1$ and the distribution of grid points labled M=38 (cf. section 3), hence $\lambda h\sim 0.48$.

Scheme	t_λ^h	t_λ	Error % in $[0,t_\lambda)$	$[t_\lambda,1]$
upwind	0.9917	0.9932	0.30	15.03
1	0.9932	0.9932	<0.01	1.56
2	0.9932	0.9932	<0.01	0.57
3	0.9932	0.9932	<0.01	0.07

For r=cos, $\lambda=1000$ and $k_i=64$ (i=1,..,31), $2k_i=k_{i-1}$ (i=32,..,36), $k_i=1$ (i=37,..,56), hence $\lambda h \sim 0.48$ and M=56, we obtain

Scheme	t_λ^h	t_λ	Error % in $[0,t_\lambda)$	$[t_\lambda,1]$
upwind	0.9912	0.9927	1.00	15.74
1	0.9927	0.9927	0.82	0.81
2	0.9927	0.9927	0.82	0.82
3	0.9927	0.9927	0.82	0.89

For r=exp, $\lambda=1000$ and the same k's (M=56) we have

Scheme	t_λ^h	t_λ	Error % in $[0,t_\lambda)$	$[t_\lambda,1]$
upwind	0.9922	0.9937	1.55	13.80
1	0.9937	0.9937	1.55	3.06
2	0.9937	0.9937	1.55	2.05
3	0.9937	0.9937	1.55	1.48

In all three examples the figures show no oscillations and the schemes are adjusted in the sense of section 3. Furthermore, the tables show that $t_\lambda^h=t_\lambda$, and (as a result) that the accuracy is satisfactory for the schemes 1, 2 and 3. Note that we need a total of only 37, 55, 55 grid points in

(0,1) for r=1, r=cos, r=exp, respectively, to achieve an error of about 1%. This is to be compared with the vast amount of grid points used in [15]. Finally, note that the schemes yield information in the boundary layer region as well as outside of it. In the examples above we have 10 to 15 grid points to represent the boundary layer.

We conclude with some theoretical results on the schemes. For simplicity we restrict ourselfs to the linear problem

$$-x'' - \lambda(p(t)x' + r(t)) = 0 \text{ on } [0,1],\ x(0)=x(1)=0 \qquad (21)$$

and we assume $p, r \in C$ as well as

$$p(t) \leq -1 \text{ on } [0,1],\ 0 < \lambda. \qquad (21a)$$

As in the case of the upwind scheme the discrete analogues given by the schemes 1, 2 and 3 have the general form (14b) where $F_h x \equiv r_h$ for all $x \in \mathbb{R}^{\Omega_h}$. Now we have

Q_h: Let h > 0 be given by (13a) and let

$$\lambda h|p(t)| \leq \begin{cases} 1 & \text{for scheme 1,2} \\ 0.1 & \text{for scheme 3} \end{cases} \text{ on } [0,1] \qquad (21b)$$

Then the properties P_h1, P_h2, P_h3, P_h5 and P_h6 hold for the schemes 1, 2 and 3.

Note that in contrast to the upwind scheme for the schemes 1, 2 and 3 the smallest step width h depends on λ and p to meet (21b). This is a restriction for the choice of the k's defining h via (13a). Indeed, if the quantity $\lambda h|p(t)|$ grows too large the resulting schemes 1, 2 or 3 produce oscillations as described in [5] for similar schemes which are not stable in λ. But we can put up with the stability in the sense of P_h3 for h and λ fixed. Note that in our examples above we have $\lambda h|p(t)|=\lambda h\sim 0.48$ which meets (12b) only for the schemes 1 and 2. Yet scheme 3 works in these cases.

5. TOWARDS THE PROOFS

For convenience we begin with the discrete operator

$$T_\lambda^h=A_h-\lambda B_h(P_hD_h+F_h) \tag{22}$$

as given by the four schemes in sections 2 and 4. If we let

$$\delta_h(t)=1,\ z_h(t)=t \text{ on } \Omega_h \tag{23}$$

then $\delta_h, z_h\in\mathbb{R}^{\Omega_h}$ and an inspection of A_h, B_h and D_h shows that

$$\begin{aligned}&A_hx(t)=x(t)\quad(t=0,1)\quad\text{for all } x\in\mathbb{R}^{\Omega_h}\\&A_h\delta_h(t)=A_hz_h(t)=0 \text{ on } \Omega_h\setminus\{0,1\}\end{aligned} \tag{24}$$

$$\begin{aligned}&D_h\delta_h=\theta,\ D_hz_h=B_h\delta_h,\ D_hx(t)=0\ (t=0,1)\\&\text{for all } x\in\mathbb{R}^{\Omega_h}\end{aligned} \tag{25}$$

hold. Let $\tau > 0$ and let $e_h = z_h + \tau\delta_h$ then from (25) we obtain

$$(A_h - \lambda B_h P_h D_h)e_h = A_h e_h - \lambda B_h P_h D_h z_h = A_h e_h - \lambda B_h P_h B_h \delta_h \geq A_h e_h + \lambda B_h \delta_h . \tag{26}$$

By (25) we may conclude

$$(A_h - \lambda B_h P_h D_h)e_h(t) \geq \begin{cases} \tau & \text{for } t=0 \\ \lambda & \text{for } t \in \Omega_h \setminus \{0,1\} \\ 1+\tau & \text{for } t=1 \end{cases} > 0 \tag{27}$$

for all $t \in \Omega_h$. For the upwind scheme $A_h - \lambda B_h P_h D_h$ is off-diagonally nonpositive and (27) therefore tells us [3a] that $A_h - B_h P_h D_h$ has a nonnegative inverse G_λ^h. The same conclusion it true for scheme 1 if we observe $\lambda h|p(t)| \leq 1$ (see (21b) in Q_h). For the schemes 2 and 3 the matrix $A_h - \lambda B_h P_h D_h$ is no longer off-diagonally nonpositive. However, it can be shown with the techniques in [4,12] that (27) yields the same conclusion as for the upwind scheme and for the scheme 1. But then (27) for $\tau=0$ says $(A_h - \lambda B_h P_h D_h)e_h \geq \lambda B_h \delta_h$ and an application of the nonnegative matrix G_λ^h shows $G_\lambda^h(B_h \delta_h) \leq \lambda^{-1} e_h \leq \lambda^{-1} \delta_h$. This implies $\|G_\lambda^h\|_{V^h} \leq \lambda^{-1}$ since $A_h - \lambda B_h P_h D_h$ leaves V^h invariant.

This already proves the statement Q_h in section 5 because under the conditions of Q_h the operators T_λ^h and $A_h - \lambda B_h P_h D_h$ differ by a constant vector $r_h \in \mathbb{R}^{\Omega_h}$ only.

If it comes to $P_h i$ (i=1,...,6) for the upwind

scheme we have to worry about F_h. But since $D_2 f(t,v) \leq 0$ on $[0,1] \times \mathbb{R}$ (cf. (16b)) and since $A_h - \lambda B_h P_h D_h$ is an M-matrix by the discussion above $P_h 1$ and $P_h 3$ follow from the corollary in section 5 of [3b] and $P_h 4$ is a consequence of section 6 in [2]. $P_h 2$ is a simple application of the mean-value-theorem and $P_h 6$ follows from $P_h 2$ since

$$T_\lambda^h \bar{x}_\lambda^h - T_\lambda^h \Theta = \lambda B_h F_h \Theta \geq \Theta, \quad T_\lambda^h \bar{z}^h - T_\lambda^h \bar{x}_\lambda^h = A_h \bar{z}^h \geq \Theta.$$

We finally turn to $P_h 5$: The existence and unicity of $\bar{z}^h \in V^h$ is trivial from an inspection of the system (14a). Now, the stability inequality of $P_h 3$ applies to $x = \bar{x}_\lambda^h$, $y = \bar{z}^h$ to yield

$$|\bar{x}_\lambda^h - \bar{z}^h| \leq G_\lambda^h |T_\lambda^h \bar{z}^h| = G_\lambda^h |A_h \bar{z}^h|.$$

This completes the proof since $|A_h \bar{z}^h| \in V^h$ and since $\|G_\lambda^h\|_{V^h} \leq \lambda^{-1}$.

We are left with the proofs Pi (i=1,..,7) in the continuous case. We first note that the inequality (1c) implies that the characteristic polynomial of $-x'' + \lambda x' - \lambda q_0 x = 0$ has two distinct real zeros. Then it is easy enough to calculate the Green's function $G_\lambda(t,s)$ mentioned in P3. This shows that (5b) holds. In particular, we have $0 \leq G_\lambda(t,s)$ on $[0,1]^2$ which is the key to the proof of the next

Lemma: Let $\bar{p}, \bar{q} \in C$ be such that

$$\bar{p}(t) \leq -\lambda, \quad \bar{q}(t) \leq \lambda q_0 \quad \text{on } [0,1]$$

If (1c) is satisfied then the Green's function $G(t,s)$ to

$$-x''-\overline{p}(t)x'-\overline{q}(t)x, \quad x(0)=x(1)=0$$

exists and the inequalities $0 \leq G(t,s) \leq G_\lambda(t,s)$ on $[0,1]^2$ hold.

By [3b] (Theorem 5) we may conclude from $\lambda D_2 f(t,v) \leq \lambda q_0$ (cf. (1b)) that

$$|x(t)-y(t)| \leq \int_0^1 K(t,s)\,|T_\lambda x(s)-T_\lambda y(s)|\,ds \qquad (t\in[0,1]) \tag{28}$$

for all $x,y\in V$ where $K(t,s)$ is the Green's function of

$$-x''-\lambda p(t)x'-\lambda q_0 x, \; x(0)=x(1)=0.$$

Since $\lambda p(t) \leq -\lambda$ we find $0 \leq K(t,s) \leq G_\lambda(t,s)$ on $[0,1]^2$ by the Lemma and (28) completes the proof of P3.

The proofs of P1, P2, P4 and P6 follow the same line as in the discrete case. For P1 and P4 we need Satz 7 in [2] and theorem 5 in [3b].

Proof of P5: Let $\overline{t}\in(0,1)$ and $y\in V$ such that $y(t)=\overline{z}(t)$ on $[0,\overline{t}]$. Then by P3 we have for $t\in[0,\overline{t}]$

$$|\overline{x}_\lambda(t)-\overline{z}(t)| = |\overline{x}_\lambda(t)-y(t)| \leq \int_0^1 G_\lambda(t,s)\,|T_\lambda y(s)|\,ds$$

$$\leq(\lambda^2-4q_0\lambda)^{-1/2}\{\exp(2q_0)\int_0^t |z''(s)|ds$$

$$+\int_t^1 \exp(-\tfrac{\lambda}{2}(s-t))|T_\lambda y(s)|ds\}$$

$$\leq(\lambda^2-4q_0\lambda)^{-1/2}\{\exp(2q_0)\int_0^1 |z''(s)|ds$$

$$+\lambda\int_t^1 \exp(-\tfrac{\lambda}{2}(s-t))|p(s)y'(s)+f(s,y(s))|ds\}.$$

The expression in the braces is uniformly bounded for $\lambda \geq 0$ and $t\in[0,1]$. This completes the proof of P5.

Proof of P7: We have $0 \leq \bar{x}_\lambda(t)$ on $[0,1]$ by P2 (note $f(t,0) \geq 0$ on $[0,1]$). This implies $f(t,\bar{x}_\lambda(t)) > 0$ in $(0,1)$ as well as $\bar{z}(t) \geq 0$ in $[0,1]$ by P5.

Let $\bar{t}\in[0,1]$ and $\bar{x}_\lambda'(\bar{t})=0$ then $\bar{t}\in(0,1)$ and

$$-\bar{x}_\lambda''(\bar{t}) = \lambda f(\bar{t},\bar{x}_\lambda(\bar{t})) > 0.$$

This implies that there is at most one $\bar{t}\in[0,1]$ such that $\bar{x}_\lambda'(\bar{t})=0$. On the other hand $\bar{x}_\lambda(0)=\bar{x}_\lambda(1)=0$ shows that there is at least one such $\bar{t}\in(0,1)$.

To prove that $t_\lambda \to 1$ as $\lambda\to\infty$ we first note that $\bar{z}(t) \geq 0$ in $[0,1]$ implies $|p(t)|\bar{z}'(t) = f(t,\bar{z}(t)) > 0$ in $(0,1)$ which means that $\bar{z}(t)$ is strongly monotone increasing in $[0,1]$. Next, let $\hat{t}\in[0,1)$ be arbitrary but fixed and assume that a sequence t_{λ_j} exists such that

$$t_{\lambda_j} \leq \hat{t} \quad (j\in\mathbb{N}) \text{ and } \lambda_j\to\infty \text{ as } j\to\infty.$$

Pick $s \in (\hat{t},1)$ and find $\bar{z}(t_{\lambda_j}) \leq \bar{z}(\hat{t}) < \bar{z}(s)$, hence there is $\varepsilon \in \mathbb{R}$ such that $0 < 2\varepsilon < \bar{z}(s) - \bar{z}(t_{\lambda_j})$ for all $j \in \mathbb{N}$. By P5 we have $|\bar{x}_{\lambda_j}(t) - \bar{z}(t)| < \varepsilon$ in $[0,s]$ for $j \geq J$.

Since $t_{\lambda_J} \in [0,s]$ the last two inequalities imply $\bar{x}_{\lambda_J}(t_{\lambda_J}) \leq \bar{z}(t_{\lambda_J}) + \varepsilon < \bar{z}(s) - \varepsilon < \bar{x}_{\lambda_J}(s)$ which contradicts the definition of t_{λ_J} and completes the proof of P7.

ACKNOWLEDGEMENT

My thanks go to Mr. J. Bigge and Miss D. Brunstering who did the computational work for the examples.

REFERENCES

1 Abrahamsson, L. R., Keller, H. B. and Kreiss, H.O. (1974). Difference Approximations for Singular Perturbations of Systems of Ordinary Differential Equations. Numer. Math. 22, 367-391.

2 Beyn, W.-J. (1976). Das Parallelenverfahren für Operatorgleichungen und seine Anwendung auf nichtlineare Randwertaufgaben. In: ISNM. Vol. 31 (J. Albrecht and L. Collatz eds.), pp 9-33. Birkhäuser Verlag, Basel, Stuttgart.

3a Bohl, E. (1974). Monotonie: Lösbarkeit und Numerik bei Operatorgleichungen. Springer Verlag, Berlin, Heidelberg, New York.

3b Bohl, E. (1978). P-boundedness of Inverses of Nonlinear Operators. ZAMM. 58, 277-287.

4 Bohl, E. and Lorenz, J. (1978). Inverse Monotonicity and Difference Schemes of Higher Order. A Summary for Two-Point Boundary Value Problems. To appear in Aequ. Math.

5 Christie, I., Griffiths, D.F., Mitchell, A.R. and Zienkiewicz, O.C. (1976). Finite Element Methods for Second Order Differential Equations with significant first Derivatives. Int. J. Numer. Meth. Engin. 10, 1389-1396.

6 Coddington, E.A. and Levinson, N. (1952). A Boundary Value Problem for a Nonlinear Differential Equation with a Small Parameter. Proc. Amer. Math. Soc. 3, 73-81.

7 Collatz, L. (1960). The Numerical Treatment of Differential Equations. Springer Verlag, Berlin.

8 Dorr, F.W., Parter, S.V. and Shampine, L.F. (1973). Applications of the Maximum Principle to Singular Perturbation Problems. SIAM Rev. 15, 43-88.

9 Heinrich, J.C., Huyakorn, P.S. and Zienkiewicz, O.C. (1977). An 'Upwind' Finite Element Scheme for Two-Dimensional Convective Transport Equation. Int. J. Numer. Meth. Engin. 11, 131-143.

10 Hemker, P. W. (1977). A Numerical Study of Stiff Two-Point Boundary Problems. Math. Cent. Tracts, Amsterdam.

11 Lentini, M. and Pereyra, V. (1975). Boundary Problem Solvers for First Order Systems Based on Deferred Corrections. In: Numerical Solutions of Boundary Value Problems for Ordinary Differential Equations. (A.K. Aziz ed.), 293-315. Academic Press, Inc., New York, San Francisco, London.

12 Lorenz, J. (1977). Zur Inversmonotonie diskreter Probleme. Numer. Math. 27, 227-238.

13 Miller, J. J.H. (1978). Sufficient Conditions for the Convergence, Uniformly in ε, of a Three Point Difference Scheme for a Singular Perturbation Problem. In: Praktische Behandlung von Differentialgleichungen in Anwendungsgebieten. Lecture Notes in Math., Springer-Verlag. To appear.

14 Mitchell, A.R. and Griffiths, D.F. (1977). Generalized Galerkin Methods for Second Order Equations with Significant First Derivative Terms. Numer. Anal. Rep. 22, Univ. of Dundee, Dep. Math..

15 Pearson, C.E. (1968). On a Differential Equation of Boundary Layer Type. J. Math. Phys. 47, 134-154.

MULTI-LEVEL ADAPTIVE TECHNIQUES (MLAT) FOR SINGULAR-PERTURBATION PROBLEMS

Achi Brandt

Weizmann Institute of Science, Rehovot, Israel

ABSTRACT

The multi-level (multi-grid) adaptive technique is a general strategy of solving continuous problems by cycling between coarser and finer levels of discretization. It provides very fast general solvers, together with adaptive, nearly optimal discretization schemes. In the process, boundary layers are automatically either resolved or skipped, depending on a control function which expresses the computational goal. The global error decreases exponentially as a function of the overall computational work, in a uniform rate independent of the magnitude (ε) of the singular-perturbation terms. The key are high-order uniformly stable difference equations, and uniformly smoothing relaxation schemes.

1. INTRODUCTION

The Multi-Level Adaptive Technique (MLAT) is a general numerical strategy for solving continuous problems such as differential and integral equations and functional minimization problems. It will be discussed here mainly in terms of the numerical solution of partial differential boundary-value problems, with special emphasis on singular-perturbation problems.

The usual approach is first to discretize the boundary-value problem in some preassigned manner (e.g., finite-element or finite-difference equations on a fixed grid), and then to submit the resulting discrete system to some numerical solution process. In MLAT, however, discretization and solution processes are intermixed with, and greatly benefit from, each other. A sequence of uniform grids (or "levels"), with geometrically decreasing mesh-sizes, participates in the process. The cooperative solution process on these grids involves relaxation sweeps over each of them, coarse-grid-to-fine-grid interpolations of corrections and fine-to-coarse transfers of residuals. This process has two important basic benefits. On one hand it acts as a very fast general solver of the discrete system of equations (including the equations on the finest grid). On the other hand it provides, in a natural way, a flexible, adaptive discretization. For convenience, we discuss these aspects one by one.

1.1. The Fast Solver

In Section 2 of this paper we portrait the multi-level process as a fast solver, i.e., regarding the coarser grids as nothing but auxiliaries for solving the finest-grid equations. The description has appeared before (Brandt (1972), (1977a), (1977b)), but here we add more detailed examples of the solution process (Sec. 2.2), and emphasize some new important aspects. In particular, the "fine-to-coarse correction function" (or the "local relative truncation error") τ_h^H is discussed, together with some of its usages.

One usage, developed in collaboration with N. Dinar, is the so called τ-*extrapolation*. It amounts to a trivial additional operation in the multi-grid program, and costs negligible amount of extra computing time. But, as shown in Sec. 2.3, it improves the solution, sometimes by very much. With it, at a total computational cost of 4 to 7 work-units (a unit being the work equivalent of one Gauss-Seidel relaxation sweep over the finest grid), a solution u^h is always obtained which is better (i.e., a closer approximation to the true *differential* solution U) than the exact solution U^h of the difference equations. Furthermore, in case some extra smoothness is present, u^h will be some orders-of-magnitude better than U^h; or, alternatively, it may be obtained at a much smaller computational cost.

As other usages of the fine-to-coarse correction function τ_h^H we list, in Section 2.4, some very efficient *methods for making nonlinear continuation (e.g., for bifurcation problems), methods for optimal-control problems, ill-posed boundary-value problems and parabolic time-dependent problems*, as well as fast solution methods that can operate with a limited computer storage. Also mentioned in Section 2.4 are new numerical experiments, made in collaboration with N. Dinar, for the *steady-state incompressible Navier-Stokes equations*, including the singular-perturbation case of large Reynolds numbers. These, and many other experiments briefly referred to, clearly indicate that the above-mentioned multi-grid efficiency (solution in just few work-units) is obtained for general elliptic and non-elliptic systems on general domains.

The fast-solver aspect of the multi-level techniques was studied by various other workers, starting, perhaps, with the "group relaxation" of Southwell (1935). See references in Brandt (1977a), and more recent references in Nicolaides (1978) and Hackbusch (1978b). Most of this work is very theoretical. That is, rigorous asymptotic bounds are derived for the multi-grid efficiency. The price of rigorosity, of course, is that the results are far from realistic: The proofs hold only for extremely small mesh sizes, and, even for those, the work estimates are orders-of-magnitude too large. (Cf. Section 10 of Brandt (1977a).) The rigorous estimates are too crude, in fact, to yield any useful information; e.g., they cannot resolve the difference between more efficient and less efficient multi-grid processes. For singular-perturbation problems, especially, the rigorous proofs hold only for mesh-sizes which are microscopic compared with the practical ones. For this reason, and since the quantity we try to estimate here is actually nothing but the computer-time, which of course we know anyway (at least aposteriori), a different type of theoretical studies are preferred by the present author. Briefly, discarding rigorosity, it is observed that the important multi-grid processes are of local nature, since long-range convergence is obtained by coarse-grid processes, which cost very little. One can therefore analyze the crucial aspects of multi-grid processes by employing a *local mode (Fourier) analysis*, ignoring for instance far boundaries and changes in the (possibly nonlinear) equations. Experiments with various types of equations (see Dinar (1978) and also Poling (1978)) shows that this analysis (which is much simpler than the rigorous theorems) precisely predicts the multi-grid efficiency. It is therefore very useful in selecting efficient algorithms (see, e.g., Appendix A in Brandt (1977a)), in understanding the numerical results, and in debugging multi-grid programs. It led, in fact, to the efficient algorithms mentioned above, which solve, for example, Navier-Stokes equations in about 7 work-units. Its mechanized use is mentioned in Section 2.4 and, for singular-perturbation problems, in Section 6.

About the *difficulties in implementing* multi-grid procedures, and what to do about them, some general comments are made toward the end of Section 2.4.

1.2. Non-Uniform, Adaptible Structures

Section 3 of this paper (following and adding to Sections 2 and 3 in Brandt (1977b)) discusses the special capability of the multi-level structure to create non-uniform, flexible discretization patterns, especially such patterns as required by singular-perturbation problems, where thin layers should sometimes be resolved, either near or away from boundaries. This capability is obtained by observing that the various grids (levels) need not all extend over the same domain. Finer levels may be confined to increasingly smaller subdomains, so as to provide higher resolution only where desired. Moreover, we may attach to each of these localized finer grids its own local system of coordinates, to fit curve boundaries or to approximate directions of interior interfaces and thin layers. Unlike global coordinate transformation, these local coordinates do not complicate the difference equations throughout the domain (hence do not turn the one-dimensional trouble of boundary approximations into a two-dimensional trouble of complicated equations). All these patches of local grids interact with each other through the multi-grid process, which, at the same time, provides fast solutions to their difference equations (an important advantage over other methods of patching grids or using transformations).

This structure, in which non-uniform discretization is produced through the sequence of uniform grids, is highly flexible. Discretization parameters, such as (finest) mesh-sizes and approximation orders, can be locally adjusted in any desired pattern, expending negligible amounts of book-keeping work and storage.

In particular, since in this structure only equidistance differencing is needed (much less expensive than differencing on variable grids), it becomes feasible to employ high-order difference approximations, even in singular-perturbation cases (see Section 5).

The discretization can thus be progressively refined and adapted. The actual adaptive solution process is governed by certain criteria, described in Section 3.6. Derived from optimization considerations, these are local criteria which automatically decide where and how to change the local discretization parameters. Furthermore, these criteria are controlled by the user through a certain function G (the error-weighting function), which, in effect, expresses the purpose of the numerical calculations, i.e., the sense (or the error norm) in which approximations to the true solution are to be measured. The resulting discretization will be of high order wherever the evolving solution is suitably smooth. Singularities of all kinds will automatically be detected and treated, usually by introducing increasingly finer levels on increasingly smaller neighborhoods of the singularity.

1.3. MLAT Solutions to Singular Perturbation Problems

The discretization patterns produced by this general adaptive process for singularly-perturbation cases are studied in Section 4. It turns out that boundary layers are sometimes resolved by the adaptive process, and in other cases they are completely "skipped", depending on the choice of the control function G. The decision whether and how to resolve the layer is automatically taken by the adaptive process itself. In any case, the convergence rate in the suitable sense (i.e., in the error norm corresponding to G) is always fast.

Rates of convergence of adaptive processes are measured by the rate of decrease of the error $E = \|u - U\|$ as a function of the computational work W, where $u = u(W)$ is the evolving numerical solution, U is the true differential solution, and $\|\cdot\|$ is the appropriate error norm. Since the grid is not uniform, nor does it have any fixed number of grid-points, the work-unit in this context must be different from the one mentioned above (which was defined in terms of the finest grid). It can be defined, for instance, as the work of applying the lowest-order difference equations at one grid-point. Thus, for example, in the conventional case, where regular grids are applied for solving a d-dimensional regular differential problem, employing $O(h^p)$ difference approximations, if a fast solver (e.g., a multi-grid algorithm) is used which solves the algebraic equations in $O(h^{-d})$ arithmetic operations, then the rate of convergence can be expressed as $E \leq C(p)W^{-p/d}$. The constant C depends not only on p but also on the solution U.

We show that, *using adaptive discretization, the same p-order convergence rate* $E = O(W^{-p/d})$ *is obtained uniformly in the size* (ε) *of the singular perturbation; that is,* $C(p)$ *does not depend on* ε. *Moreover, if the order-of-approximation* p *is adaptible too, the rate of convergence is uniformly exponential.* More precisely, for cases requiring boundary-layer resolution it is shown that $E \approx cW^{\alpha}\exp(-cW^{\alpha})$, with α and c independent of ε. We assume, of course, that the convergence rate for the reduced problem would not be slower. In cases the boundary-layer is skipped, the rate depends solely on the rate of the reduced problem. We also show that for this type of results it is not necessary to adapt p locally; p *may be an adaptible constant.*

These convergence rates are *uniform*; they are obtained for any ε, small, moderate or large. Nothing actually should be known in advance about the value of ε. It is not even required at all to know that this is a singular-perturbation problem. Most other solution methods, by contrast, solve either the regular case ($\varepsilon = 0(1)$) or the asymptotic case (ε very small), but not the whole range. (Quite often, intermediate values of ε are the most difficult to solve). Here, no special analyses are required, no need to separate the reduced problem from the singular-perturbation, and, in particular, no need to compute the proper reduced boundary conditions. No matching procedures are employed. The method works similarly for interior singular layers, e.g., for ODE problems with turning points, even when (as in nonlinear problems) the location of the singularity is not known in advance.

Although no apriori knowledge is needed about the size of the singular-perturbation, some rules should be kept in dealing with potentially singularly-perturbed problems. As is well known, difference schemes should be constructed which are uniformly stable. This aspect is discussed in Section 5, including the construction of *high-order uniformly-stable difference operators*. Similarly, in the multigrid processing of potentially singular problems, relaxation schemes with *uniform smoothing rates* should be employed. Such schemes are described in Section 6.

Remark. Sections 5 and 6 are abbreviated here. For their full text, see Brandt (1978b). It will include further discussion of ellipticity and uniform well-posedness of finite-difference systems, as well as remarks on the uniformly well-posed approximation of singular-perturbation problems with highly-oscillating solutions.

1.4. Table of Contents

2. MULTI-GRID FAST SOLVERS

To understand the basic numerical processes of MLAT, consider first the usual situation where a partial differential problem

$$LU(x) = F(x), \quad \text{for} \quad x = (x_1,\dots,x_d) \quad \text{in a domain} \quad \Omega \subseteq \mathbb{R}^d, \tag{2.1a}$$

$$\Lambda U(x) = \Phi(x), \quad \text{for} \quad x \quad \text{on the boundary of} \quad \Omega, \tag{2.1b}$$

is discretized in a preassigned manner on a given uniform grid G^h, with mesh-size h, yielding the finite-difference equations

$$L^h U^h(x^h) = F^h(x^h), \qquad (x^h \in G^h). \tag{2.2}$$

Here $U = (U_1, U_2, \dots, U_q)$ and its discrete approximation U^h are q-dimensional vectors of unknown functions, L and Λ are linear or nonlinear differential operators and $L^h U^h(x^h)$ is, correspondingly, a linear or nonlinear expression involving values of U^h at x^h and at neighboring grid points. At various instances of the solution process, we have on G^h an approximation to U^h, which we will generally denote by u^h.

In this section multi-level techniques for the fast solution of (2.2), with coarser grids serving as auxiliaries, will be described. In this context the term multi-grid, rather than multi-level, can be used. The difference between "grid" and "level" arises only in the more general situation (see Section 3 below).

2.1. Basic Multi-Grid Processes

To obtain a fast solution to equation (2.2) via the multi-grid method, we add to G^h a sequence of coarser uniform grids. Let G^H be such a coarser grid; e.g., let the grid-lines of G^H be every other grid-line of G^h, so that its meshsize is $H = 2h$.

One way of inexpensively obtaining an approximate solution u^h to (2.2) is first to obtain an (approximate) solution u^H to the corresponding coarser problem

$$L^H U^H(x^H) = F^H(x^H), \qquad (x^H \in G^H) , \tag{2.3}$$

(which is much less expensive to solve since it contains far fewer unknowns) and then to interpolate u^H to the fine grid:

$$u^h = I_h^H u^H . \tag{2.4}$$

The symbol I_h^H stands for the operation of interpolating from G^H to G^h. Polynomial interpolations of any order can be used. (The optimal order is discussed in Section A.2 of Brandt (1977a). Generally, if m_j is the highest order of derivatives of U_j in L and p is the order of approximation, then an interpolation of order at least $m_j + p$ (i.e., polynomials of degree at least $m_j + p - 1$) should be used for u_j^H to ensure full multi-grid efficiency. In some particular situations, even greater efficiency can be achieved by still higher interpolation; see footnotes 1 and 5 below.)

How good the approximation (2.4) is depends on the smoothness of the solution U^h. In some cases U^h is so smooth that, if the interpolation I_h^H and the coarse grid operator L^H are of order high enough to exploit that smoothness, then u^h obtained by (2.4) satisfies

$$\|u^h - U\| \lesssim \|U^h - U\| , \tag{2.5}$$

in some suitable norm. This means that u^h solves (2.2) "to the level of the truncation error", which is all we can meaningfully require in solving (2.2). In such cases, however, the fine grid G^h is not really needed: the coarser grid G^H already yields a solution with the required accuracy. If G^h is at all needed, our first approximation (2.4) will require a considerable improvement.

Can we compute a correction to u^h again by some interpolation from the coarse grid G^H? Namely, can we somehow approximate the error $V^h = U^h - u^h$ by some V^H computed on G^H? Normally[1)], the answer is no. If u^H in (2.4) is a good enough approximation to U^H, then V^h will be a rapidly oscillating function that cannot meaningfully be described on the coarse grid G^H. Therefore, before we can reuse coarse grids, the error V^h should be smoothed out.

An efficient smoothing is obtained by *relaxation sweeps*. A standard example is the Gauss-Seidel relaxation sweep. This is a process in which all points x^h of G^h are scanned one by one in some prescribed order. At each point the old value $u^h(x^h)$ is replaced by a new value, which is computed so that (2.2) is satisfied at that particular point x^h (or nearly satisfied, in case (2.2) is nonlinear at that point; one Newton step of changing $u^h(x)$ is enough). Having completed such a sweep, the system (2.2) is not yet solved, because its equations are coupled to each other; but the new approximation u^h is hopefully "better" than the old one.

In fact, a well known, and extensively used, method for solving (2.2) is by a long sequence of relaxation sweeps. When the system (2.2) is linear, convergence of u^h to U^h is obtained by a sequence of relaxation sweeps if and only if the system is definite. But the rate of convergence is very slow. Typically, if m is the order of L and N_i is the number of grid intervals in the x_i direction, then the number of sweeps required for convergence is proportional to $(\min[N_1,\ldots,N_d])^m$.

A closer examination, e.g., by Fourier analysis, of the error V^h, shows that the components slow to converge are those whose wavelength is large compared with the mesh-size of h. The high-frequency components, however, converge very fast; they practically disappear after a few sweeps. For example, in Gauss-Seidel relaxation for the 5-point Laplace operator

$$L^hU^h(x,y) \equiv \Delta_h U^h(x,y) \equiv \frac{1}{h^2}\{U^h(x+h,y) + U^h(x-h,y)$$

$$+ U^h(x,y+h) + U^h(x,y-h) - 4U^h(x,y)\}\ , \qquad (2.6)$$

the convergence factor of the Fourier component $\exp[i(\theta_1 x + \theta_2 y)/h]$ (i.e., the factor by which the magnitude of its amplitude in the error expansion is reduced by one sweep) is

$$\mu(\theta_1,\theta_2) = \left|\frac{e^{i\theta_1} + e^{i\theta_2}}{4 - e^{-i\theta_1} - e^{-i\theta_2}}\right| . \qquad (2.7)$$

For the longest components $\theta_j = O(N_j^{-1})$, and hence $\mu = 1 - O(N_1^{-2} + N_2^{-2})$. But for high-frequency components, say with $\max|\theta_j| \geq \frac{\pi}{2}$, we have $\mu \leq .5$, so that in three relaxation sweeps these components are reduced by almost an order of magnitude.

This means that *relaxation sweeps, inefficient as they are in solving problems, are very efficient in smoothing out the error* V^h. This property, which is extensively used in multi-level algorithms, is very general. It holds for

Gauss-Seidel relaxation of any uniformly elliptic scalar ($q = 1$) difference operator, whether linear or nonlinear. For elliptic *systems* ($q > 1$), efficient smoothing is obtained by suitable variants of the Gauss-Seidel relaxation. Even degenerate and singularly-perturbed elliptic operators are smoothed out with similar efficiency, provided more sophisticated variants are used, such as line relaxations in suitable directions, or "distributed" Gauss-Seidel relaxations. Moreover, some of these variants are very efficient even for non-elliptic systems. (See Section 3 in Brandt (1976), Section 3 in Brandt (1977a), Lectures 5, 6 and 7 in Brandt (1978a) and Section 6 below.) It is also important to note that fortunately the smoothing efficiency does not depend on some sensitive relaxation parameters. Such parameters are sometimes needed (e.g., a relaxation factor is required in simultaneous-displacement relaxation schemes, which are used in conjunction with vector or parallel processing); but since smoothing is a local process, the optimal values of the parameters depend on the local operator only, and can easily be calculated by local Fourier analysis. Large deviations from the optimal values have only mild effect.

Thus, after a couple of relaxation sweeps, the error V^h is smooth, and a good approximation to it can inexpensively be computed on the coarser grid G^H. To see how this is done in the general nonlinear case, observe that on G^h the equation satisfied by V^h is the "*residual equation*"

$$\hat{L}^h V^h(x^h) = r^h(x^h), \qquad (x^h \in G^h) \,, \tag{2.8a}$$

where [10]

$$\hat{L}^h V^h \equiv L^h(u^h + V^h) - L^h u^h \,, \tag{2.8b}$$

$$r^h \equiv F^h - L^h u^h \,. \tag{2.8c}$$

In the Linear case $\hat{L}^h = L^h$, and on first reading one may like to keep this case in mind. (2.8) is of course fully equivalent to (2.2), but we are interested in this form because V^h, not U^h, is the smooth function which we like to approximate on the coarser grid G^H. r^h is the *"residual function"*, and, like V^h, it is smoothed-out by relaxation. The approximation to (2.8) on the coarse grid is

$$L^H(I_h^H u^h + V^H)(x^H) - L^H(I_h^H u^h)(x^H) = \bar{I}_h^H r^h(x^H), \quad (x^H \in G^H), \qquad (2.9)$$

where V^H is designed to be the coarse-grid approximation to V^h, and I_h^H and $\bar{I}_h^H$ are interpolation operators (not necessarily the same) from G^h to G^H. Since the points of G^h are often a subset of the points of G^H, one can actually use direct "injection", i.e., $I_h^H u^h(x^H) = u^h(x^H)$. In many cases, however, it is preferrable to use "full weighting", i.e., to use $I_h^H u^h(x^H)$ which is a weighted average of values $u^h(x^h)$ at points $x^h \in G^h$ near x^H, in such a way that all values $u^h(x^h)$ equally contribute to the coarse-grid values.

Observe that at this stage we could *not* approximate the equation $L^h(u^h + V^h) = f^h$ on the coarse grid by the simpler approximation

$$L^H(I_h^H u^h + V^H) = I_h^H f^h , \qquad (2.9')$$

since the error of this approximation depends on the rapidly-oscillating part of u^h, which may be large compared with the function V^h we seek to approximate. In (2.8), by contrast, even if V^h is small, the left-hand side is still approximately a linear operator in V^h, and the left-hand side of (2.9) nicely approximates[2)] that linear operator. In fact, the coefficients of that quasi-linear operator on the fine grid depend on u^h, while on the coarse grid they have similar dependence on $I_h^H u^h$. Hence, if I_h^H is a proper averaging operator, the coarse grid coefficients will automatically be averages of the fine grid coefficients, so that even if u^h is highly oscillatory, the coarse-grid equation is a proper "homogenization" of the fine-grid equation. (For discussions of homogenization see, e.g., Babuška (1975) and Spagnolo (1975).)

Observe also that residuals r^h are defined, and are transferred (with some averaging) to the coarser grid, not only with respect to the interior equations, but also with respect to the boundary conditions. In order that such transfers are done in the right scale, it is important that (i) the difference equations (2.2), and similarly (2.3), approximate (2.1) without change of scale (e.g., without multiplying through by h^m. Equations (2.8)-(2.9) refer to the *divided* form of the difference approximations. Keeping this in mind, one can of course write the *program* with differently scaled equations, provided r^h is multiplied by a suitable constant when transferred to the coarser grid.) (ii) Difference-equations approximating different differential equations should be clearly separated. For example, do not scramble together equations approximating (2.1a) with equations approximating (2.1b). Do not incorporate the boundary condition into the neighboring interior equation. Failure to observe these rules is a common error in multilevel programming.

To avoid complicated linearization in solving (2.9), a new unknown function

$$U_h^H = I_h^H u^h + V^H \tag{2.10}$$

is introduced (instead of V^H) on G^H, in terms of which

(2.9) becomes

$$L^H U_h^H(x^H) = F_h^H(x^H), \qquad (x^H \in G^H) , \qquad (2.11a)$$

where[10)]

$$F_h^H = L^H(I_h^H u^h) + \bar{I}_h^H r^h . \qquad (2.11b)$$

The advantage of this new form is that it is the same equation as (2.3), except for a different right-hand side. (The difference between the two right-hand sides is an important quantity which will be exploited below. See Section 2.3.) Moreover, (2.11) and (2.3) has the same form as (2.2). Hence, the same routines (e.g., the same relaxation routine) can be used in treating all of them. (See for example the simple sample program in Appendix B of Brandt (1977a).)

It is also worth noting that our new unknown (2.10) represents, on the coarse grid, the sum of the basic approximation u^h and its correction V^h. Thus, u_h^H is the full current approximation, represented on G^H. The scheme of using u_h^H is therefore called the *Full Approximation Scheme (FAS)*. To be distinguished from the Correction Scheme (CS), which directly uses V^H. (The Correction Scheme is messy in nonlinear problems, and cannot be applied on composite grids (see Section 3). We therefore continue our discussion here only in terms of the more general scheme FAS.) At convergence, when $u^h = U^h$ and $V^h = 0$, we have $U_h^H = I_h^H U^h$. Thus U_h^H is a coarse-grid function which coincide with the fine-grid solution - a fact which will also be very useful below (Sections 2.3 and 2.4).

Once (2.11) is solved (or approximately solved), we want to use its solution to correct the basic approximation u^h. In doing so we should keep in mind that $V^H = U_h^H - I_h^H u^h$, and not u_h^H itself, is the function which approximates a smooth fine-grid function, and hence V^H (or its computed approximation) is what we should interpolate to the fine grid and use it there as a correction to u^h. Thus, denoting our approximate solution to (2.11) by u_h^H, the corrected approximation on the fine grid should be

$$u_{NEW}^h = u_{OLD}^h + I_H^h(u_h^H - I_h^H u_{OLD}^h) \ . \tag{2.12}$$

Observe that this interpolation is *not* equivalent to

$$\bar{u}_{NEW}^h = I_H^h u_h^H \ , \tag{2.13}$$

since $I_H^h I_h^H$ is not the identity operator. The important difference is that (2.12) preserves the high-frequency information contained in u_{OLD}^h, while (2.13) does not use u_{OLD}^h, and thus loses this information. Interpolation of the type (2.12) is called *FAS interpolation*.

The order of the interpolation I_H^h in (2.12) need not be as high as in the first coarse-to-fine interpolation (2.4). Order m_j is enough (cf. the discussion following (2.4)). For example, if the differential equation is of second-order, then linear interpolation is enough.

We summarize the basic processes above: To solve the fine-grid equations (2.2), an initial approximation (2.4) is obtained from an approximate solution u^H *to the coarse grid equation (2.3). Then the approximation is improved by a "multi-grid cycle". This cycle includes a couple of relaxation sweeps followed by the "coarse-grid correction" (2.12), in which* u_h^H *is an approximate solution to the coarse-grid correction equations (2.11).*

In most cases, at the end of the multi-grid cycle the approximation u^h *will satisfy (2.5) and therefore require no further improvement.* This is because the relaxation sweeps effectively liquidate the high-frequency components of the error, while the coarse-grid correction liquidates the lower components. In fact, we will see below (Section 2.3) that, with some modification in the algorithm, at the end of one multi-grid cycle $||u^h - U||$ may be much smaller than $||U^h - U||$. If, however, for any reason, a greater accuracy in solving (2.2) is desired, additional multi-grid cycles can be performed. Typically, each cycle which includes three sweeps of a suitable relaxation scheme will reduce $||u^h - U^h||$ by a factor of .2 to .08.

We still have to specify how *the coarse-grid equations*, first (2.3) and later (2.11a), are actually solved. They *are solved in the same way that (2.2) is solved, namely, by a combination of relaxation sweeps and coarse-grid corrections, using a grid still coarser than* G^H. More precisely, (2.3) is solved by a first approximation obtained from a still-coarser grid (grid G^{2H}, for example), and then a multi-grid improvement cycle (using G^{2H} again). For solving equation (2.11a) the first approximation is $I_h^H u^h$; one multi-grid cycle (using G^{2H}) is enough for improving this approximation to the required level of accuracy.

The full algorithm has several variations. One is flow charted here as Figure 1. This is essentially the same algorithm as described in Section 1.3 of Brandt (1977b). Sample runs of it can be produced by the MUGTAPE (1978a) program FASFMG. It contains three *switching parameters*: α, δ and η. Usually $\alpha = 2^{-p}$, where p is the approximation order. Optimal values for δ and η are discussed in Sections A.6 and A.7 of Brandt (1977a). In practice the precise optimization is not important. One can take $\eta = \bar{\mu}$ and $\delta = \bar{\mu}^r$, where $\bar{\mu}$ is an estimate for the smoothing factor (computed for example by SMORATE; cf. Section 6),

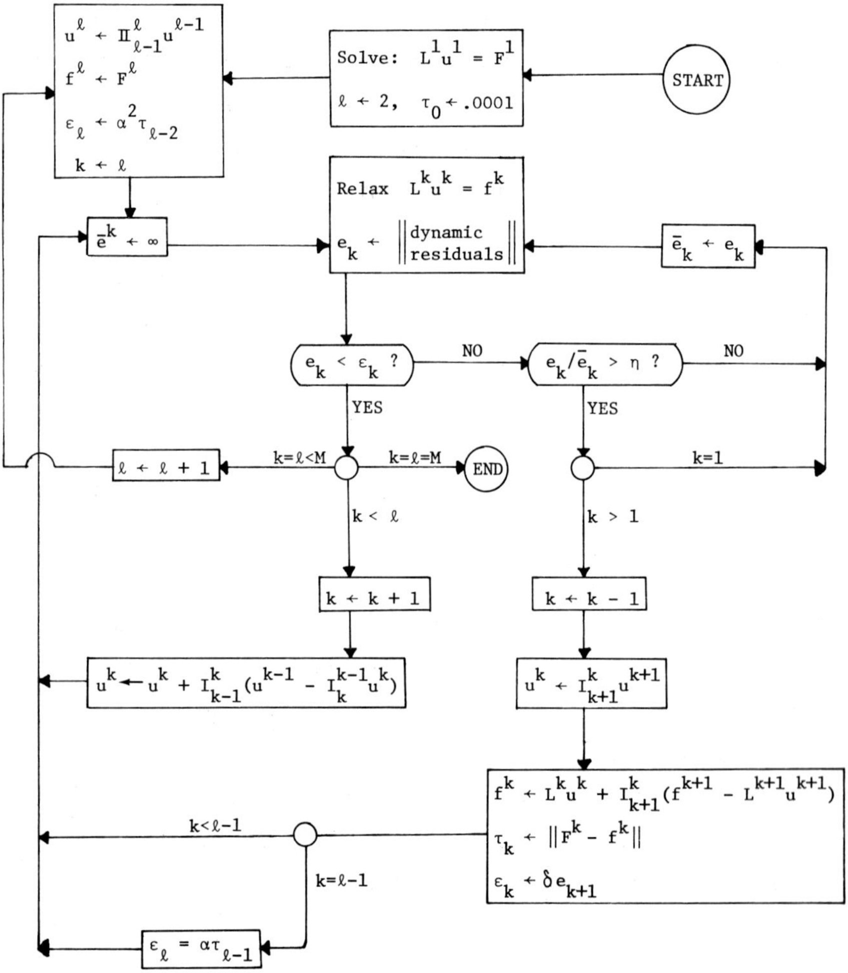

Fig. 1. FAS Full Multi-Grid (FAS FMG) Algorithm.

(See Legend on next page.)

Fig. 1. FAS Full Multi-Grid (FAS FMG) Algorithm. In this flowchart, as in the program itself, the different levels (grids) are not labelled by their mesh-size, as in the text, but by a positive integer k. $k = 1$ *is the coarsest level,* $k = M$ *is the ultimately finest level, and* $k = \ell$ *is the currently finest one (i.e., the finest so far used by the algorithm).*

Thus, the original finite-difference equation on level k *(before being changed to serve as a correction equation like (2.11)) is* $L^k U^k = F^k$. u^k *denotes the current approximation, and* f^k *the current right-hand side, on level* k. I^k_{k-1} *denotes interpolation of order* m *(the order of the differential equation) from level* $k - 1$ *to the next finer level* k. $\Pi^{\ell}_{\ell-1}$ *denotes higher-order* ($m + p$ *order) interpolation.* I^k_{k+1} *denotes transfer (by some averaging) from level* $k + 1$ *to the next coarser level* k. τ_k *denotes an approximate measure of the local truncation errors on level* k *(cf. Section 2.3).* e_k *is a measure of the residuals, taken during the relaxation sweep on level* k. $\bar{e}_k$ *is the value of* e_k *at the previous sweep, so that* $e_k / \bar{e}_k > \eta$ *signals slow convergence.* ε_k *is a tolerance designed so that* $e_k < \varepsilon_k$ *signals convergence of the current k-level problem. For* $k < \ell$ *the k-level problem is the correction problem to level* $k + 1$, *hence* $\varepsilon_k = \delta e_{k+1}$. *On the currently finest level* ($k = \ell$) *we need convergence to within the estimated size of the truncation error. Hence* $\varepsilon_\ell = \alpha \tau_{\ell-1}$, *and before* $\tau_{\ell-1}$ *is known* $\varepsilon_\ell = \alpha^2 \tau_{\ell-2}$. *The parameters* η, δ *and* α *are discussed in the text.*

and r is the number of relaxation sweeps per multi-grid cycle. With good relaxation schemes $\bar{\mu} \approx .5$ and $r \approx 3$.

A slightly different algorithm is shown step-by-step in Table 1, in the next section. That algorithm can also be run through program FASFMG (using its FASFIX subroutine). The difference between the two is that the algorithm in Fig. 1 is "accommodative", its flow depends on some internal checks, while that in Table 1 is "fixed", its entire flow is prescribed in advance, depending only on some input parameters.

2.2. *Full Multi-Grid Run: An Example*

Table 1 shows the steps and the results of a multi-grid solution process for the 5-point Poisson equation (cf. (2.6))

$$\Delta_h U^h(x^h) = F(x^h), \qquad (x^h \in G^h) , \tag{2.14}$$

where G^h is a 97×65 grid with mesh-size $h = 1/32$, covering the rectangle $\{0 \leq x_1 \leq 3,\ 0 \leq x_2 \leq 2\}$. Dirichlet boundary conditions are given on the boundary of the rectangle. In this particular run the boundary conditions and $F = \Delta U$ were chosen so that the exact solution to the differential problem is $U = \sin(3x_1 + 3x_2)$.

The flow of the algorithm can be seen from the first column of the table. It tells us the mesh-size H of the grid on which a Gauss-Seidel (GS) relaxation sweep is made. Thus, the process starts with 5 sweeps on G^{16h}, a 7×5 grid with mesh-size $H = 16h = 1/2$, starting with the approximation $u^{16h} \equiv 0$. Since the grid is very coarse, after 5 sweeps u^{16h} solves (2.3) well below the truncation level. This last u^{16h} is then *cubic*-interpolated (i.e., interpolation of order 4) to the finer grid G^{8h} to serve there as the first approximation u^{8h}, as indicated

TABLE I

Output of Multigrid Runs for a Poisson Problem, Produced by the MUGTAPE (1978) Program FASFMG

Level H	$\|r^H\|_G$	Relaxation Work	$\|u^H - U\|_\infty$ Usual	$\|u^H - U\|_\infty$ τ-extrap.	$\|U^H - U\|$
16h	2.11	.0039			
16h	.750	.0078			
16h	.270	.0117			
16h	.104	.0156			
16h	.0417	.0195	.25	.25	.249
Cubic interpolation $u^{8h} = I_{16h}^{8h} u^{16h}$					
8h	3.73	.0351			
8h	1.34	.0508	.150	.150	
				$\tau_{8h}^{16h} \leftarrow 4\tau_{8h}^{16h}/3$	
16h	.570	.0547			
16h	.269	.0589	.0611	.0600	
32h	.203	.0596			
32h	.0262	.0605			
32h	.00164	.0615	.0586	.0538	
16h	.145	.0654	.0765	.0600	
8h	.492	.0811	.0798	.0499	.0572
Cubic interpolation $u^{4h} = I_{8h}^{4h} u^{8h}$					
4h	1.01	.144			
4h	.602	.206	.0645	.0407	
				$\tau_{4h}^{8h} \leftarrow 4\tau_{4h}^{8h}/3$	
8h	.418	.222			
8h	.288	.237	.0398	.0246	
16h	.150	.241			
16h	.0473	.245	.0117	.0517	
32h	.0114	.246			
32h	.000095	.247			

TABLE I - Continued

32h	.000006	.248	.00713	.00458	
16h	.0145	.252	.0123	.00565	
8h	.0988	.266	.0162	.00731	
4h	.201	.330	.0180	.00525	.0138
Cubic interpolation $u^{2h} = I_{4h}^{2h} u^{4h}$					
2h	.267	.580			
2h	.205	.830	.0168	.00467	
				$\tau_{2h}^{4h} \leftarrow 4\tau_{2h}^{4h}/3$	
4h	.177	.893			
4h	.158	.955	.0146	.00387	
8h	.109	.971			
8h	.0752	.986	.00893	.00242	
16h	.0378	.990			
16h	.0139	.994	.00206	.000870	
32h	.00551	.995			
32h	.000715	.996			
32h	.000045	.997	.00223	.000109	
16h	.00435	1.00	.00254	.000253	
8h	.0274	1.02	.00364	.000226	
4h	.0504	1.08	.00404	.000675	
2h	.0665	1.33	.00419	.000472	.00344
Cubic interpolation $u^{h} = I_{2h}^{h} u^{h}$					
h	.0680	2.33			
h	.0569	3.33	.00412	.00448	
				$\tau_{h}^{2h} \leftarrow 4\tau_{h}^{2h}/3$	
2h	.0540	3.58			
2h	.0513	3.83	.00394	.000411	
4h	.0455	3.89			
4h	.0393	3.95	.00340	.000313	
8h	.0278	3.97			
8h	.0192	3.99	.00206	.000167	
16h	.00945	3.99			
16h	.00355	3.99	.000562	.0000594	
32h	.00138	3.99			
32h	.000123	3.99			
32h	.000008	4.00	.000540	.0000231	

TABLE I - Continued

16h	.00113	4.00	.000625	.0000261	
8h	.00684	4.02	.000879	.0000176	
4h	.0127	4.08	.000984	.0000337	
2h	.0163	4.33	.00102	.0000438	
h	.0182	5.33	.00103	.0000518	.00086

in the table. Two GS sweeps are then made on G^{8h}. Then, the table shows, a switch is made back to the coarser grid G^{16h}. Such a "switch to the coarser grid" means that coarse-grid correction equations such as (2.11) are set up on G^{16h}, namely, their right-hand side is calculated and an initial approximation $u^{16h} = I_{8h}^{16h} u^{8h}$ is introduced. Then, as shown, 2 relaxation sweeps are made on those G^{16h} equations starting with that approximation. Then a switch is made to the still-coarser grid G^{32h} (our coarsest grid here, which is 4×3 and has therefore only 2 interior points), where 3 sweeps are made. Then a switch back to the finer grid G^{16h} is made. Such a "switch to the finer grid" means that a FAS interpolation, like (2.12), is made from G^{32h} to G^{16h}. These interpolations are all of order 2, that is, *linear* interpolations. One relaxation on G^{16h}, then switch back to G^{8h} and one sweep on it, and a multi-grid cycle for G^{8h} has been completed. At this point u^{8h} solves (2.3) to the level of its truncation error, so we can already use it, with *cubic* interpolation, as the initial approximation on G^{4h}. Etc., etc., the first column in Table 1 shows the flow all the way to a final approximation on the finest grid $(H = h)$.

At each relaxation sweep over G^H the residuals $r^H(x^H)$ are measured and their norm is accumulated. This norm is shown on the second column of Table 1. The precise definition of the measured quantity $\|r^H\|_G$ is not important at this point. In this specific run we chose the following definition: r^H are the *dynamic* residuals, i.e., $r^H(x^H)$, as defined by (2.8c), is calculated using values of u^H as they stand immediately before the point x^H is scanned by the relaxation sweep, that is, immediately before the value $u^H(x^H)$ is changed. This type of residual is least expensive to calculate, since it is (almost) calculated anyhow in the course of computing the new value of $u^H(x^H)$. The norm $\|\cdot\|_G$ used in the table is the grid analog of the continuous norm

$$\|r\|_G = \frac{\int G(x)\,|r(x)|\,dx_1 dx_2}{\int G(x)\,dx_1 dx_2} \tag{2.15}$$

where G is some function related to the error functional we are interested in (see Section 3.6). In this particular run we had $G(x) = C_x^2$, where C_x is the distance of x from the nearest corner.

The third column of Table 1 shows the accumulated work, measured in *work units*. A work unit is the work of one relaxation sweep on the finest grid G^h. Hence, a relaxation sweep on G^{Nh} is counted as N^{-2} work units, since it contains about N^{-2} the number of grid-points contained in G^h. This work count neglects any other work except that of relaxation sweeps, because (i) Relaxation sweeps account for at least 70% of the total actual work. (ii) The other work, mostly that of interpolations, is not directly expressible in terms of work units. We could measure running time, but this depends too much on the particular hardware, software and program being used. (iii) The theoretical prediction of convergence rates is made, too, in terms of these work units, so it is convenient to use them for comparing experiments with theory.

Observe how little is the work accumulated on the coarser grids. In sum, this solution algorithm requires only 5.33 work units. A more precise count of *all* operations (including interpolations) shows that, if $\|r^H\|$ is not computed[3)], *this algorithm requires less than 42*n *arithmetic operations, where* n *is the total number of points in the finest grid.*

Incidentally, in this particular problem (2.14), most of these arithmetic operations are additions. In fact, one can arrange it so that the only multiplications (and divisions) involved are by factors 4, 2, 1/2 or 1/4, which in binary floating-point arithmetic can be performed as additions. (For this purpose, cubic interpolation should be performed through the difference operator itself, as for example in Hyman (1977).) Thus, *no multiplications or divisions are required.*

In experiments at ICASE made by Craig T. Poling, the time measured for this algorithm on a CDC 6600 was .083 seconds for a 33 × 33 grid, and .303 for a 65 × 65 grid. A similar algorithm (using another kind of relaxation) on CDC STAR-100 computer required .011 seconds for a 65 × 65 grid and .0347 seconds for a 129 × 129 grid, i.e., about 2 microseconds per grid point.

The first three columns in Table 1 exhibit the standard output of a multi-grid run. The last three columns can be produced only in experimental runs made for cases in which the exact differential solution U is known. The purpose of these columns is to show the quality of the computed approximations u^H. Here, u^H denotes the current approximation on G^H. Thus, in the coarse of a correction cycle for level $h_1 < H$, u^H denotes the full approximation on level G^H, similar to (2.10). The exact solution of the difference equations (2.3) on grid G^H is denoted by U^H.

The fourth column of Table 1 shows, at various stages, the maximum difference between the exact differential solution U and the computed u^H, where the values of u^H at each stage are those obtained at the *end* of the G^H relaxation sweep corresponding to that row of the table. The fifth column of the table will be explained in Section 2.3. The sixth column gives, for each level H, the maximum discretization error $|U^H - U|$. It is shown on the row corresponding to the end of the multi-grid correction cycle for level H. The main observation is, of course, that at this stage $\|u^H - U\|$ is not considerably bigger that $\|U^H - U\|$, so *the algorithm indeed solves the discrete problem to the level of the discretization error.*

This performance is predictable, and will be the same for any other data. The local mode analysis shows that the multi-grid cycle used here reduces $\|u^h - U^h\|$ by a factor .08. A factor .25 would in fact suffice, since all we need in this cycle is to reduce the error from approximately $\|U^{2h} - U\|$ to approximately $\|U^h - U\|$.

Moreover, observe the row one before the last in Table 1. It shows that already at this stage u^{2h} solves the problem to the level of the h discretization error, i.e., $\|u^{2h} - U\|$ is not much bigger than $\|U^h - U\|$. Hence, *the level of the* h *discretization error is actually obtained in only* 4.3 *work-units. The total number of arithmetic operations (additions and subtractions) required is only about* 35n. The full 42n operations are required only if we need the solution, to the same accuracy, at all points of G^h, not only at points of G^{2h}.

Another interesting comment which follows concerns the storage requirement for this algorithm. If the algorithm indeed terminates at 4.33 work-units, no FAS interpolation to G^h is made. Hence the values of u^h need not be stored. All the operations made on G^h are the initial cubic interpolation from G^{2h}, followed by two relaxation sweeps and the calculation of F_h^{2h}. All these operations can be made by one pass over G^h, requiring in storage only 5 columns of the grid at a time. (The first sweep of relaxation is made on the fourth of these columns, then the second sweep can already be made for the third column and the residual transfer can then be made for the second column.) Thus, *the algorithm requires no storage (not even external storage) for the finest grid.* The storage required for the coarser grids is only $\frac{1}{4} + \frac{1}{16} + \cdots \leq \frac{1}{3}$ that of the finest grid. A modified multi-grid algorithm can work with even smaller storage (see Section 2.4).

2.3. Relative Local Truncation Errors and τ Extrapolation

To realize further aspects of the multi-grid method, we now slightly shift our point of view. Going back to the coarse-grid correction equations (2.11), we rewrite them, using (2.8c), in the form

$$L^H U_h^H = F^H + \tau_h^H \tag{2.16}$$

where $F^H = \bar{I}_h^H F^h$ and

$$\tau_h^H = L^H(I_h^H u^h) - \bar{I}_h^H(L^h u^h) \ . \tag{2.17}$$

At convergence, where $u^h = U^h$, we have $V^h = 0$ and, by (2.10), $U_h^H = I_h^H U^h$. Hence, τ_h^H actually represents the *local truncation error of* G^H *relative to* G^h, i.e., the error which arises when the fine-grid solution U^h is substituted in the coarse-grid equation (2.3). Compare this to the usual local truncation error, i.e., to the error τ^h which arises when the true *differential* solution U is substituted into the G^H equations (2.3); namely

$$\tau^H = L^H U - LU . \tag{2.18}$$

We can now reverse our viewpoint. Instead of regarding the coarse-grid as a devise for calculating the correction (2.12) to the fine grid solution, we can view the fine grid as a devise for calculating the correction τ_h^H to the coarse-grid equations, a correction which will make the solution of these coarse-grid equations to coincide (up to interpolation) with the fine-grid solution. τ_h^H is therefore called the *fine-to-coarse correction function*. It is a kind of defect correction (cf. Sec. 3.4). This point of view open many more algorithmic possibilities, such as the multi-grid method for non-uniform discretization (Section 3.3 below), continuation techniques, methods for ill-posed, optimal-control, and time-dependent problems, and small-storage procedures (Section 2.4).

The quantity τ_h^H itself is very useful. First, it is an approximation to the local truncation error τ^H. More precisely, we see from (2.17)-(2.18) that

$$\tau_h^H \approx \tau^H - \tau^h . \tag{2.19}$$

This relation will be used in the adaptive processes (Section 3.6). Here we show how to use it to improve the multi-grid solution.

The local truncation error can be expanded in Taylor series, yielding always a relation of the form

$$\tau^h(x) = C(x)h^p + O(h^{\bar{p}}), \qquad (\bar{p} > p) , \tag{2.20}$$

where $C(x)$ depends on the (unknown) solution, but does not depend on h. Applying (2.20) also to τ^H, and using

(2.19), we find

$$\tau^H \approx \frac{1}{1-\eta^p}\tau_h^H + O(H^{\bar{p}}), \qquad (\eta = \frac{h}{H}, \text{ hence usually } \eta = \frac{1}{2}) . \quad (2.21)$$

Thus, if we replace τ_h^H in the coarse-grid correction equations (2.16) by $\tau_h^H/(1-\eta^p)$, we effectively introduce the *true* truncation error (up to order $H^{\bar{p}}$) instead of the *relative* one, and the solution U_h^H should become a much better approximation to the true differential solution U. We call this replacement a *local truncation extrapolation*, or, briefly, τ-*extrapolation*.

Note that τ_h^H is defined with respect to both the interior difference equations and the boundary conditions, hence τ-extrapolation can be applied for both.

The τ-extrapolation costs very little. Only one operation (multiplication by $(1-\eta^p)^{-1}$) is added, and only at coarser - grid points, so it amounts to less than $\frac{1}{3}$ operation per fine-grid-point. The stages in the algorithm where this extrapolation is made are shown in the fifth column of Table 1. Also shown in that column are the results of this operation in terms of the error $||u^H - U||_\infty$ at various stages. We see that with exactly the same flow[4], the algorithm produces now much smaller errors; in fact u^h *now, after* 4.3 *work-units, is a much better approximation (to* U*) than the exact finite-difference solution* U^h.
With more work and some changes in the algorithm (quintic instead of cubic interpolations, and two instead of one multi-grid cycles for each level, τ-extrapolation being made on the second of the two), $||u^h - U||_\infty$ could be made much smaller yet[5].

The impressive improvement depends of course on the smoothness of the solution. Similar improvements could be obtained, in some such cases, by using Richardson extrapolation (extrapolating from the *solutions* u^H and u^h). Indeed, in case the solution is *not* smooth, e.g., if it oscillates wildly on the scale of G^h, the τ-extrapolation does not considerably improve u^H. But exactly in this case the one multi-grid cycle is enough to reduce $\|u^h - U^h\|$ well below $\|U^h - U\|$, since, exactly in this case, τ^h is not considerably smaller than τ^H. So the point of the τ-extrapolation is that it can always be used, for negligible extra cost in either programming or computer time, and it produces u^h that, at 4.3-5.3 work-units, is guaranteed to be no worse than U^h, the full solution of the difference equations, with the nice additional feature that any available smoothness is automatically exploited to improve u^h even further.

It should also be pointed out that the τ-extrapolation can improve the solution in many cases where the Richardson extrapolation cannot. The τ-extrapolation depends on Taylor expansion for the *local* truncation error, while the Richardson extrapolation requires such an expansion for the *global* discretization error $U^h - U$. In many cases the later expansion does not exist; for example, no such expansion is possible when the local truncation error is not uniform. Another nice feature of the τ-extrapolation procedure is that it produces the improved solution at all points of the fine-grid, while Richardson extrapolation gives it only at points common to the fine and the coarse grid.

A remark on both types of extrapolations: When the extrapolated solution is considerably better than the fine-grid solution, its accuracy is actually comparable to the accuracy of a higher-order solution on the *coarse* grid. (Because the coarse-grid higher-order error term is not removed by the extrapolation.) That higher-order coarse-grid solution is in principle less expensive in computer resources than the extrapolated solution, since it does not use the fine grid at all. A fully adaptive procedure (cf. Section 3.6) would probably prefer in such a situation to use the higher-order scheme on the coarser grid.

2.4. General Properties of Multi-Grid Solutions (Advertisement)

We have discussed at length the multi-grid solution of one particular problem. The algorithm, however, does not depend on any of the particular features of that problem.

The *shape of the domain* is immaterial. Only low-frequency error components are affected by it, and such components are always liquidated on the coarsest grids for negligible amount of work. The only effect of complicated boundaries is to complicate the difference equations at adjacent points and thus to make each relaxation sweep somewhat more expensive. In terms of work-units as defined above, the efficiency remains the same. Many experiments (reported in Brandt (1972), with many more examples in Shiftan (1972)) confirm this.

Variations in the coefficients, or nonlinearities, in the differential equations usually also affect only low-frequency components, and are therefore still treated at the same multi-grid efficiency. When these variations are wild, i.e., when the coefficients change significantly over the scale of the grid, attention should be given to the proper choice of residual-weighting ($\bar{I}_h^H$ in (2.9) or (2.11)), since the residual function r^h is considerably less smooth than the error V^h. (See discussions in Section A.4 of Brandt (1977a). More precise rules of residual-weighting are given in Lecture 17 of Brandt (1978a). The weights near boundaries depend on the type of boundary conditions.) It is also important for such cases that the coefficients of the coarse-grid difference equations represent local averages of the fine-grid coefficients. This, in fact, is automatically obtained if (for variable-coefficients linear problems) the difference equations on each grid are derived by suitably averaging the differential equations, or if (for nonlinear problems, where the coefficients depend on the solution) the solution-weighting (I_h^H in (2.9) or (2.11)) is a full weighting operator.

Numerical experiments (Brandt (1978a) p. 17-7, and more details in Ophir (1978)) were conducted with difference equations of the form

$$L^h U^h_{\alpha,\beta} \equiv a^h_{\alpha\beta} \frac{U^h_{\alpha-1,\beta} - 2U^h_{\alpha,\beta} + U^h_{\alpha+1,\beta}}{h^2}$$

$$+ c^h_{\alpha\beta} \frac{U^h_{\alpha,\beta-1} - 2U^h_{\alpha,\beta} + U^h_{\alpha,\beta+1}}{h^2} = F^h_{\alpha,\beta} , \qquad (2.22)$$

where *the coefficients* a^h *and* c^h *vary wildly*; e.g., a^h and c^h are random; or $a^h_{\alpha\beta} = c^h_{\alpha\beta} = 1$ at even points ($\alpha + \beta$ even) but $a = .01$ and $c = 1$ otherwise; etc. When, and only when, the algorithm (with proper line relaxation) used full weighting in the transfer to the coarse grid of both the residuals and the coefficients, the solution was almost as efficient as in corresponding constant-coefficient cases.

Many experiments were also made for various nonlinear problems. In 1975-76 Multi-grid programs were developed for the steady-state small-disturbance *transonic flow* problems (see South and Brandt (1976), and Section 6.5 in Brandt (1977a)), in which the differential equations are mixed elliptic-*hyperbolic*, and the solutions contain shock discontinuities. Although these programs are somewhat obsolete (with the present stage of multi-grid experience and knowledge they could be improved in various ways), they do clearly show that the typical multi-grid efficiency can be obtained for this type of problems.

Recently, multi-grid codes for *steady-state incompressible Navier-Stokes problems* have been developed (see Lecture 7 in Brandt (1978a), Brandt (1978c) and Dinar (1978)). The algorithm is similar to the one shown in Section 2.2 above, except for the more elaborated "distributed" relaxation scheme which is required here (a) because it is an elliptic *system*, not a scalar equation, and (b) because, for large Reynolds numbers R the system is *singularly perturbed*. Cavity and pipe problems were solved, with R ranging from 0 to large values (but below the values causing instabilities). For any such R, the process required 6 to 7 work-units, and produced solutions closer to the true differential solutions (in cases those were known) than the exact solutions of the finite-difference equations.

Successful multi-grid applications are also reported for the *minimal-surface equations* (D. J. Jones, Lecture 15 in Brandt (1978a)), for the *pressure iteration in Eulerian and Lagrangian hydrohynamics* (Brandt, Dendi and Ruppel (1978)), and for some simple problems in *finite-element formulations* (Nicolaides (1978) and Poling (1978)). Not listed here are some other multi-grid applications, which seem not to have realized the true multi-grid efficiency[6)].

In some nonlinear problems *a continuation (or embedding) process* should be made, either because there are several solutions to the differential equations or because an initial approximation to the solution cannot otherwise be obtained. In such cases a certain problem parameter, γ say, is introduced, so that instead of a single isolated problem we consider a continuum of problems, one problem $P(\gamma)$ for each value of γ in an interval $\gamma_0 \leq \gamma \leq \gamma_*$, where $P(\gamma_0)$ is easily solvable (e.g., it is linear) and $P(\gamma_*)$ is the target (the given) problem. The continuation method is to advance γ from γ_0 to γ_* in steps $\delta\gamma$. At each step we use the final solution of the previous step $P(\gamma - \delta\gamma)$ (or extrapolation from several previous steps) as an initial approximation in an iterative process for solving $P(\gamma)$. The main purpose is to ensure that the approximations we use are all "close enough" to the respective solution, so that some desirable properties are maintained and convergence is ensured. The process should of course be made carefully in case of *bifurcation*, when several solutions branch off at a certain value of γ. (See, e.g., Keller (1977).)

With the multi-grid solution method the continuation process may cost very little. First, because it can be made on coarser grids. Sometimes, however, too coarse grids do not retain enough of the solution features, and the continuation may not accomplish its purpose. This means that oscillations on the scale of a certain mesh-size h cannot be ignored. These oscillations, however, do not usually change much at each $\delta\gamma$ step. The trick then is to use form (2.16) of the coarse-grid equations. We can make several continuation steps on the coarse grid G^H only, *retaining the values of* τ_h^H *fixed*. By doing this we effectively freeze the high-frequency part of the solution, and retain its influence on the coarse-grid equations. Only once in several (sometimes many) steps we need to calculate on the finer grid too, in order to update the fine-to-coarse correction τ_h^H. This of course may be carried further: the G^h equations themselves may include a $\tau^h_{h/2}$ correction term, and once in several visits to G^h we may go to calculate on $G^{h/2}$ in order to update that term; etc.

A similar technique can be used in *optimal control problems*. Here, some parameters controlling the partial-differential problem should iteratively be adjusted so as to achieve some optimal condition (minimal "cost"). In such problems we can again make most of the iterations on coarse grids, with frozen values of the fine-to-coarse corrections τ^H_h, making only infrequent visits to finer grids for updating τ^H_h. (Together with τ^H_h we should also freeze the fine-to-coarse correction of the cost functional.) If, for example, the control itself is a grid function (e.g., a term in the right-hand side of the equations), a change in the control value at a point will introduce only smooth changes in the solution in regions away from that point. It is therefore enough to keep refined only a small portion of the domain at a time, while at other regions the coarse level is used with frozen fine-grid corrections. This technique should combine with the usual multi-grid approach of obtaining the first approximation on each finer grid by interpolating from a solution to a similar problem on the next coarser grid.

Such techniques can also serve some *ill-posed problems*, where the solution should fit some data which are not the normal, well-posed boundary conditions. In such situations only smooth components of the solution can be meaningfully fitted to the data (or to smooth averages of data). The data-fitting can therefore be made on a coarse grid, G^H say, but the coarse-grid equations should have the fine-to-coarse correction (2.17), so that they represent a reasonable approximation to the differential equations.

Small-storage multi-grid algorithms are also based on the form (2.16) of the coarse-grid equations. Observe that, once τ_h^H is known, the fine grid is no longer needed. But τ_h^H depends mainly on high-frequency components of u^h, which can be computed by having in storage only a small segment of G^h. An algorithm based on this idea ("segmental refinement", Section 7.5 in Brandt (1977a)) requires in principle only some $15^d \log n$ storage locations, where d is the problem dimension and n is the number of points in the finest grid. No external memory is assumed.

Time-dependent problems, especially of parabolic type, usually require the use of implicit difference equations. The system of equations to be solved at each time step is similar to the steady-state equations, and can be solved usually by one multi-grid cycle, starting from the previous-time solution as the first approximation. Moreover, in some important cases (e.g., the heat equation) after a short time $t_0 = O(h^2)$ the high-frequency components of the solution practically reach their steady state, and further changes occur only in the smooth components. Hence, for many time-steps, the values of τ_h^H hardly change. Freezing τ_h^H for several time-steps we can then solve the coarse-grid equation (2.16) without using the fine grid at all. Only infrequently should we make a time-step with fine grid calculations, to update the high-frequencies of the solution and the values of τ_h^H. In this way *we retain fine-grid accuracy but each implicit time step of this kind costs on the average much less than an explicit time step on the finest grid*. This kind of techniques for parabolic time-dependent problems, are discussed in Lecture 9 of Brandt (1978a), and some preliminary experiments are reported by Dinar (1978).

Parallel or vector processing can be fully exploited by the multi-grid algorithm. The main processes, namely, interpolations and relaxation sweeps, are completely parallelizable, although it requires the use of a slightly different type of relaxation, with smoothing rates (per sweep) somewhat slower than the (sequential) Gauss-Seidel relaxation. (See Section 3.3 in Brandt (1977a).)

Implementation difficulties. A fair amount of knowledge is required in implementing multi-grid algorithms, including some general knowledge common to all multi-grid programs, plus particular expertise related to the specific type of problem at hand.

Generally, one has to be familiar with the basic rules of interpolation and residual-weighting, with the normal flow of multi-grid runs[7)], and, last but not least, with the *data structure* used in multi-grid programs. Without a suitable data structure the program will become complicated, with many unnecessary repetitions. It is advisable to follow the programming techniques exhibited in the simple Sample Program (Appendix B of Brandt (1977a)) and in the programs of MUGTAPE (1978). With this technique, each operation (such as relaxation, coarse-to-fine interpolation, fine-to-coarse residual weighting) is written once for all grids. Moreover, most operations can be written once for all programs; that is, the same code can be used by all the programs which use the same data structure. This includes the coarse-to-fine and fine-to-coarse transfer operations (e.g., interpolation), the operation of introducing the values of a given function into a given grid, operations of generating and manipulating grids (e.g., augmenting a grid, coarsening a grid, generating the interior part of a grid, or transposing a grid from row structure to column structure), displays of grids or grid-functions, etc. Three different data structures are used in existing programs: One for rectangular grids, one for single-string grids (where the domain can be defined as $\{(x,y)|x_1 \leq x \leq x_2, f_1(x) \leq y \leq f_2(x)\}$), and one for general grids. The latter is called the QUAD structure, and is described in Brandt (1977b) and in Lectures 12 and 13 of Brandt (1978a). With these techniques, *the programming of a multi-grid solution for a new problem is essentially reduced to the programming of the relaxation routine.* (The residual-weighting routine should also be programmed anew for each problem, but its part that depends on the problem is a simple modification of a similar part in the relaxation routine.)

Particular expertise is required in designing the relaxation sweeps. For a uniformly elliptic scalar equations the simplest Gauss-Seidel is the best scheme (on sequential machines), but suitable modifications are required for degenerate-elliptic, non-elliptic, indefinite or singular-perturbation problems, as well as for cases of parallel or vector processing, and for problems involving more than one unknown function. Generally speaking, the particular knowledge required for designing relaxation is similar in each case to the specific knowledge required in discretizing the differential equations. Similar - but not identical. As in learning discretization methods, one should learn relaxation methods gradually, starting from simplest models, gaining some basic insights, and only then proceeding to complex real-world problems.

The design of relaxation is much facilitated by a *standard gauge we have for apriori measuring the relaxation efficiency.* The only role of relaxation in multi-grid programs is to smooth the errors. The efficiency of the entire algorithm depends on (and can be approximately predicted by) the "smoothing factor" of the relaxation sweeps. Since smoothing is a local process, the smoothing factor can be calculated by a local mode analysis (cf. Section 6). For simple equations this can be done by hand (see examples in Sections 3.1 and 6.2 of Brandt (1977a), and Section 3 of Brandt (1976).). For general equations it is done by using the computer routine SMORATE, available on MUGTAPE (1978). The user of this routine supplies a description of the relaxation scheme and other parameters (in a format explained by comment cards in the routine). The output contains the smoothing factor and other useful information, including an estimate of the multi-grid convergence factor per work-unit. The routine can therefore be used to *optimize relaxation*, i.e., to select the best relaxation type and parameters from a given set of possibilities.

Some general orientation concerning the relaxation of singular-perturbation problems is given in Section 6 below.

A major advantage of the multi-grid solution process, in particular for singular-perturbation and other irregular problems, is its *full compatibility with adaptive processes*. The reason is that the multi-grid process in itself is adaptive: in adaptive processes mesh sizes are adapted to the computed solution; the multi-grid process goes one step further and employ mesh sizes adapted to the *error* in the computed solution. Let us now turn to these adaptive aspects of the multi-level techniques.

3. NON-UNIFORM AND ADAPTIVE DISCRETIZATIONS

3.1. Non-Uniformity Organized by Uniform Levels

In principle, the multi-grid solution process described in Section 2 could work equally well when the finest grid G^h is non-uniform and even non-rectangular, with grid points at arbitrary locations. A relaxation with good smoothing rates on a general grid is obtained by employing all line and marching directions. The main difficulty with general grids, however, is practical: Merely to formulate and use the difference equations, let alone solve them, is complicated. It requires lengthy calculations and large memories for storing geometrical information, such as the location of each grid point, its neighbors, the coefficients of its difference equation, etc. The multi-grid processing of such arbitrary grids generates additional practical difficulties since it requires the introduction of coarser grids and the grouping of grid points in grid lines (for line relaxation).

These arbitrary general grids, however, are not really needed. We will show below a method of organization which is less general but in which any desired refinement pattern can still be obtained, and easily changed, with negligible book-keeping and with difference equations always defined on equi-distant points. This flexible organization will naturally lend itself to multi-grid processing and to local transformations, and will lay the groundwork for efficient adaptation.

It is proposed to organize non-uniform grids as "*composite grids*". A composite grid is a union of uniform subgrids

$$\ldots, G^{4h}, G^{2h}, G^{h}, G^{h/2}, \ldots \qquad (3.1)$$

where the superscripts denote the mesh-size. The grids are usually positioned so that every other grid-line of G^h is a grid line of G^{2h}. Unlike the description in Section 2, however, the subgrids are not necessarily extended over the entire domain Ω: The domain of G^h may be only a proper part of the domain of G^{2h}, so that different degrees of refinement can be created at different subdomains. See Figure 2. Each G^h is extended, as a rule, over those subdomains where the desired mesh-size is roughly 1.5h or less. G^h may be thus disconnected, but its domain is always a subdomain of G^{2h}. The effective mesh-size at each neighborhood will be that of the finest grid covering that neighborhood. Clearly, any desired mesh-size $\bar{h}$ can be approximated by some effective mesh-size h', where $0.75\bar{h} \leq h' \leq 1.5\bar{h}$. Mesh-sizes never require better approximation.

The composite grid is very flexible, since local grid refinement (or coarsening) is done in terms of extending (or contracting) uniform subgrids, which is relatively easy and inexpensive to implement. A scheme for constructing, extending and contracting uniform grids, together with various service routines for such grids (efficient sweeping aids, interpolations, displays, etc.) is described in Brandt (1977b) and is partly available on MUGTAPE (1978a). One of its advantages is the efficient storage. The amount of logical information (pointers) for describing a uniform grid is proportional to the number of strings of points, and is therefore usually small compared with the number of points on the grid. Similarly, the amount of logical operations for sweeping over a grid is only proportional to the number of strings. Changing a grid is inexpensive, too.

Moreover, this composite structure will at the same time provide a very efficient solution process to its difference equations, by using its levels (3.1) also as the multi-grid sequence (as in Section 2). Each G^h will automatically

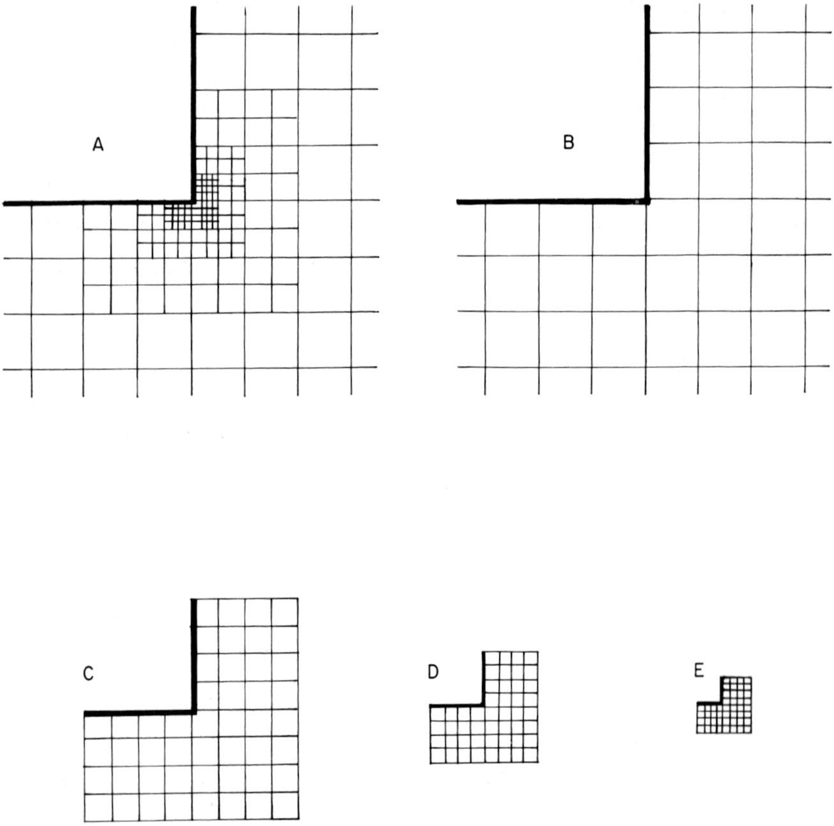

Fig. 2. A piece of non-uniform grid (A) and the uniform subgrids it is made of (B, C, D, E).

play the role of the correcting coarse-grid whenever the finer subgrid $G^{h/2}$ is present (see Section 3.3).

In adaptive procedures, the sequence of subgrids will be kept open-ended, so that we can add increasingly finer or coarser levels, as needed. (Increasingly coarser levels may be needed if the problem's domain Ω is unbounded and the bounded computational domain is chosen adaptively).

The coarsest subgrid should of course be kept coarse enough to have its system of difference equations relatively inexpensive to solve. Hence, there will usually be several coarse subgrids extending over the entire domain Ω. That is, they will not serve to produce different levels of refinement, but they are kept in the system for serving in its multi-grid processing.

There seems to be certain waste in the proposed system, because one function value may be stored several times when its geometrical point belongs to several subgrids G^k. This is not really the case: The extra values are exactly those needed for the multi-grid processing. In the process, the different subgrids may have different values at the same geometrical point. Moreover, it is only a small fraction (2^{-d}) of the points that are actually being repeated.

3.2. Unisotropic Refinements

For singular-perturbation and other problems, it is sometimes desired to have a grid which resolves a certain thin layer, such as a boundary layer. Very fine mesh-sizes are then needed in one direction, namely, across the layer, to resolve its thin width. Even when the required mesh-size is extremely small, not many grid points are needed, since the layer is correspondingly extremely thin. (See Section 4.)

Provided, of course, that the *fine mesh-size is used only in that one direction*. We need therefore a structure for mesh-sizes which get finer in one direction only.

In case the thin layer is along coordinate lines (x_j = constant), we resolve it again by using a sequence of uniform grids, except that their meshes are no longer square; h_i, the mesh-size in the x_i direction, may be very different from h_j. See Figure 3. We still require all mesh-sizes to be binary multiples of some basic size h_0, that is, on the k-th subgrid the mesh-size in the x_i direction is

$$h_i^k = 2^{n_{ik}} h_0, \qquad (n_{ik} \text{ integer}, \quad i = 1,\dots,d) \ . \tag{3.2}$$

For the multi-grid processing we require that for each such subgrid k, except for the very coarsest $(k = 1)$, there exists in the scheme a coarser subgrid, number $\ell = \ell(k)$ say, such that for each $1 \le i \le d$ either $n_{i\ell} = n_{ik}$ or $n_{i\ell} = n_{ik} + 1$, and $\Sigma\, n_{i\ell} > \Sigma\, n_{ik}$. Grid ℓ will be the grid from which corrections are interpolated to grid k, and to which residuals from grid k are transferred. We call ℓ "the *predecessor* of k", and k "a *successor* of ℓ". Each subgrid, except for the coarsest, has exactly one subgrid defined as its predecessor, and may have any number of successors. The domain on which each subgrid is defined is always contained in, or coincide with, the domain of its predecessor. Thus, the set of subgrids is arranged logically in a tree, instead of the linear ordering we had before.

All these subgrids are still uniform, and can still easily be handled (created, extended, displayed, etc.) by the system mentioned above. Except that some of the interpolation routines required for this more general situation

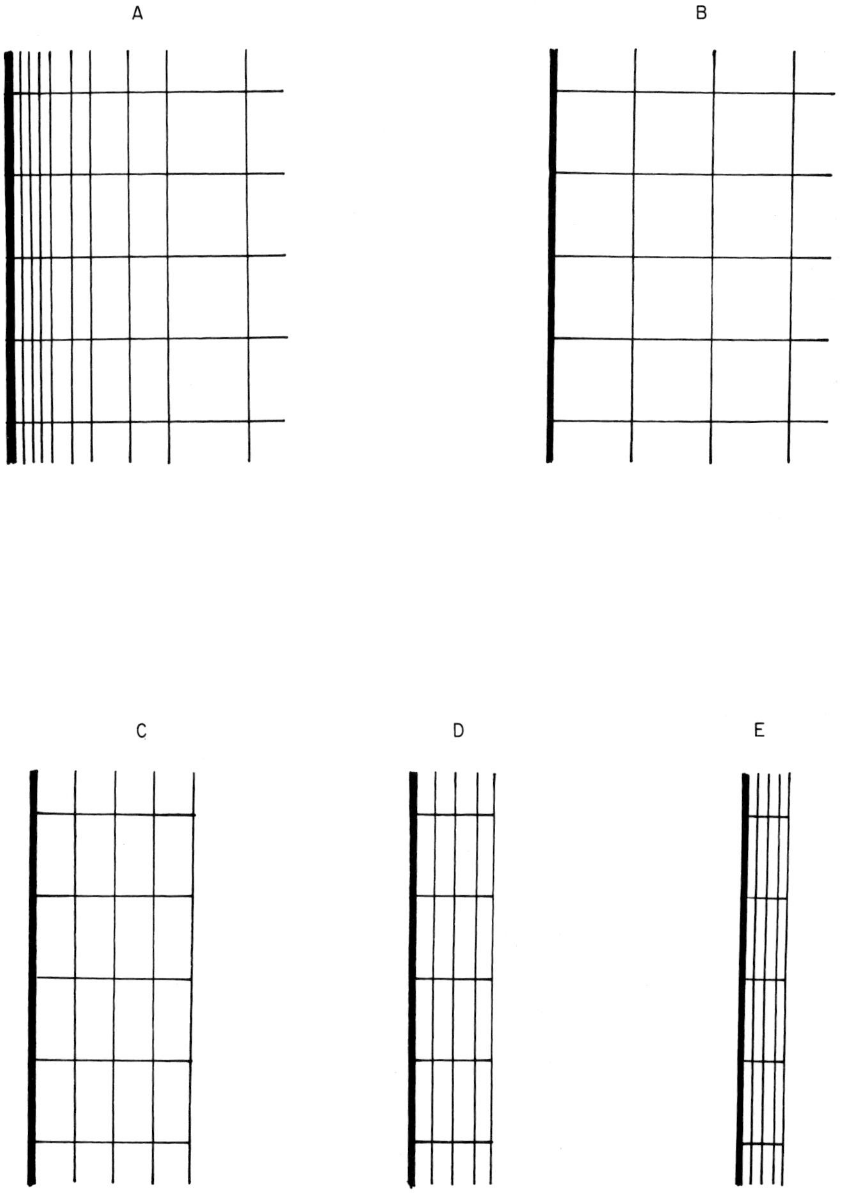

Fig. 3. A piece of non-uniform, boundary-layer type grid (A) and the uniform rectangular subgrids it is made of (B, C, D, E).

are still missing in MUGTAPE (1978a); they have been prepared for MUGTAPE (1978b).

In case the thin layer is not along coordinate lines, some local coordinate transformation is required. This technique is explained in Section 3.5 below.

3.3. Difference Equations and Multi-Grid Processing

The difference equations and their multi-grid solution for the composite structure are explained in Section 2.2 of Brandt (1977b). The main idea is that a fine-to-coarse correction function τ_h^H, correcting the G^H equations, (see (2.16)-(2.17) and the subsequent discussion) can be computed wherever the finer grid G^h exists, or, more precisely, at any interior point of G^H which is also an interior point of its successor G^h. At all other points the original coarse grid equation (2.3) can be used. From this point of view it is clear that we can have various patches of finer subgrids ("successors") thrown over various desired subdomains of G^H; the finer-subgrid accuracy will be established in the equations of such a subdomain via the τ_h^H correction. On any part of such a subdomain a patch of still-finer subgrids can be defined, etc.

The multi-grid process proceeds essentially as before. If, for example, we regard the coarsest subgrid as level 1, and all the successors of level j as level $j + 1$, we could use exactly the same algorithm; e.g., the one shown in Table 1 above. Except that now, each operation (relaxation, fine-to-coarse residual transfer, etc.) at each level is performed not on one subgrid, but on the sequence of all subgrids of that level in some preassigned order. Important improvement: Relaxing each cycle progressively smaller <u>parts</u> of the coarser grids.

Note that successors of a given subgrid G^H may geometrically overlap. All is needed in such a case is to set priority relations between the successors, to tell which correction τ_h^H applies at those G^H points which are interior points of more than one successor. Such priority relations are automatically implied by the ordering of subgrids within each level.

An important advantage of this structure is its flexibility. One can add more and more patches of increasingly finer subgrids, where and when they are needed. One can also discard some such patches. Notice, however, that even when a piece of finer grid, G^h say, is discarded one can still retain its correction τ_h^H in the G^H equations. (This leads to multi-grid procedures which require only a small memory. See Section 2.4 in Brandt (1977b).)

Another important advantage of the outlined structure is that our *difference equations are defined on uniform grids only* (patched together by the usual multi-grid interpolations). Such difference equations on equidistant points are simple and can be read from small standard tables, while on general grids their weights would have to be recomputed (or stored) separately for each point, entailing very lengthy calculations especially for high-order approximations. Thus, our system *facilitates the use of high and variable (adaptive) order of approximation*.

Still another advantage is that relaxation sweeps, too, are done on uniform grids only. This simplifies the sweeping and is particularly essential where symmetric and/or alternating-direction line relaxations are required for obtaining high smoothing rates.

3.4. Remark on High-Order Approximations

An efficient and convenient way of using high-order difference equations, especially when the order is adaptible (see Section 3.6), is by the well-known technique (suggested by L. Fox) of "deferred correction" (see, e.g., Pereyra (1968), Lentini and Pereyra (1975) and Stetter (1978)). Simply, before starting a multi-grid cycle for improving an approximation u^h, add to the right-hand side of the fine-grid equations (2.2) the correction

$$\sigma_p^h(x^h) = L^h u^h(x^h) - L_p^h u^h(x^h), \qquad (x^h \in G^h) , \qquad (3.3)$$

where L_p^h is the higher-order (order p) operator. Then proceed with the multi-grid cycle as usual. The roll of σ_p^h is similar to the roll of the fine-to-coarse correction τ_h^H. We can thus call σ_p^h the *high-order-to-low-order ("deferred", or "defect") correction.*

A certain amount of work is saved if p is advanced in steps; e.g. each multi-grid cycle advance p by 1 or 2. In the adaptive procedures described below, p is always advanced gradually.

Note that, since the multi-grid cycle operates with the original operator L^h, no new routines (such as relaxation routine) should be added, and the same multi-grid efficiency is obtained as in solving low-order equations. Except that the number of multi-grid cycles may increase linearly with p, since $\tau^{2h}/\tau^h \approx 2^p$.

3.5. Local Coordinate Transformations

Another dimension of flexibility and versatility can be added to the above system by allowing each subgrid to be defined in terms of its own set of coordinates.

Near a boundary or an interface, for example, the most effective local discretizations are made in terms of local coordinates in which the boundary (or interface) is a coordinate line. In such coordinates it is easy to formulate high-order approximations near the boundary; or to introduce mesh-sizes which are much smaller across than along the boundary layer (see Section 3.2); etc.

Usually it is easy to define suitable *local* coordinates (see below), and uniformly discretize them, but it is more difficult to patch together all these local transformations, especially in an adaptible way. In the above structure, however, this difficulty does not arise, since we can introduce independent and overlapping patches of "successor" grids.

Each set of coordinates will generally have more than one subgrid defined on it, so that (i) local refinement, in the style of Figure 2 and/or Figure 3 above, can be made within each set of coordinates; and (ii) the multi-grid processing retains its full efficiency by keeping the mesh-size ratio between any subgrid and its predecessor properly bounded (e.g., $\geq \frac{1}{2}$).

Since local refinement can be made within each set of coordinates, the *only* purpose of the coordinate transformation is to provide a certain grid direction, i.e., to have a given manifold (e.g., a piece of boundary) coincide with a grid hyperplane. We can therefore limit ourselves to simple forms of transformations. For example, in 2-dimensional problems, let a curve (a boundary, an interface, etc.) be given in the general parametric form

$$x = x_0(s), \quad y = y_0(s), \qquad (0 \leq s \leq s_1) \tag{3.4}$$

where s is the arclength, i.e.,

$$x_0'(s)^2 + y_0'(s)^2 = 1 \ . \tag{3.5}$$

To get a coordinate system (r,s) in which such a curve will be a grid line, we can always use the transformation

$$x(r,x) = x_0(s) - ry_0'(s), \quad y(r,s) = y_0(s) + rx_0'(s) \ . \tag{3.6}$$

Near the given curve $(r = 0)$ this transformation (a special case of transformations discussed in Starius (1977a)) is orthogonal, owing to (3.5), and transforms any small $h \times h$ square to another $h \times h$ square.

The main advantage of this transformation is that it is fully characterized by the single-variable functions $x_0(s)$, $y_0(s)$. These functions (together with $x_0'(s)$, $y_0'(s)$ and $q(s) = x_0''/y_0' = -y_0''/x_0'$) can be stored as one-dimensional arrays, in terms of which efficient interpolation routines from (x,y) grids to (r,s) grids, and vice versa, can be programmed once for all. (Such a general routine, however, is not easy to program, and is still missing in MUGTAPE (1978).) The difference equations in (r,s) coordinates are also simple enough in terms of these arrays. For example, by (3.5-6),

$$\frac{\partial}{\partial x} = -y_0' \frac{\partial}{\partial r} + \frac{x_0'}{1+rq} \frac{\partial}{\partial s}, \quad \frac{\partial}{\partial y} = x_0' \frac{\partial}{\partial r} + \frac{y_0'}{1+rq} \frac{\partial}{\partial s} \ . \tag{3.7}$$

Hence we can easily approximate the (x,y) derivatives by (r,s) finite-differences, with numerical values of $x_0'(s)$, $y_0'(s)$ and $q(s)$ directly read from their stored tables. (No interpolation is needed if the tables contain values for s points which correspond to grid lines and half-way between grid lines.)

Such a system offers much flexibility. Precise treatment of boundaries and interfaces by the global coordinates is not required, since along boundaries the global grids are only correction grids to the local ones. The local coordinates are easily changeable (changing only the one-dimensional tables of x_0, y_0, x_0', y_0', q) and can therefore be adapted to a moving interface.

The main difference between this structure and the one used by Starius (1977a), (1977b) and (1978) is that the boundary grids are completely embedded in the global grids (their predecessors), allowing a fast multi-grid solution of the equations. Also, since we have the multi-grid method for local refinement, the coordinate transformation is used only for orienting the grid, hence only the simpler transformation (3.6) is needed, allowing simpler differencing and interpolations.

Another variant of this procedure is required in case the location of the curve (interface, shock, etc.) is not fully defined. For example, a solution may include many shocks, some weaker and some stronger, and it is hopeless to try to recognize where a shock occurs, let along determine its exact curve. The usual procedure is to let the shocks develop by themselves, e.g., by adding some artificial viscosity which spread shocks over several mesh-sizes. Sometimes, however, this procedure is unacceptable because too much artificial viscosity is used near strong shock (and because of other reasons). We like to have a procedure which will automatically use smaller mesh-sizes near stronger shocks. This will be done by the general adaptive procedure (Section 3.6 below) if we choose the error functional E so that it contains some measure of the artificial viscosity. In order to obtain full efficiency, however, we like the procedure to be able to produce mesh-sizes which are much smaller in one direction (the direction perpendicular to the shock) than in the other. We therefore need a structure for adapting the grid *orientation*, too. Notice that on some coarse grids the orientation is immaterial; the finer the grid the more precisely its orientation should be chosen. Hence, the direction can be refined successively, together with the mesh-sizes. An example is drawn and explained in Figure 4. We see that in this method the more general transformation (3.6) is not needed; only rotations are used. Hence the difference equations may be as simple as in the original (e.g., cartesian) coordinates. This method may therefore be preferable even in cases the curve (e.g., boundary) is known.

3.6. Adaptation Techniques

The flexible organization and solution process, described above, facilitate the implementation of variable mesh-size $h(x)$ and the employment of high and variable approximation order $p(x)$. How, then, are mesh-sizes and approximation-orders to be chosen? Should boundary layers, for example, be resolved by the grid? What is their proper resolution? Should high-order approximation be used at such layers? How does one detect such layers automatically? In this section we survey (for more details, see Brandt (1977a), Chapters 8 and 9) a general framework for automatic selection of $h(x)$, $p(x)$ and other discretization parameters in a (nearly) optimal way. This system automatically resolves or avoids from resolving thin layers, depending on the goal of the computations, which can be stated through a simple function.

As our directive for sensible discretization we consider the problem of minimizing a certain error estimator E subject to a given amount of solution work W (or minimizing W for a given E. Actually, the control quantity will

Figure 4A

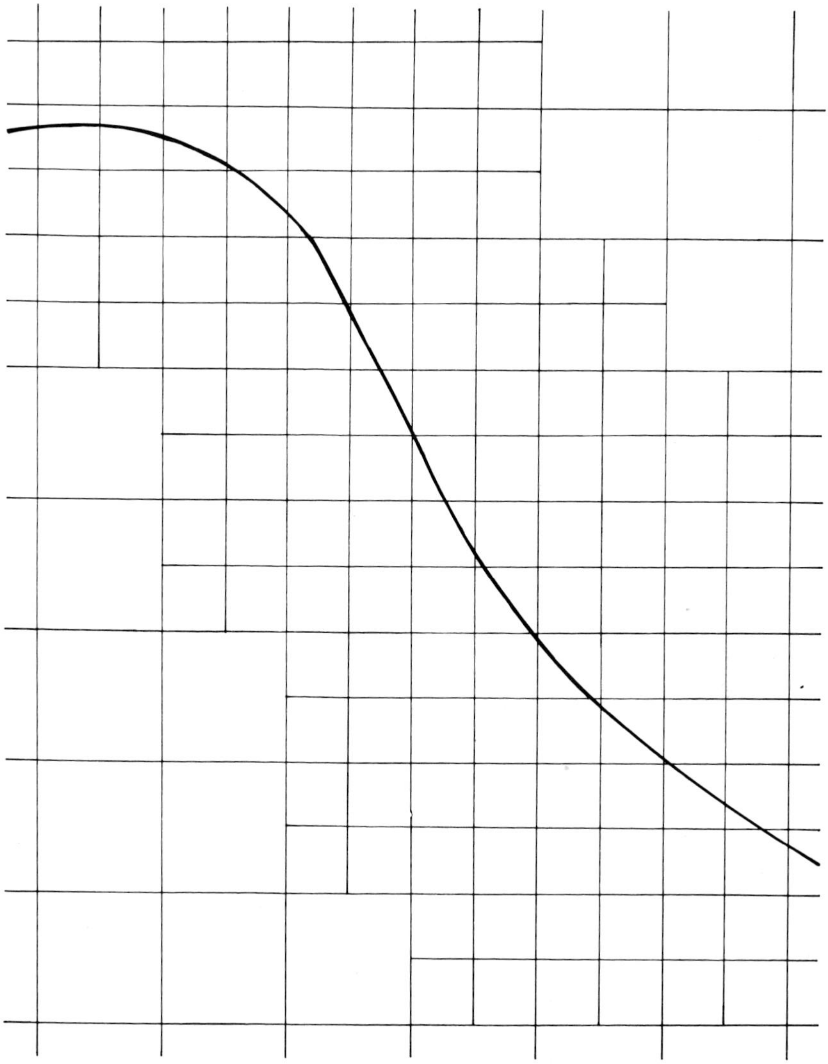

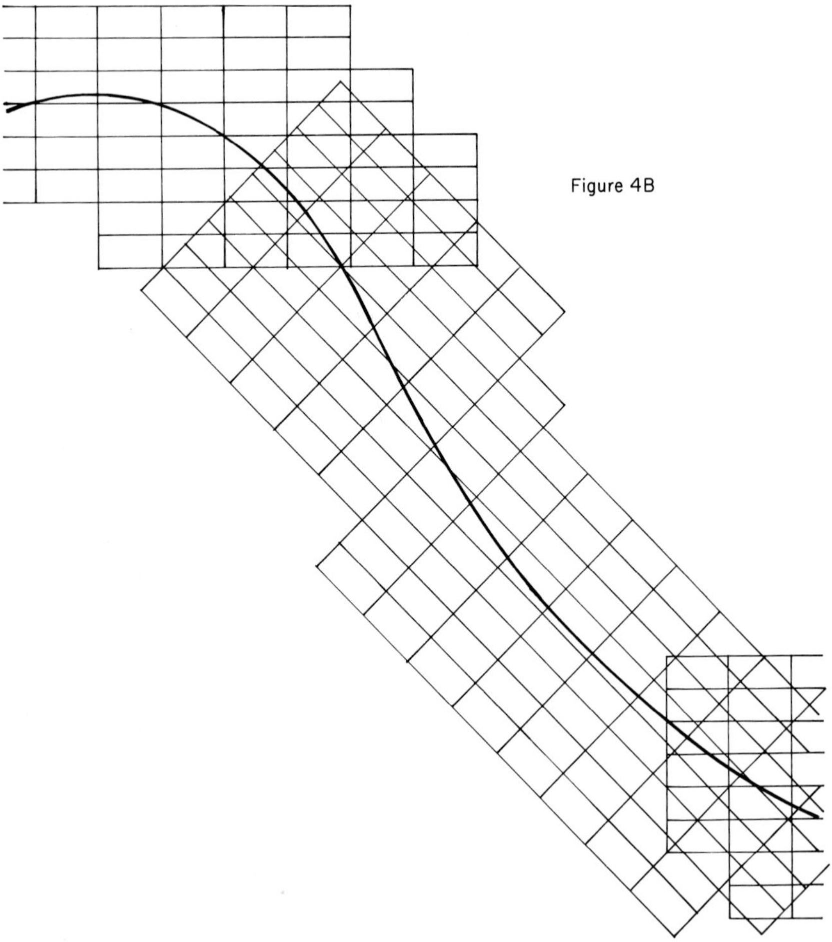

Figs. 4A and 4B. Grid orientation around an interior thin layer. The two coarsest levels (A) have the usual orientation 0. The next level (B) has 3 orientations: 0, $\frac{\pi}{4}$ and $-\frac{\pi}{4}$ (the later is not applied here). The next level (not shown) would have 7 orientations: 0, $\pm\frac{\pi}{8}$, $\pm\frac{\pi}{4}$, $\pm\frac{3\pi}{4}$; etc. The successors (refinements) of a grid will always have either the same orientation or one of the two closest ones (e.g., each successor of the $\frac{\pi}{4}$-oriented grid in B will have orientation $\frac{\pi}{4}$, $\frac{\pi}{8}$ or $\frac{3\pi}{8}$).

be neither E nor W, but their rate of exchange). This optimization problem should of course be taken quite loosely, since full optimization would require too much control work and would thus defeat its own purpose.

The error estimator E has generally the form

$$E = \int_{\Omega} G(x)\tau(x)dx , \tag{3.8}$$

where $\tau(x) = \tau(x,h,p)$ is the magnitude of the local truncation error at x (see (2.18)). $G(x) \geq 0$ is the error-weighting function. It should in principle be imposed by the user, thus defining his goal in solving the problem. In practice $G(x)$ serves as a convenient control. It is only the relative orders of magnitude of $G(x)$ at different points x that really matter, and therefore it can be chosen by some simple rules. For example, if it is desired to compute ℓ-order derivatives of the solution up to the boundary then

$$G(x) \approx d_x^{m-1-\ell} , \tag{3.9}$$

where d_x is the distance of x from the boundary, and m is the order of the differential equation.

The work functional W is roughly given by

$$W = \int_{\Omega} \frac{w(p(x))}{h(x)^d} dx , \tag{3.10}$$

where d is the dimension and h^{-d} is therefore the number of grid points per unit volume. $w = w(p)$ is the solution work per grid-point. In multi-grid processing, for $p \leq 6$ the work depends mainly on the number of cycles (cf. Section 3.4), hence $w \approx w_0 p$. For (unusually) large p, $w = O(p^3)$ since evaluating L_p^h at each cycle involves $O(p)$ terms and $O(p)$ arithmetic precision.

Treating $h(x)$ and $p(x)$ as continuous variables, the Euler equations of minimizing (3.8) and (3.10) can be written as

$$G \frac{\partial \tau}{\partial h} - \frac{\lambda d w(p)}{h^{d+1}} = 0 , \qquad (3.11a)$$

$$G \frac{\partial \tau}{\partial p} + \frac{\lambda w'(p)}{h^{d}} = 0 , \qquad (3.11b)$$

where λ is a constant (the Lagrange multiplier), representing the marginal rate of exchanging optimal accuracy for work: $\lambda = -dE/dW$. The = sign in (3.11b) should be replaced by $\geq$ at points x where p attains its minimal allowable value p_0 (usually 1 or 2). In case we use fixed-order difference equations (adapting $h(x)$ only, p is fixed in advance), equation (3.11b) should be omitted. If constant order is used (p is constant over the domain, but instead of being fixed in advance this constant is to be optimized) equation (3.11b) is replaced by

$$\frac{\partial E}{\partial p} + \lambda \frac{\partial W}{\partial p} = 0 . \qquad (3.12)$$

In principle, once λ is specified, equations (3.11) determine, for each $x \in \Omega$, the local optimal values of $h(x)$ and $p(x)$, provided the truncation function $\tau(x,h(x),p(x))$ is fully known. In some problems the general behavior of $\tau(x,h,p)$ near singularities or in singular layers is known in advance by some asymptotic analysis, so that approximate formulae for $h(x)$ and $p(x)$ can apriori be derived from (3.11). More generally, however, equations (3.11) are coupled with, and should therefore be solved together with, the given differential equations. Except that (3.11) is solved to a cruder approximation. This is done in the following way:

In the multi-grid solution process (possibly incorporating a continuation process), incidentially to the stage of computing f^{2h} from u^{2h}, u^h and f^h (see Figure 1 above, $2h$ corresponding to level k and h to $k+1$) we can get an estimate of the decrease in the error estimator E introduced by the present discretization parameters. For example, the quantity

$$\begin{aligned} -\Delta E(x) &= G(x)\,|\tau(x,2h,p) - \tau(x,h,p)| \\ &\approx G(x)\,|\tau_h^{2h}| \\ &= G(x)\,|f^{2h}(x) - F^{2h}(x)| \end{aligned} \tag{3.13}$$

(cf. (2.17)-(2.19) above) may serve as a local estimate for the decrease in E per unit volume owing to the refinement from $2h$ to h (cf. (3.8)). Each such decrease in E is related to some additional work W (per unit volume). For example, that refinement from $2h$ to h required the additional work (per unit volume; cf. (3.10))

$$\Delta W = \frac{w(p)}{h^d} - \frac{w(p)}{(2h)^d} \,. \tag{3.14}$$

We say that the present parameter (h in the example) is highly profitable if the local rate of exchanging accuracy for work $Q = -\Delta E/\Delta W$ is much larger than the control parameter λ.

More sophisticated tests may be based on assuming τ to have some form of dependence on h and p. Instead of calculating Q for the previous change (from $2h$ to h in the example) we can then extrapolate and estimate the rate $\bar{Q}$ for the next change (from h to $h/2$), which is the more appropriate rate in testing whether to make that next change. The "extrapolated test" is not, however, much different in practice, and may actually be equivalent to testing the former Q against another constant λ.

In deciding whether and where to change the discretization, we adopt rules which stablize the adaptive process. For example, a change (e.g., refinement from h to $h/2$, which in practice (see Sec. 2.2) means an extension of the uniform grid with mesh-size $h/2$) is introduced only if there is a point where the change is "overdue" (e.g., a point where $\bar{Q} > 15\lambda$). But, together with each point where the change is introduced, it is also introduced at all neighboring points where the change is just "due" (e.g., where $\bar{Q} > 3\lambda$).

We can use the Q vs λ test to decide on all kinds of other possible changes, such as: Changing the order p to $p + 1$ (or $p - 1$); or changing the computational boundaries (when the physical domain is unbounded); or we can use such a test to decide whether to discard some terms from the difference operator (such as the highest order terms in some regions of a singular-perturbation problem); or to decide on unisotropic changes in h and p (e.g., changing Δx to $\Delta x/2$ without changing Δy); etc.

In case of optimizing a global quantity (such as constant approximation order; cf. (3.12)) similar tests can still be applied, except that ΔE and ΔW should of course be measured globally (summing over the entire region) instead of locally.

The computer work invested in the tests is negligible compared with the solution work itself. The measure (3.13) is taken only on $G^{2h} \cap G^h$, and the stage of computing it occur only once per several relaxation sweeps on G^h.

4. MULTI-LEVEL ADAPTIVE SOLUTIONS TO SINGULAR-PERTURBATION PROBLEMS: CASE STUDIES

To get a transparent view of the discretization patterns and the accuracy-work relations typical to the adaptive procedures proposed above when applied to singular perturbation problems, we consider now several cases which are simple enough to be fully analyzed. The simplicity of the solution, it should be emphasized, is not used in the solution process itself.

4.1. Optimal Discretization of One-Dimensional Case

Consider the 2-point boundary-value problem

$$\varepsilon \frac{d^2U}{dx^2} + \frac{dU}{dx} = 0 \quad \text{in} \quad 0 < x < 1 , \tag{4.1}$$

with constant $\varepsilon > 0$ and with boundary conditions $U(0)$ and $U(1)$ such that the solution is $U = e^{-x/\varepsilon}$. An elliptic (stable) difference approximation to such an equation can be central for $\varepsilon \geq 2h$ but should be properly directed for small ε. (See Section 5 below.) In either case, the truncation error behaves like

$$\tau(x,h,p) = \left(\frac{h}{\gamma\varepsilon}\right)^p \frac{e^{-x/\varepsilon}}{\varepsilon} , \tag{4.2}$$

where γ is a constant close to 2. (Actually, γ slightly depends on p; see (5.14). For simplicity, we neglect this dependence.) For the error weighting function we choose

$$G(x) \equiv 1 , \tag{4.3}$$

which would be the choice (see (3.9)) when one is interested in computing boundary first-order derivatives (corresponding, e.g., to boundary pressure or drag, in some physical models). Then, assuming $w(p) = w_0 p$, the optimization equation (3.11a) yields

$$\tau = \frac{\lambda w_0}{h} \quad , \tag{4.4}$$

and (3.11b) therefore becomes

$$h = \frac{\gamma}{e}\,\varepsilon \quad \text{if} \quad p > p_0; \quad h \geq \frac{\gamma}{e}\,\varepsilon \quad \text{if} \quad p = p_0 \ . \tag{4.5}$$

From (4.2), (4.4) and (4.5) it follows that

$$h = \frac{\gamma}{e}\,\varepsilon, \quad p = \log \frac{\gamma}{\lambda w_0} - 1 - \frac{x}{\varepsilon} \quad \text{for} \quad 0 < x \leq x_0 \ , \tag{4.6}$$

$$h = \frac{\gamma}{e}\,\varepsilon\, e^{(x-x_0)/(\varepsilon p_0+\varepsilon)} \ , \ p = p_0 \quad \text{for} \quad x_0 \leq x < 1 \ , \tag{4.7}$$

where

$$x_0 = \varepsilon(\log \frac{\gamma}{\lambda w_0} - p_0 - 1) \ . \tag{4.8}$$

If $x_0 \geq 1$ then (4.6) applies throughout, hence

$$W = \int_0^1 \frac{w_0 p}{h}\,dx = \frac{w_0 e}{\gamma\varepsilon}\,(\log \frac{\gamma}{\lambda w_0} - 1 - \frac{1}{2\varepsilon}) \ , \tag{4.9}$$

$$E = \int_0^1 \tau dx = \frac{\lambda w_0 e}{\gamma\varepsilon} = \frac{1}{\varepsilon}\exp(-\frac{\gamma\varepsilon}{w_0 e}\,W - \frac{1}{2\varepsilon}) \ , \tag{4.10}$$

and the condition $x_0 \geq 1$ itself becomes, by (4.8)-(4.9),

$$W \geq (p_0 + \frac{1}{2\varepsilon})\,\frac{w_0 e}{\gamma\varepsilon} \ . \tag{4.11}$$

Thus, if W satisfies (4.11), E converges like (4.10). That is, *for large values of* ε, *the total error* E *decreases exponentially as a function of the overall work* W.

Notice that when ε is large no boundary-layer is formed, and the mesh is uniform. Note also that the optimal mesh-size $h = \gamma\varepsilon/e$ is independent of the work W (or the exchange rate λ). That is, when more work can be afforded, it should *not* be invested in refining the grid, but in increasing the approximation-order p. In the MLAT processes described above p will automatically increase by 2 (or by 1, if non-central approximations are used) every multi-grid cycle, until the desired accuracy (or the work limit, or the prescribed exchange rate λ) is reached.

4.2. Boundary-Layer Resolution

For small ε, however, the mesh-size $h \approx \varepsilon$ is impractical. Indeed, in the optimal discretization (4.6)-(4.7), for small ε we get small x_0, and an "external region" $x_0 \le x < 1$ is formed where the mesh-size grows exponentially. *The small mesh-size is used only to resolve the boundary layer.* In this simplified problem the solution away from the boundary layer (i.e., for $x \gg \varepsilon$) is practically constant, so that indefinitely large h is suitable. Usually h will grow exponentially, as in (4.7), from $h = \gamma\varepsilon/e$ to some definite value suitable for the external region (the optimal mesh-size of the reduced problem). In the transition region we have $p = p_0$, i.e., the minimal order of differencing is used in the region where h changes. This may be useful in practical implementations.

From (4.6)-(4.8) and (4.2)-(4.4) we get, for small ε,

$$W = \int_0^1 \frac{w_0 p}{h} = \frac{w_0 e}{2\gamma}\left[\left(\log\frac{\gamma}{\lambda w_0} - 1\right)^2 + 2p_0 + p_0^2\right], \tag{4.12}$$

$$E = \int_0^1 \tau dx = \frac{\lambda w_0 e}{\gamma}\log\frac{\gamma}{\lambda w_0}, \tag{4.13}$$

where $\exp(-1/\varepsilon)$ and similar terms are neglected. Using (4.12) we can express λ in terms of W and substitute that expression is (4.13). In reasonable calculations $W \gg w_0$, and then the relation simplifies to

$$E \approx \left(\frac{2\gamma e}{w_0} W\right)^{1/2} \exp\left[-\left(\frac{2\gamma}{w_0 e} W\right)^{1/2}\right]. \tag{4.14}$$

Thus, essentially -

For small values of ε, *the total error* E *decreases exponentially as a function of* $W^{1/2}$, *where this rate is independent of* ε *and does not deteriorate as* $\varepsilon \to 0$. In principle, this rate is better than $O(h^p)$, for any fixed p.

Notice that here h does depend on W (or λ), but only in some transitional layer. In the inner part of the boundary layer $h = \gamma\varepsilon/e$ still holds, while away from that layer h tends to the optimal mesh-size of the reduced problem. (If the reduced problem is itself regular, its optimal mesh-size will be determined by the "local scale" of that problem. This scale is independent of W, as it is for example in Section 4.1 above for the case of moderate ε. That scale is too small to resolve only when the reduced problem is singular). *What depends on the total computational work is the distance* x_0 *from the wall at which the meshsize starts to grow exponentially.* In fact, from (4.8) and (4.12) we see that

$$x_0 \approx \varepsilon\left(\frac{2\gamma}{w_0 e} W\right)^{1/2} . \tag{4.15}$$

Defining *the computational boundary layer* as the region where $h < h_0$ for some h_0 independent of ε, the width w_{CBL} of the layer is, by (4.7),

$$w_{CBL} \approx x_0 + (p_0 + 1)\varepsilon \log \frac{1}{\varepsilon} . \tag{4.16}$$

Another, quite obvious but interesting observation can be made at this junction, based on (4.11) above. Even for small ε, if W is sufficiently large then the exponential relation (4.10) holds. Hence, in an asymptotic theory for $W \to \infty$ (corresponding to asymptotic theory for $h \to 0$, which is so common in numerical analysis) the relations of this section, which undoubtedly dominate the numerical process at reasonable values of W, would not be seen. What we should be interested in are *values of* W *which are large but independent of* ε, and therefore not large compared with negative powers of ε.

4.3. Fixed-Order and Constant-Order Discretizations

The optimal approximation-order p calculated above varies with the location x. This is not essential. Indeed, if p is fixed then (3.11b) is omitted, but (3.11a) and (4.2)-(4.3) still imply $\tau = \lambda w(p)/(hp)$, and hence

$$h = \gamma \left(\frac{\lambda w}{\gamma p}\right)^{\frac{1}{p+1}} \varepsilon e^{x/(\varepsilon p+\varepsilon)} \quad . \tag{4.17}$$

Hence, for small ε,

$$W = w \int_0^1 \frac{dx}{h} = \frac{w(p+1)}{\gamma} \left(\frac{\gamma p}{\lambda w}\right)^{\frac{1}{p+1}} , \tag{4.18}$$

$$E = \frac{\gamma w}{p} \int_0^1 \frac{dx}{h} = \frac{\lambda}{p} W = (p+1)\left(\frac{\gamma W}{(p+1)w}\right)^{-p} . \tag{4.19}$$

Thus, $E = CW^{-p}$ with C independent of ε, so that the convergence order is p (analogous to error $E = O(h^p)$ when h is constant). *The variable mesh-size* (4.17) *keeps the convergence rate essentially unimpaired by the singular perturbation, even though convergence is considered of the first derivative up to the boundary.*

The constant approximation-order p need not of course be fixed in advance. It may be optimized just as well, using global tests as mentioned above. From (4.19) it follows that the minimal E for a given W is obtained when p satisfies

$$1 + \frac{pw'(p)}{w(p)} = \log \frac{\gamma W}{(p+1)w(p)} \quad , \tag{4.20}$$

and for $w = w_0 p^{\ell}$ the total error will be

$$E \approx (\gamma W)^{1/(\ell+1)} \exp\left[- \frac{1+\ell}{e} \left(\frac{\gamma W}{w_0}\right)^{1/(\ell+1)} \right] . \tag{4.21}$$

For $\ell = 1$ this rate is almost the same as (4.14). Thus, *we do not lose much by using constant, optimized* p, which, on the other hand, may be considerably simpler to program.

From (4.17)-(4.18) and (4.20) we see that the width of the computational boundary layer is now

$$w_{CBL} \approx \varepsilon(p+1) \log \frac{1}{\varepsilon} \quad . \tag{4.22}$$

For small ε this is $(p+1)/(p_0+1)$ times wider than the variable-order case (4.16).

4.4. A Case of Skipping the Boundary-Layer

To see the effect of choosing different error-weighting functions, consider again problem (4.1), but with the choice

$$G(x) \equiv x \quad . \tag{4.23}$$

This will be the choice *in case one is interested in approximating* U *only, not its derivative, and to approximate it in the* L_1 *sense.* By substituting (4.2) and (4.23) into (3.11), and solving for p, we would obtain

$$p = \log \frac{x\gamma}{\lambda w_0} - 1 - \frac{x}{\varepsilon} \tag{4.24}$$

$$\leq \log \frac{\varepsilon\gamma}{\lambda w_0} - 2 \quad .$$

For bounded (independent of ε) λ and sufficiently small ε, this p is smaller than p_0. Hence $p = p_0$ should replace (3.11b) and substituting (4.2) into (3.11a) we actually get

$$\left(\frac{h}{\gamma\varepsilon}\right)^{p_0+1} = \frac{\lambda w_0 e^{x/\varepsilon}}{x\gamma} \geq \frac{\lambda w_0 e}{\varepsilon\gamma} . \qquad (4.25)$$

Thus, *for bounded* λ *and sufficiently small* ε, $h >> \varepsilon$ *everywhere, so that the boundary-layer is not resolved by the grid.*

In fact, since $h >> \varepsilon$, all interior grid points lie in a region where the rate of convergence would normally be determined by the reduced equation (see Section 4.5). In our simple example (4.1), the reduced equation has the trivial solution $U \equiv 0$, and accordingly $E \to 0$ as $\varepsilon \to 0$, for any fixed W (or λ). This can be verified from (3.8), (4.2) and (4.25).

In case one is interested in *approximating* U *in the* L_∞ *sense*, a precise choice of the error-weighting function is

$$G(x) = 1 - e^{-x/\varepsilon} . \qquad (4.26)$$

With this function, solving (3.11) for p we would get

$$p(x) = \log \frac{\gamma(1 - e^{-x/\varepsilon})}{\lambda w_0} - 1 - \frac{x}{\varepsilon} , \qquad (4.27)$$

$$\max p(x) = p(\varepsilon \log 2) = \log \frac{\gamma}{4\lambda w_0} - 1 ,$$

and hence, for λ reasonably small, $p(\varepsilon \log 2) > p_0$. Therefore, around $x = \varepsilon \log 2$, we again have $h = \frac{\gamma}{e}\varepsilon$. Beyond this point (for $x > \varepsilon \log 2$) the discretization pattern is essentially the same as in Section 4.2 above (since $G(x)$ is essentially the same). Before this point $(x \leq \varepsilon \log 2)$ we have $h(x) > x$, so that in practice we do not have there more grid points. Thus, the grid throughout is essentially as in Section 4.2. Similarly, for constant p the mesh-size distribution will be as in Section 4.3. The accuracy-work relation, too, is essentially as before.

4.5. *Remarks on More General Problems*

For general problems it is of course impossible to find *apriori* the relation between work and accuracy that would result from multi-level adaptive solution processes. In fact, in most non-trivial cases, an optimal (or nearly optimal) choice of discretization parameters ($h(x)$, p, etc.) is not known in advance, since it depends on the particular solution. This is exactly why adaptive techniques are needed. Nevertheless, in this section we will try to indicate that the simple relations described in Section 4.2-4.4 are typical to many, perhaps most, singular-perturbation problems, even in complicated, high-dimensional problems.

Consider first a more general one-dimensional, constant-coefficient equation of the form

$$\varepsilon^{m-n}U^{(m)} + a_{m-1}\varepsilon^{m-n-1}U^{(m-1)} + \cdots + a_n U^{(n)} + a_{n-1}U^{(n-1)} + \cdots + U^{(0)} = 0 , \qquad (4.28)$$

normalized so that

$$U(x) = e^{-x/\varepsilon} \qquad (4.29)$$

is a solution. And assume the boundary conditions are such that (4.29) is actually *the* solution. The truncation error is then approximately (see (5.14))

$$\tau(x,h,p) = \left(\frac{h}{\gamma\varepsilon}\right)^p \frac{e^{-x/\varepsilon}}{\varepsilon^n} , \qquad (4.30)$$

where γ is again a constant (slightly different than in (4.2), but still $\gamma \to 2$ for larger p. We again neglect the changes in γ).

If we are interested in computing $U^{(j)}(x)$ near $x = 0$, the error-weighting function for small x should behave like

$$G(x) \approx \varepsilon^{j-m+n} x^{m-1-j} . \tag{4.31}$$

For the adaptive process, the multiplicative constant in G is immaterial. For our convergence estimates, however, the correct order of ε should be used. The behavior (4.31) results from the observation that if $\varepsilon^{m-n} V^{(m)}(x) = \delta_\xi(x)$ in $(0,1)$ and $V(0) = V(1) = 0$, then, for $0 < x < \xi << 1$, $u^{(j)}(x) = O(\varepsilon^{-m+n} \xi^{m-1-j})$. An additional ε^j factor appears in (4.31) since we are interested in measuring *relative* errors in $u^{(j)}(x)$, and by (4.29), $u^{(j)}(x) = O(\varepsilon^{-j})$ for $x \leq O(\varepsilon)$.

For $j = m - 1$, $G\tau$ is the same as in the special case discussed in Sections 4.2 and 4.3 above. Therefore exactly the same discretization parameters and the same accuracy-work relations will follow. For smaller j, the accuracy-work relation cannot get worse; it may even improve, depending on the norm used (cf. Section 4.4).

In more general singular-perturbation problems, the solution $U(x)$ can be written as a sum of a function $\tilde{U}(x)$ which tends uniformly to the solution U_0 of the reduced problem, and boundary-layer terms, each of which behaves like (4.29) above for some suitable ε. (See O'Malley (1974).) Consider the case of a fixed order p, as in Section 4.3 above. Let $h_0(x)$ be the mesh-size distribution optimal (at some given λ) for the reduced problem, and $h_i(x)$ the optimal distribution in case the solution contain only the i-th boundary term $(i = 1,2,\ldots,\sigma)$. Let E_i and W_i denote the corresponding total error and overall work $(i = 0,1,2,\ldots,\sigma)$. For any given solution (containing all terms) choose

$$h(x) = \min_{0\le i\le\sigma} h_i(x) \ . \qquad (4.32)$$

Then clearly $E \le \Sigma E_i$ and $W \le \Sigma W_i$. In an optimal choice of h, E will be even smaller (for the same value of W; or vice versa). Hence, essentially,

The convergence rate behaves either like one of those described above for the boundary terms (Section 4.3), *or like the convergence rate of the reduced problem, whichever is slower.*

The situation is a little more complicated for variable p, but we saw before that we don't loose much by using a constant p. Moreover, the optimal p (4.20) does not depend on ε and can therefore serve uniformly for all the boundary-layer terms.

Similar convergence rates should be expected in *higher-dimensional problems*, too. To see this, examine the behavior near some portion of the boundary. Assuming our computations use boundary coordinates (see Remark below), we can regard the boundary as $x_1 = 0$. Using Fourier transformation in all but the x_1 coordinate, the solution u can again be written, in many cases, as a sum of $\tilde{u}$ (tending asymptotically to the reduced solution U_0) and boundary-layer terms behaving like (4.29). Assuming also that the only significant adaptation of mesh-size is needed in the x direction (i.e., perpendicular to the boundary), we may repeat the above argument, using (4.32), and arrive at the same conclusion.

Remark about boundary coordinates: "Boundary Coordinates" is a coordinate system in which the boundary is contained, at least locally, in a coordinate hyperplane (e.g., $\{x_1 = 0\}$). In Section 3.5 above it is explained how to construct and use such a coordinate system in multi-grid processes. For the finite-difference equations it is important to use a grid along such boundary coordinates. Otherwise it is impossible to simultaneously use small mesh-sizes in the direction perpendicular to the boundary and large ones in the other direction(s), as required for obtaining efficiencies similar to the one-dimensional ones.

Thin transition layers not on the boundary, such as turning points in ordinary differential equations or contact-discontinuities and shocks in higher dimensions, are likely to be treated by multi-level adaptive techniques as efficiently as the boundary-layer cases analyzed above, since the procedures did not assume any apriori knowledge concerning the location of the layer. The layer is discovered, and if necessary resolved, by the numerical process, using general and automatic criteria. The only difficulty is, in higher dimensional problems, to get a coordinate system in which the internal layer is a coordinate hyperplane. To a suitable approximation, however, this can be done, using the procedure described in Figure 4 above.

Not all singular-perturbation problems can efficiently be solved by the above techniques, of course. For example: problems with highly oscillatory solutions, such as the Helmholtz equation

$$\varepsilon^2 \Delta u + u = 0 \, . \tag{4.33}$$

In usual norms, this problem is *not uniformly well-posed.*

That is, the change in the solution caused by a certain change in the data is not uniformly bounded: it may increase indefinitely as $\varepsilon \to 0$. Such problems should be reformulated, using other variables and norms, so as to make them uniformly well-posed. (See Section 5 in Brandt (1978b).)

5. UNIFORMLY WELL-POSED HIGH-ORDER DIFFERENCE EQUATIONS

An extended version of this section appears as Section 5 in Brandt (1978b). It discusses the concepts of well-posedness and uniform well-posedness, ellipticity and uniform ellipticity, and their significance for singular-perturbation problems in general, and for their multi-level solutions in particular. Closely related are the extensive theoretical investigations of Frank (1978 and references therein). Related preliminary observations were made in Brandt (1976).

Here we summarize some of the more practical aspects.

5.1. General Remarks

In approximating potentially singular-perturbation equations it is essential to ensure that the discrete problem is uniformly well-posed (uniformly stable) not only with respect to the mesh-size (h), but also with respect to the singular-perturbation size (ε). That is, in suitable norms, a small change in data should cause a small change in the solution, *uniformly in both* h *and* ε. For this to be possible, the original differential problem should be uniformly stable (in ε). This, however, is not sufficient. Innocent-looking difference approximations L^h may easily be uniformly stable in h (i.e., for any fixed ε), but not jointly in h and ε. In such cases satisfactory approximations will still be obtained by sufficiently small h, but that h will have to be small *compared with* ε (or some power of ε), and hence too small to be practical. In particular such mesh-sizes are unacceptable (even for moderate ε) for the coarse grids of a multi-level structure.

There are no general procedures to construct uniformly stable difference approximation; nor even general procedures to check uniform stability of given difference schemes. This is in fact already true for the differential equations. But there are some important classes of uniformly stable operators and some practical ways of constructions.

For various boundary-value problems to be well posed it is required that the partial-differential operator (2.1a) is *elliptic*, i.e., that the homogeneous system of equations has no non-constant periodic solution. This, together with appropriate boundary conditions, ensures well-posedness. Similarly, the difference operator (2.2) can be defined as elliptic if there is no periodic U^h such that $L^h U^h = 0$. For scalar operators $(q = 1)$ such a definition was introduced by Thomée (1964), and various results related to the stability of such operators were proved by him and by Thomée and Westergren (1968). Many more results were published in conjunction with finite-element formulations, which yield scalar or vectorial elliptic difference operators. (See Ciartet (1978).). Most of these results, however, hold only for sufficiently small mesh-sizes, and are therefore not directly applicable in the present context (where "sufficient-small" means smaller than ε). Slightly different notions of ellipticity are needed. The most useful for applications is, perhaps, the following.

R-Ellipticity. Assume the $q \times q$ difference operator L^h in (2.2) has constant coefficients, and let

$h = (h_1,\ldots,d_d)$, where h_j is the mesh-size in the x_j coordinate. Denote

$$\theta = (\theta_1,\ldots,\theta_d), \quad \theta\cdot x/h = \theta_1 \frac{x_1}{h_1} + \cdots + \theta_d \frac{x_d}{h_d}, \tag{5.1}$$

$$|\theta| = \max(|\theta_1|,\ldots,|\theta_d|) . \tag{5.2}$$

Then, for any constant q-vector V

$$L^h e^{i\theta\cdot x/h} V = B(\theta,h) e^{i\theta\cdot x/h} V , \tag{5.3}$$

where A is a $q \times q$ matrix, easily obtained by replacing in the matrix L^h each h_j-translation with the complex function $\exp(i\theta_j)$. B is called the *matrix-symbol* of L^h. The difference operator L^h is called R-elliptic if

$$\mathcal{R}e\ V^T B(\theta,h) V > 0 \quad \text{for all} \quad 0 < |\theta| \leq \pi \quad \text{and all} \tag{5.4}$$

$$\text{real q-vectors} \quad V \neq 0$$

An operator L^h with variable coefficients is called R-elliptic if the frozen-coefficients operator at every point is R-elliptic. A nonlinear difference operator is called R-elliptic if the corresponding linearized operator is R-elliptic (which may depend on the solution around which the linearization is taken).

This is not a complete definition of ellipticity. For example, if L_1^h and L_2^h are R-elliptic, then $L_1^h L_2^h$ is not necessarily also R-elliptic. But the definition gives, on one hand, a concept much more general than the special case of positive-type operators (which trivially satisfy (5.4)); in fact, a definition general enough for almost all scalar equations. On the other hand, the definition has some nice properties. One property is that it restricts the location of the operator, while Thomée's definition allows any translation to be added to the operator (which of course cannot be permitted in discussing finite mesh-sizes, because for example, it allows the two difference equations at two neighboring points to coincide, i.e., to be just one equation). This restriction is essential in discussing relaxation schemes, where a relation is required between each difference equation and the point at which it is relaxed. Another nice property is that the sum of R-elliptic operators is clearly also R-elliptic. One can therefore construct R-elliptic operators one term at a time.

We can use this property for singular-perturbation operators. If both the perturbed and the reduced equation are elliptic, the required uniform stability is obtained by constructing a difference approximation which is *uniformly* elliptic. A simple way to achieve this is to construct R-elliptic approximations to the various terms in the equation, so that R-elliptic approximation is obtained, in particular, for the reduced equation.

5.2. Examples

R-elliptic approximations, of arbitrary order, will be constructed in this section for the basic one-dimensional operators. Since this construction do not use any relation between terms, these approximations can be used as building blocks for approximating many ordinary and partial differential operators. The approximations are constructed on uniform grids only. As shown in Section 3, this is all we need in a multi-grid environment.

Using the operator notation

$$\delta u(x) = u(x+\tfrac{h}{2}) - u(x-\tfrac{h}{2}), \quad \nabla u(x) = u(x) - u(x-h),$$

$$\Delta u(x) = u(x+h) - u(x),$$

$$\mu u(x) = \frac{1}{2}\,[u(x+\tfrac{h}{2}) + u(x-\tfrac{h}{2})], \tag{5.5}$$

$$Du = \frac{du}{dx},$$

and the calculus of such operators (see, e.g., Dahlquist and Björk (1974) p. 311) one can derive the expansions

$$hD = \mu\delta \sum_{q=0}^{\infty} a_q(-\delta^2)^q \tag{5.6}$$

$$(hD)^2 = \delta^2 \sum_{q=0}^{\infty} \frac{a_q}{q+1}(-\delta^2)^q \tag{5.7}$$

where

$$a_0 = 1, \quad a_q = \frac{a_{q-1}}{4+\frac{2}{q}} \quad \text{and hence} \quad (a_q)^{1/q} \to \frac{1}{4}\,. \tag{5.8}$$

From this we find the following expressions for the 2s-order central approximations to the first and the second derivatives, and for the corresponding local truncation errors:

$$u'(x) = \frac{1}{h}\,\mu\delta \sum_{q=0}^{s-1} a_q(-\delta^2)^q\, u(x) + (-1)^s a_s\, h^{2s}\, u^{(2s+1)}(\xi_1)\,, \tag{5.9}$$

$$-u''(x) = \frac{1}{h^2}(-\delta^2) \sum_{q=0}^{s-1} \frac{a_q}{q+1}(-\delta^2)^q\, u(x) - (-1)^s \frac{a_s}{s+1}\, h^{2s}\, u^{(2s+2)}(\xi_2)\,, \tag{5.10}$$

where ξ_1 and ξ_2 are some intermediate points.

Let us now check these difference approximations for R-ellipticity. It is easy to see that the symbols corresponding to $\mu\delta$ and to $-\delta^2$ are, respectively, $i\sin\theta$ and $2(1 - \cos\theta)$. The latter is positive for all $0 < |\theta| \leq \pi$. Hence all the above approximations to $-u''$ (note the sign!) are R-elliptic. On the other hand any central approximation to either u' or $-u'$ has purely imaginary symbol, and is therefore never R-elliptic.

In various elliptic equations, $-u''$ is added to au', where a may have any sign. We therefore need to construct R-elliptic approximations to both u' and $-u'$. This is done by adding to (5.9) an R-elliptic term of order high enough: To obtain an approximation $O(h^{2s-1})$, add any positive multiple of the term $\frac{1}{h}(-\delta^2)^s$; to retain the $O(h^{2s})$ order, add any positive multiple of $\frac{1}{h}\nabla(-\delta^2)^s$ or $-\frac{1}{h}\Delta(-\delta^2)^s$. These terms are R-elliptic since the symbol of ∇ and $-\Delta$ are $1 - e^{-i\theta}$ and $1 + e^{i\theta}$, respectively, so that their real part is positive for $0 < |\theta| \leq \pi$. The values of the positive multiples can be chosen so that the $O(h^\ell)$ approximation uses exactly $\ell + 1$ points ($\ell = 2s - 1, s$). This gives the following R-elliptic approximations and truncation errors:

$$\pm u'(x) = \left\{\pm \frac{1}{h}\mu\delta \sum_{q=0}^{s-1} a_q(-\delta^2)^q + \frac{a_{s-1}}{2h}(-\delta^2)^s\right\}u(x) \qquad (5.11)$$

$$+ (-1)^s \frac{a_{s-1}}{2} h^{2s-1} u^{(2s)}(\xi_3)$$

$$u'(x) = \left\{\frac{1}{h}\mu\delta \sum_{q=0}^{s-1} a_q(-\delta^2)^q + \frac{a_{s-1}}{2h}\nabla(-\delta^2)^s\right\}u(x) \qquad (5.12)$$

$$+ (-1)^s \left\{\frac{a_{s-1}}{2} + a_s\right\} h^{2s} u^{(2s+1)}(\xi_4) ,$$

$$-u'(x) = \left\{-\frac{1}{h}\mu\delta \sum_{q=0}^{s-1} a_q(-\delta^2)^q - \frac{a_{s-1}}{2h}\Delta(-\delta^2)^s\right\}u(x) \quad (5.13)$$

$$- (-1)^s \left\{\frac{a_{s-1}}{2} + a_s\right\} h^{2s} u^{(2s+1)}(\xi_5) .$$

Observe that these operators are completely one-sided (so called "upwind") only for the $O(h)$ and $O(h^2)$ approximations. One can describe the above formulae as the non-central operators closest to the central among all operators which use the minimal number of points. The one-sided operators using the same number of points are not R-elliptic (for orders higher than 2).

The error term. For theoretical purposes (as in Section 4.1 above) it is more convenient to express the magnitude of the error terms in (5.10) and (5.13) in the forms

$$\left(\frac{h}{\gamma_s^2}\right)^{2s} u^{(2s+2)}(\xi_2) \quad \text{and} \quad \left(\frac{h}{\gamma_s^1}\right)^{2s} u^{(2s+1)}(\xi_5) , \qquad (5.14)$$

respectively, and similarly for (5.11) and (5.12). It is clear from (5.8) that both $\gamma_s^2 \to 2$ and $\gamma_s^1 \to 2$ as s grows. In fact, as shown in Table 2, each γ_s^i does not change much with s, and for theoretical convenience we treat them as constants.

TABLE 2

s	1	2	3	4	5
a_s^{-1}	6	30	140	630	2772
γ_s^1	1.22	1.71	1.86	1.93	1.97
γ_s^2	3.46	3.08	2.87	2.73	2.64

High-order approximations near boundaries may pose a problem, since the above difference operators may need function values at points which are not on the grid. One way out is to impose this technical restriction on the adaptive process (Section 3.6) which will, as a result, choose to further refine toward the boundary. The refinement will be geometric, so that without using too many points (their number is proportional to the high approximation order desired in the interior), the mesh-size near the boundary will be small enough to allow low-order approximation. In this respect the boundary behaves like a singular curve. Incidentally, for certain error norms (correspondingly: for certain functions G), lower order can be used (correspondingly: will be affected by the adaptive process) near the boundary without spoiling the global order of approximation. (Cf. Bramble and Hubbard (1962).)

6. RELAXATIONS WITH UNIFORM SMOOTHING RATES

A full version of this section appears as Section 6 in Brandt (1978b). Here we summarize the main points through simple examples.

The role of relaxation sweeps in multi-grid algorithms is to smooth the error (Section 2.1). The efficiency of relaxation is therefore measured by its "smoothing factor" $\bar{\mu}$ and the corresponding "smoothing rate" $\bar{\nu} = |\log \bar{\mu}|^{-1}$. The smoothing factor is defined in terms of the local mode analysis. Namely, if $\mu(\theta)$ is the convergence factor per relaxation sweep of the θ Fourier-component (see for example (2.7)) then,

$$\bar{\mu} = \max_{\frac{\pi}{2} \le |\theta| \le \pi} \mu(\theta), \quad \text{where} \quad |\theta| = \max |\theta_i| \ . \tag{6.1}$$

The range $\frac{\pi}{2} \le |\theta| \le \pi$, which is somewhat arbitrary (see Section A.3 in Brandt (1977a)), is chosen because these are the components which are too high to be seen on the coarser (2h) grid, so they cannot normally be reduced by the coarse-grid corrections. This definition seems to assume an infinite domain (where the Fourier expansion is made), but the behavior of such high-frequencies is practically independent of the domain. Thus, the smoothing rate $\bar{\nu}$ is, roughly, the number of relaxation sweeps required to reduce all high-frequency components by the factor $1/e$. $\bar{\mu}$ and $\bar{\nu}$ can be calculated for any relaxation scheme by the MUGTAPE (1978a) routine SMORATE. A table of representative values is given in Brandt (1977a), pp. 351-352.

For uniformly elliptic operators all (reasonable) relaxation schemes have bounded smoothing rates. (See the general theorems in Section 3.1 of Brandt (1976).) For singular-perturbation problems, however, many relaxation schemes will have $\bar{\nu}$ which grows indefinitely as the size of the perturbation decreases $(\varepsilon \to 0)$. That is, the convergence rates of some components θ will not be bounded uniformly in ε. One may sometimes still get a nice multi-grid process if those bad components have only a small contribution to the error norms (see Poling (1978)), but it is better and safer to use other relaxation schemes, with smoothing rates which are bounded *uniformly in* ε.

Two kinds of degeneracies will usually occur in relaxing singular-perturbation problems. One kind occurs in the boundary layer, in dimension $d \ge 2$, when a highly stretched grid (as in Figure 3E) is used. To see the problem, consider the usual, pointwise Gauss-Seidel relaxation for the 5-point Laplace operator on a grid with aspect-ratio

$\alpha = h_1/h_2 << 1$. Examining the component $\theta = (0,\frac{\pi}{2})$ for example, it is easy to see that $\bar{\nu} > .5\alpha^{-2}$. Since α may be comparable to ε, this smoothing rate may be extremely slow. The same slow rate will occur for *any* point-wise relaxation. If, in addition to the Laplace operator, the differential equation has also lower-order terms (as it does, of course, in singular-perturbation problems), the trouble still occurs, since on the finest grid the higher-order term (the perturbation) is dominant.

This kind of trouble can always be avoided by using Gauss-Seidel *line* relaxation. This means a relaxation in which we scan G^h (cf. Section 2.1) not point by point, but line by line. At each line, all the values $u^h(x^h)$ associated with that line are simultaneously replaced by new values which are computed so that they simultaneously satisfy all the difference equations associated with that line. In the above example the lines should be horizontal lines (x_2 = const.), and the resulting smoothing rate will uniformly be $\bar{\nu} = 2/\log 5$, no matter how small α is. The same smoothing rate will be obtained generally in the boundary layer, provided line relaxation is employed with *lines perpendicular to the boundary*.

Another type of degeneracy occurs on the coarser grids, where the reduced part of the equation dominates the smoothing process. The relaxation there should be one which is suitable for the reduced equation. Still more difficult may be the case of intermediate grids, where both the reduced and the perturbation parts interact with the smoothing process. Consider for example the ordinary differential operator

$$LU \equiv \varepsilon \frac{d^2u}{dx^2} + a\frac{du}{dx} , \tag{6.2}$$

and its lowest-order[8)] stable approximations (see Section 5.2)

$$L^hU^h \equiv \frac{\varepsilon}{h^2}\{U^h(x-h) - (2+\eta)U^h(x) + (1+\eta)U^h(x+h)\} \tag{6.3a}$$

for $a \geq 0$,

$$L^hU^h \equiv \frac{\varepsilon}{h^2}\{(1-\eta)U^h(x-h) - (2-\eta)U^h(x) + U^h(x+h)\} \tag{6.3b}$$

for $a \leq 0$,

where $\eta = ah/\varepsilon$ may be moderate or large. Denote by $\bar{\mu}_+(\eta)$ and $\bar{\mu}_-(\eta)$, respectively, the smoothing factors for the forward and backward Gauss-Seidel relaxation of (6.3). Forward and backward refer to the marching direction, i.e., to the order in which the points x are relaxed. A straightforward calculation gives, for $\eta \geq 0$,

$$\bar{\mu}_+(\eta) = \bar{\mu}_-(-\eta) = \left|\frac{1+\eta}{2+\eta+i}\right| \tag{6.4}$$

$$\bar{\mu}_-(\eta) = \bar{\mu}_+(-\eta) = \left|\frac{1}{2+\eta+i(1+\eta)}\right| . \tag{6.5}$$

Observe that $\bar{\mu}_+(\eta) \to 1$ as $\eta \to \infty$, which means that the forward relaxation is not uniformly smoothing for $a > 0$, and should not be used. No "relaxation parameter" will help here. On the other hand in this case $(a > 0)$ we have $\bar{\mu}_-(\eta) < 5^{-1/2}$, so the backward relaxation has a very good uniform smoothing rate. The backward direction corresponds to the *direction of convection, or the down stream direction*, in physical problems modelled by (6.2). Generally, physical insight is an invaluable source for devising successful relaxation schemes.

For $a < 0$ the situation is reversed: Backward relaxation is useless at small ε, but the forward relaxation has excellent (very small) smoothing rates. Slightly more difficult is the case where $a = a(x)$ changes sign in the domain. In that case each relaxation direction will have slow smoothing at some part of the domain. The good scheme then is *symmetric relaxation*, i.e., sweeping forward and then backward. The smoothing factor (per single sweep) of this is

$$\bar{\mu}_s(\eta) = [\bar{\mu}_+(\eta)\bar{\mu}_-(\eta)]^{1/2} = \left|\frac{1+\eta}{3+3\eta+\eta^2+i(2+\eta)^2}\right|^{1/2} \leq 3^{-1/2} . \tag{6.6}$$

Hence $\bar{\nu}_s \leq 2/\log 3$, uniformly bounded for all values of η, positive or negative. Observe that, in fact, the larger is $|\eta|$ the better is the smoothing rate.

The same holds for singular-perturbation equations in higher dimensions: Very good smoothing rates are obtained by a proper choice of the relaxation marching direction. In some situations *all* marching directions should be employed successively if a uniform smoothing is to be achieved. This may require more sweeps per multi-grid cycle, which one would like to avoid. We can, in fact, construct relaxation schemes which have *bounded smoothing rates even when marching against the local convection direction*. These schemes necessarily belong to the following class.

Distributed Relaxation. In classical relaxation we relate the unknown at a grid point to the difference equation at that same point. That is to say, we change that unknown to satisfy the corresponding equation; or, as in line relaxation, we change simultaneously a set of unknowns to satisfy the corresponding set of equations. This "marriage" between the unknown and the equation at the same point is not always natural. In many cases, especially in solving a *system* of differential equations $(q > 1)$, the natural thing is to change *several* unknowns in order to satisfy just *one* difference equation. Such a scheme is called distributed relaxation. (See Lecture 7 in Brandt (1978a).) A special case of such a relaxation was suggested by Kaczmarz (1937) and analyzed by Tanabe (1971)[9]. Various cases of distributed relaxation for singular-perturbation problems are analyzed by Dinar (1978).

Let us show how distributed-relaxation yields uniformly bounded smoothing rates even when the marching direction is upstream, i.e., against the direction of convection. Take again the operator (6.3a) and assume forward relaxation. Instead of changing only the approximation $u^h(x)$ to satisfy $L^h u^h(x) = F(x)$, change now both $u^h(x)$ and $u^h(x + h)$: Change $u^h(x)$ by adding to it δ, and $u^h(x + h)$ by adding to it $-w\delta$, where w is a fixed coefficient (see below) and δ is calculated so that the equation $L^h u^h(x) = F(x)$ is satisfied after these changes. This marching process is stable for $w < (1 + \eta)/2$. The larger w the better is the smoothing rate. By taking w not far from the critical value $(1 + \eta)/2$, we can get smoothing rates $\bar{\nu}$ which are less than 1 for all η, and $\bar{\nu} = O(\eta^{-1})$ for large η.

w is called the distribution coefficient, and should not be confused with the familiar "relaxation parameter". The above scheme is called Distributed Gauss-Seidel (DGS) relaxation, because, as in the Gauss-Seidel scheme, each difference equation in its turn is fully satisfied by the changes. For all problems examined, including incompressible Navier-Stokes equations, DGS was found to be the best smoother.

One final remark: The difference operator must be uniformly stable (see Section 5), otherwise no relaxation scheme can have uniformly bounded smoothing rates. For example, if the central difference approximation is used instead of (6.3a), even backward relaxation would have

$$\bar{\mu}_-(\eta) = \left|\frac{1 - \eta/2}{2 + i(1 + \eta/2)}\right| \to 1 \quad \text{as} \quad \eta \to \infty .$$

FOOTNOTES

1) An exception is the case when the coarse-grid difference operator L^H does not fully use the smoothness of the solution. In that case, if I_h^H in (2.4) is of sufficiently high order, then V^h will be smooth enough to be approximated by some V^H. This situation is related, however, to the use of an inappropriate approximation order, and will therefore not arise in a fully adaptive procedure.

2) Provided the two $I_h^H u^h$ appearing in (2.9) are identically the same. A common programming error is that they differ at some special points.

3) It is not necessary to compute the residual norm, since this particular algorithm is "fixed", its flow does not depend on the internal measures, and the number of sweeps made at each stage is prescribed in advance. For more complicated equations an "accommodative" algorithm, with internal switching criteria (e.g., the algorithm in Figure 1 above), may be desired. But, for more complicated equations, relaxation is more expensive, so that the extra work in computing $\|r^H\|$ is relatively small.

[4] τ-extrapolation is produced by the same FASFMG program of MUGTAPE (1978) through simple changes shown there by Comment cards.

[5] Alternatively, extra smoothness on the scale of the finest grid G^h can be used to produce a solution with errors smaller than the truncation errors *in very little work*. Indeed, if the difference equations do not exploit all the smoothness in the solution, an approximation to the level of the G^h truncation errors is obtained (with τ-extrapolation) already on one of the coarser grids. All is needed then is to interpolate from that grid to G^h, with high enough order of interpolation.

[6] For example, in the approach taken by Hackbusch (1978), the solution of a coupled pair of elliptic equations requires work equivalent to too many (at least 28/3, instead of just 2) solutions of a single equation.

[7] An abnormal run can usually be detected by examining the condensed output (output similar to the first three columns in Table 1). See Debugging Techniques, Lecture 18 in Brandt (1978a).

[8] It is enough to study relaxation schemes for the lowest order operator, because (i) one can compute higher-order approximations via the lower-order ones (see Section 3.4). (ii) For any relaxation scheme, the smoothing-rate dependence on the approximation-order is not very significant.

[9] References due to Blair Swartz and Gene Golub.

[10] F^h in the definition of r^h should later be understood as the <u>current</u> right-hand side $f^h = F^h_{h/2}$ (see Fig. 1).

ACKNOWLEDGEMENT

The work reported here was performed under NASA Contract No. NAS1-14101 while the author was in residence at ICASE, NASA Langley Research Center, Hampton, VA 23665.

REFERENCES

Babuska, I. (1975). Homogenization and its Application. Mathematical and Computational Problems. In: *Numerical Solution of Partial Differential Equations III* (SYNSPADE 1975), (B. Hubbard, ed.). Academic Press, New York.

Bakhvalov, N.S. (1969). The Optimization of Methods of Solving Boundary Value Problems with a Boundary Layer. *Zurnal Vycislitel'noi Matematiki i Matematiceskoi Fiziki* 9, 841-859.

Bramble, J.H. and Hubbard, B.E. (1962). On the Formulation of Finite Difference Analogues of the Dirichlet Problem for Poisson's Equation. *Numerische Mathematik* 4, 313-327.

Brandt, A. (1973). Multi-Level Adaptive Technique (MLAT) for Fast Numerical Solution to Boundary Value Problems. *Proceedings of the 3rd International Conference on Numerical Methods in Fluid Mechanics* (Paris, 1972), *Lecture Notes in Physics* 18, pp. 82-89, Springer-Verlag, Berlin and New York.

Brandt, A. (1976). Multi-Level Adaptive Techniques, IBM Research Report RC6026.

Brandt, A. (1977a). Multi-Level Adaptive Solutions to Boundary-Value Problems. *Mathematics of Computation* 31, 333-390.

Brandt, A. (1977b). Multi-Level Adaptive Solutions to Partial Differential Equations - Ideas and Software. *Proceedings of Symposium on Mathematical Software* (Mathematics Research Center, University of Wisconsin, March 1977), (John Rice, ed.), pp. 277-318. Academic Press, New York.

Brandt, A. (1978a). Lecture Notes of the ICASE Workshop on Multi-Grid Methods. With contributions also by J.C. South (Lecture 8), J. Oliger (10), F. Gustavson (13), C.E. Grosch (14), D.J. Jones (15) and T.C. Poling (16). ICASE, NASA Langley Research Center, Hampton, Virginia.

Brandt, A. (1978b). Multi-Level Adaptive Techniques (MLAT) and Singular-Perturbation Problems, Mathematics Research Center Report, University of Wisconsin, Madison. To appear.

Brandt, A. (1978c). Multi-Grid Solutions to Flow Problems, Numerical Methods for Partial Differential Equations. *Proceedings of Advanced Seminar*, Mathematics Research Center, University of Wisconsin, Madison, October 1978. (S. Parter, ed.). To appear.

Brandt, A., Dendi, J.E., Jr. and Ruppel, H. (1978). The Multi-Grid for the Pressure Iteration in Eulerian and Lagrangian Hydrodynamics, LA-UR 77-2995 report of Los Alamos Scientific Laboratory, Los Alamos, New Mexico.

Ciarlet, P.G. (1978). *The Finite Element Method for Elliptic Problems*. Studies in Mathematics and its Applications. (J.L. Lions, G. Papanicolaou and R.T. Rockafellar, eds.). North-Holland Publishing Co., New York.

Dahlquist, G. and Björck, A. (1974). *Numerical Methods*. Prentice-Hall, Inc., Englewood Cliffs, New Jersey.

Dinar, N. (1978). Fast Methods for the Numerical Solution of Boundary-Value Problems. Ph.D. Thesis, Weizmann Institute of Science, Rehovot, Israel.

Frank, L.S. (1978). Coercive Singular Perturbations: Stability and Convergence. This Proceedings.

Hackbusch, W. (1978a). On the Fast Solution of Parabolic Boundary Control Problems, to appear in *SIAM Journal on Control and Optimization*.

Hackbusch, W. (1978b). On the Fast Solving of Elliptic Control Problems. Report 78-14, Angewandte Mathematik, Universität zu Köln.

Hyman, J.M. (1977). Mesh Refinement and Local Inversion of Elliptic Partial Differential Equations. *Journal of Computational Physics* 23, 124-134.

Kaczmarz, S. (1937). Angenaherte Auflosung von Systemen Linearer Gleichungen. *Bulletin de l'Académie Polonaise des Sciences et Lettres* A, 355-357.

Keller, H.B. (1977). Numerical Solution of Bifurcation and Nonlinear Eigenvalue Problems. In: *Applications of Bifurcation Theory*. (P. Rabinowitz, ed.), pp359-384. Academic Press, New York.

Lentini, M. and Pereyra, V.L. (1975). Boundary Problem Solvers for First Order Systems Based on Deferred Corrections. In: *Numerical Solutions of Boundary Value Problems for Ordinary Differential Equations*. (A.K. Aziz, ed.). Academic Press, New York.

MUGTAPE (1978a). A Tape of Multi-Grid Software and Programs. Distributed at the ICASE Workshop on Multi-Grid Methods. Contributions by A. Brandt, N. Dinar, F. Gustavson and D. Ophir.

MUGTAPE (1978b). Updated Version of the Tape, prepared by D. Ophir, Mathematics Department, Weizmann Institute of Science, Rehovot, Israel.

Nicolaides, R.A. (1978). On Finite-Element Multi-Grid Algorithms and Their Use. ICASE report 78-8, ICASE, NASA Langley Research Center, Hampton, Virginia.

O'Malley, R.E., Jr. (1974). Boundary Layer Methods for Ordinary Differential Equations with Small Coefficients Multiplying the Highest Derivatives. *Proceedings of Symposium on Constructive and Computational Methods for Differential and Integral Equations. Lecture Notes in Mathematics* 430, 363-389. Springer-Verlag, New York.

Ophir, D. (1978). Language for Processes of Numerical Solutions to Differential Equations. Ph.D. Thesis, Mathematics Department, Weizmann Institute of Science, Rehovot, Israel.

Pereyra, V.L. (1968). Iterated Deferred Corrections for Nonlinear Boundary Value Problems. *Numerische Mathematik* 11, 111-125.

Poling, T.C. (1978). Numerical Experiments with Multi-Grid Methods. M.A. Thesis, Department of Mathematics, The College of William and Mary, Williamsburg, Virginia.

Shiftan, Y. (1972). Multi-Grid Method for Solving Elliptic Difference Equations. M.Sc. Thesis (in Hebrew), Weizmann Institute of Science, Rehovot, Israel.

South, J.C., Jr. and Brandt, A. (1976). Application of a Multi-Level Grid Method to Transonic Flow Calculations, ICASE Report 76-8, NASA Langley Research Center, Hampton, Virginia.

Southwell, R.V. (1935). Stress Calculation in Frameworks by the Method of Systematic Relaxation of Constraints. I, II. *Proceedings of the Royal Society London Series A* 151, 56-95.

Spagnolo, S. (1975). *Convergence in Energy for Elliptic Operators, Numerical Solution of Partial Differential Equations III* (SYNSPADE 1975). (B. Hubbard, ed.). Academic Press, New York.

Starius, G.C. (1977a). Constructing Orthogonal Curvilinear Meshes by Solving Initial Value Problems. *Numerische Mathematik* 28, 25-48.

Starius, G.C. (1977b). Composite Mesh Difference Methods for Elliptic Boundary Value Problems. *Numerische Mathematik* 28, 242-258.

Starius, G. (1978). Numerical Treatment of Boundary Layers for Perturbed Hyperbolic Equations. Report No. 69, Department of Computer Sciences, Uppsala University, Uppsala, Sweden.

Stetter, H.J. (1978). The Defect Correction Principle and Discretization Methods. *Numerische Mathematik* 29, 425-443.

Tanabe, K. (1971). Projection Method for Solving a Singular System of Linear Equations and its Applications. *Numerische Mathematik* 17, 203-214.

Thomée, V. (1964). Elliptic Difference Operators and Dirichlet Problem. *Contributions to Differential Equations* 3, 301-324.

Thomée, V. and Westergren, B. (1968). Elliptic Difference Equations and Interior Regularity. *Numerische Mathematik* 11, 196-210.

THE NUMERICAL SOLUTION OF SINGULAR SINGULARLY-PERTURBED INITIAL VALUE PROBLEMS

J.E. Flaherty
Department of Mathematical Sciences, Rensselaer Polytechnic Institute, Troy, New York 12181, U.S.A.

R.E. O'Malley, Jr.
Department of Mathematics and Program in Applied Mathematics, University of Arizona, Tucson, Arizona 85721, U.S.A.

ABSTRACT

We consider the vector initial value problem $\varepsilon\dot{y} = f(y,t,\varepsilon)$, $y(0) = y^0(\varepsilon)$ in the situation when the $m \times m$ matrix $f_y(y,t,0)$ is singular with constant rank $k < m$ and has k stable eigenvalues. We show how to determine the unique limiting solution Y_0 of the reduced problem $f(Y_0,t,0) = 0$ and how to obtain a uniform asymptotic expansion of the solution which is valid for small values of ε on finite t intervals. A numerical technique is developed to calculate the limiting solution and the results of some examples are compared with an existing code for stiff differential equations.

1. INTRODUCTION

We consider the initial value problem

$$\varepsilon\dot{y} = f(y,t,\varepsilon) \quad , \; y(0,\varepsilon) = y^0(\varepsilon) \tag{1.1}$$

for m nonlinear differential equations on a finite interval $0 \le t \le T$ in the limit as the small positive parameter ε

p. 60.

Theorem 1.

The following inequality holds

$$\|\hat{u}^n - \hat{u}^o\|_B \le \frac{L}{\kappa} \inf_{u^n \in U_n} \|u^n - \hat{u}^o\|_B .$$

In Dontchev *et al* (1977) a more general class of problems is analysed and, on some additional assumptions, a similar estimation is obtained.

In the present paper we apply this result to an optimal control problem with a parameter ε in the derivatives of a part of the state, which may be small. We show, that the general Ritz's computational scheme for this problem is converging uniformly in ε.

2. We shall consider the following singularly perturbed optimal control system for a finite time interval $[0,T]$:

$$\begin{aligned} \dot{x} &= A_{11}(t)x + A_{12}(t)y + B_1(t)u , \\ \varepsilon\dot{y} &= A_{21}(t)x + A_{22}(t)y + B_2(t)u, \\ & x \in R^n , \ y \in R^m , \ u \in R^r , \end{aligned} \tag{2}$$

with given initial state $x(0) = x_o$, $y(0) = y_o$. The matrices $A_{ij}(t)$, $B_i(t)$; $i, j \in \{1,2\}$ are continuous, $\varepsilon \in [0,\varepsilon_o]$ is a singular perturbation parameter. We assume that the eigenvalues of the matrix $A_{22}(t)$ have negative real parts.

The following performance index

$$J_\varepsilon = g(x(T)) + \int_0^T f(x(t),u(t),t)dt \tag{3}$$

is to be minimized over the space $L_2^{(r)}[0,T]$. We assume that the function $g(x)$ is convex and differentiable and its derivative is Lipschitz-continuous. The function $f(x,u,t)$ is continuous with respect to the triple (x,u,t) and differentiable with respect to x and u. Its derivatives are continuous functions and satisfy the following Lipschitz-conditions uniformly in [0,T]:

$$\left\| \frac{\partial f(x+\Delta x,u+h,t)}{\partial x} - \frac{\partial f(x,u,t)}{\partial x} \right\|_{R^n} \leq \rho(\|\Delta x\|_{R^n} + \|h\|_{R^r}), \qquad (4a)$$

$$\left\| \frac{\partial f(x+\Delta x,u+h,t)}{\partial u} - \frac{\partial f(x,u,t)}{\partial u} \right\|_{R^r} \leq \rho(\|\Delta x\|_{R^n} + \|h\|_{R^r}), \qquad (4b)$$

where the Lipschitz's constants ρ are denoted in the same way for these two inequalities.

We assume that there exists a positive number κ so that for every $t \in [0,T]$; $x, z \in R^n$; $u, v \in R^r$; $\alpha \in [0,1]$:

$$f(\alpha x+(1-\alpha)z,\alpha u+(1-\alpha)v,t) \leq$$

$$\alpha f(x,u,t)=(1-\alpha)f(z,v,t)-\kappa\alpha(1-\alpha)\|u-v\|_{R^r}.$$

It is known that, see e.g. Vasil'ev (1974), on the above hypotheses, the functional J(u) is strongly convex with the same constant κ (uniformly in ε) and Fréchet-differentiable in $L_2^{(r)}[0,T]$. Its derivative has the form of

$$(J'_\varepsilon(u))(t) = -\frac{\partial f(x(t),u(t),t)}{\partial u} + B_1^T(t)\eta(t) + B_2^T(t)\sigma(t), \qquad (5)$$

where $\eta(t)$, $\sigma(t)$ satisfy the following adjoint equations

$$\begin{aligned} \dot\eta &= -A_{11}^T(t)\eta - A_{21}^T(t)\sigma - \frac{\partial f(x(t),u(t),t)}{\partial x}, \\ \varepsilon\dot\sigma &= -A_{12}^T(t)\eta - A_{22}^T(t)\sigma, \\ \eta(T) &= -\frac{dg(x(T))}{dx}, \quad \sigma(T) = 0. \end{aligned} \qquad (6)$$

Lemma.

There exists a constant L so that for every $u,v \in L_2^{(r)}[0,T]$ and for every $\varepsilon \in [0,\varepsilon_0]$

$$\|J'_\varepsilon(v) - J'_\varepsilon(u)\|_{L_2} \leq L\|v - u\|_{L_2}.$$

Proof: Let the functions u(t), h(t) be arbitrarily chosen in

$L_2^{(r)}[0,T]$ and $v(t) = u(t)+h(t)$,
$\Delta y(t)=y(v,t)-y(u,t)$, $\Delta\eta(t)=\eta(v,t)-\eta(u,t)$, $\Delta\sigma(t)=\sigma(v,t)-\sigma(u,t)$.
We show that there exist constants d , d , d so that

$$\|\Delta x\|_{C[0,T]} \le d_1 \|h\|_{L_2}, \tag{7}$$

$$\|\Delta \eta\|_{C[0,T]} \le d_2 \|h\|_{L_2}, \tag{8}$$

$$\|\Delta \sigma\|_{C[0,T]} \le d_3 \|h\|_{L_2}, \tag{9}$$

where

$$\|\Delta x\|_{C[t_1,t_2]} = \max_{t_1 \le t \le t_2} \|\Delta x(t)\|_{R^n}$$

etc. Then, applying the formulae (4b), (5) and Hoelder's inequality one can easily complete the proof.

From the assumption for the matrix $A_{22}(t)$ it follows (Vasil'eva and Butuzov (1975)), that there exist positive constants μ_o, μ so that

$$\|\phi(t,\tau,\varepsilon)\| \le \mu_o \exp(-\mu\frac{t-\tau}{\varepsilon}) , \tag{10}$$

where $\phi(t,\tau,\varepsilon)$ is the fundamental solution of the equation $\varepsilon\dot{z} = A_{22}(t)z$ and where the usual matrix norm is used. Let $a_{1i} = \|A_{1i}\|_{C[0,T]}$, $a_{2i} = \mu_o\|A_{2i}\|_{C\ 0,T}$, $i=1,2$, $b = \|B_1\|_{C[0,T]}$, $b = \mu_o\|B_2\|_{C[0,T]}$. From Cauchy's formula and (10) we get

$$\|\Delta x(t)\| \le \int_0^t (a_{11}\|\Delta x(\tau)\| + a_{12}\|\Delta y(\tau)\| + b_1\|h(\tau)\|)d\tau = \alpha(\tau),$$

$$\|\Delta y(t)\| \le \frac{1}{\varepsilon}\int_0^t \exp(-\mu\frac{t-\tau}{\varepsilon})(a_{21}\|\Delta x(\tau)\| + b_2\|h(\tau)\|)d\tau = \beta(t).$$

Hence

$$\begin{aligned}\dot{\beta} &= -\frac{\mu}{\varepsilon}\beta + \frac{a_{21}}{\varepsilon}\|\Delta x(t)\| + \frac{b_2}{\varepsilon}\| h(t) \| \\ &\le -\frac{\mu}{\varepsilon}\beta + \frac{a_{21}}{\varepsilon}\alpha(t) + \frac{b_2}{\varepsilon}\| h(t) \|, \quad \beta(0) = 0\end{aligned}$$

and

$$\beta(t) \le (a_{21}\alpha(t) + b_2 \parallel h(t) \parallel - \varepsilon\dot{\beta}(t))/\mu$$

We obtain

$$\alpha(t) \le \int_0^t (a_{11}\alpha(\tau) + a_{12}\beta(\tau) + b_1 \parallel h(\tau) \parallel)\, d\tau$$

$$\le \int_0^t (a^*\alpha(\tau) + b^* \parallel h(\tau) \parallel - \frac{a_{12}\varepsilon}{\mu}\dot{\beta}(\tau))\, d\tau$$

$$\le \int_0^t (a^*\alpha(\tau) + b^* \parallel h(\tau) \parallel)\, d\tau - \frac{a_{12}\varepsilon}{\mu}\beta(t)$$

$$\le \int_0^t (a^*\alpha(\tau) + b^* \parallel h(\tau) \parallel)\, d\tau,$$

where $a^* = a_{11} + a_{12}a_{21}/\mu$, $b^* = b_1 + a_{12}b_2/\mu$, all the norms are in the corresponding euclidean spaces and the positiveness of $\beta(t)$ is used. Then, applying Gronwall's inequality and Hoelder's inequality we get (7).

Using Lipschitz-continuity of $\frac{dg}{dx}$, (4a) and the same construction as above we can easily obtain (8). The estimation (9) follows from (8) and the inequality

$$\parallel \Delta\sigma(T-t) \parallel_{R^m} \le \frac{a_{12}}{\varepsilon}\int_0^t \exp(-\mu\frac{t-\tau}{\varepsilon}) \parallel \Delta\eta \parallel_{C[0,T]} d\tau$$

$$\mu_0 \frac{a_{12}}{\mu} \parallel \Delta\eta \parallel_{C[0,T]}.$$

Thus, the proof is completed.

3. Let S^n be a finite-dimensional subspace of the space $L_2^{(r)}[0,T]$. We apply the so-called primal Ritz's method to

the problem (2,3). We solve this problem on the assumption that the admissible set of controls is S^n and, in such a way, we transform the problem to a finite-dimensional one. Let $\hat{u}_\varepsilon$ be the solution of the original problem and $\hat{u}^n_\varepsilon$ is the solution, obtained by Ritz's method, $\varepsilon\in[0,\varepsilon_0]$ is fixed. From the results given in the first two parts of the paper, the following estimate can be obtained.

Theorem 2.

There exists a constant C so that for every $\varepsilon\in[0,\varepsilon_0]$

$$\|\hat{u}^n_\varepsilon - \hat{u}_\varepsilon\|_{L_2} \leq C \inf_{u^n\in S^n} \|u^n - \hat{u}_\varepsilon\|_{L_2}.$$

Let $f(x,u,t) = p(x,t) + q(u,t)$ and the assumptions, given in section 2, hold. Let $p(x,t) \geqq 0$ and there exists a number $\gamma>0$ so that for every $u\in R^r$, $q(u,t) \geq \gamma\|u\|^2$. Then the optimal control $\hat{u}_\varepsilon(t)$ is a continuous function with respect to the time t for every fixed $\varepsilon\in[0,\varepsilon_0]$ (Gičev (1976)). Moreover, in Gičev and Dontchev (1978), it is shown that, if $\varepsilon_1\in[0,\varepsilon_0]$ and $\hat{u}_1$ corresponds to ε_1 then, for every T* (0,T)

$$\lim_{\varepsilon\to\varepsilon_1} \|\hat{u}_\varepsilon - \hat{u}_1\|_{C[0,T^*]} = 0$$

and $\|\hat{u}_\varepsilon\|_{C[0,T]}$ is bounded in $+0,\varepsilon_0]$.

Combining this result with Theorem 2 we get that there exists a constant C so that for every $\varepsilon_1,\varepsilon_2 \in [0,\varepsilon_0]$

$$\|\hat{u}^n_1 - \hat{u}_2\|_{L_2} \leq C_1 \inf_{u^n\in S^n} \|u^n - \hat{u}_2\|_{L_2} + O(|\varepsilon_1 - \varepsilon_2|), \qquad (11)$$

where $\hat{u}_i$ corresponds to ε_i, i=1,2 and $\hat{u}^n_1$ is the approximate solution for $\varepsilon=\varepsilon_1$.

Note, that for $\varepsilon = 0$ the order of the system decreases. This may extremely simplify the problem (Dontchev (1977)).

If S^n is the space of polynomials on [0,T], then according to the classical results of the approximation theory (Jackson (1930)) and (11) we get

$$\|\hat{u}_1^n - \hat{u}_2\|_{L_2} \leq C_2\omega(\tfrac{1}{n}) + O(|\varepsilon_1 - \varepsilon_2|) ,$$

where ω is the modulus of continuity of the function $\hat{u}_2$.

Consider now a more simple problem. Assume $q(u,t) = u^T R(t)u$ where $R(t)$ is a continuously differentiable and positively definite matrix. Let $B_i(t)$, $i=1,2$ be continuously differentiable. Then the optimal control $u_\varepsilon(t)$ is a continuously differentiable function in time t for ε fixed Let u_o correspond to $\varepsilon = 0$. From (11) an important estimate follows:

$$\max \{ \|\hat{u}_\varepsilon^n - \hat{u}_o\|_{L_2}, \|\hat{u}_o^n - \hat{u}_\varepsilon\|_{L_2} \} \leq \frac{C_3}{n} + O(\varepsilon).$$

At the end we note that the presented approach can be applied to other optimal control problems with regular and singular perturbations.

REFERENCES

Dontchev, A.L. and Ivanov, R.P. (1977). Estimations of the solutions of the constrained extremal problem by means of its best approximation. *Serdika*, 3, No. 3 236-241. (In Russian).

Vasil'ev, F.P. (1974). Lecture notes on methods for solving of extremal problems, Moscow University Press, M. (In Russian).

Vasil'eva, A.B. and Butuzov, W.F. (1975). Asymptotic solutions of singularly perturbed equations. *Nauka, M.* (In Russian).

Gičev, T.R. (1976). Correctness of an optimal control problem with an integral convex performance index. *Serdika*, 2, No. 4, 334-342. (In Russian).

Gičev, T.R. and Dontchev, A.L. (1978). Linear optimal control system with singular perturbation and convex performance index. *Serdika*, 4, No. 1. (To appear).

Dontchev, A.L. (1977). Linear model simplification for singularly perturbed optimal control systems. *Compt. Rend. Acad. Bulg. Sc.*, 30, No. 4, 449-502.

Jackson, D. (1930). The theory of approximation. N.J.

CONVERGENCE OF RITZ'S METHOD, UNIFORMLY IN A SINGULAR PERTURBATION, FOR OPTIMAL CONTROL PROBLEMS

Asen L. Dontchev
Institute of Mathematics, Bulgarian Academy of Sciences, Sofia, P.O. Box 373, Bulgaria

ABSTRACT

We consider a linear control system with a small parameter in the derivatives of a part of the state variables and a convex performance index. We give sufficient conditions for the convergence of the general Ritz's computational method uniformly in the singular perturbation. The presented approach is based on some previous author's results.

1. Consider the following family of extremal problems

$$\min J(u) \ , \ u \in U_n \ , \quad n = 0,1, \ldots, \tag{1}$$

where $\{U_n\}_1^\infty$ is a sequence of closed and convex sets U_n in a reflexive Banach space B. Let the functional J(u) be Fréchet-differentiable and its derivative be Lipschitz-continuous with a constant L. We assume also that J(u) is strongly convex, that is, there exists a positive constant κ so that for every u,v B and for every $\alpha \in [0,1]$

$$J(\alpha u + (1-\alpha)v) \leq \alpha\ J(u) + (1-\alpha)J(v) - \kappa\alpha(1-\alpha)\|u-v\|_B^2 .$$

From the usual arguments it follows that for every n = 0,1, ... there exists an unique solution $\hat{u}^n$ of the problem (1). Let $U_0 = B$.

The following result can be immediately obtained from Dontchev and Ivanov (1977) (Theorem 3) and from Vasil'ev (1974), tends to zero. Familiarity with singular perturbation problems leads one to expect that (under appropriate stability conditions) the solution of (1.1) would converge to a limiting solution Y_0 of the reduced system

$$f_0(Y_0,t) \equiv f(Y_0,t,0) = 0 \tag{1.2}$$

as $\varepsilon \to 0$, at least away from an initial boundary layer region of nonuniform convergence. For example, in the classical Tikhonov problem (cf. Wasow (1976)), when the Jacobian F_y $f_y(y,t,0)$ has stable eigenvalues for all y and t (the region of stability can be further restricted), then (1.2) has a unique solution $Y_0(t)$ which is the limiting solution of (1.1) for $t > 0$. The solution generally converges nonuniformly at $t = 0$ since there is no reason to expect that $Y_0(0) = y^0(0)$. Indeed, if f is infinitely differentiable in y and t and has an asymptotic expansion in ε then the solution $y(t,\varepsilon)$ of (1.1) can be represented asymptotically in the form

$$y(t,\varepsilon) = Y(t,\varepsilon) + \Pi(\tau,\varepsilon), \tag{1.3}$$

throughout $0 \le t \le T$. The outer solution Y and the boundary layer correction Π both have asymptotic expansions in ε, and Π tends to zero as the stretched (or boundary layer) variable

$$\tau = t/\varepsilon \tag{1.4}$$

tends to infinity.

We wish to consider (1.1) when matrix $f_y(y,t,0)$ is singular, and in particular satisfies:

Hypothesis (H): $f_y(y,t,0)$ *has constant rank k,* $0 \le k < m$ *for all t in* $0 \le t \le T$ *and all y; its nonzero eigenvalues have negative real parts there; and its null space is spanned by* $m - k$ *linearly independent eigenvectors.*

In this case we will find that the asymptotic solution of

(1.1) still has the form (1.3) whenever the reduced system (1.2) is consistent and solvable, i.e., whenever (1.2) has at least one solution. However, because f_{0y} is singular, (1.2) no longer has a unique solution and additional analysis is necessary to determine the unique limiting solution for $t > 0$. We call such problems "singular singularly-perturbed problems". Two simple scalar examples illustrating some of the possibilities are (i) $\varepsilon\dot{y} = 1$, $y(0) = 0$ and (ii) $\varepsilon\dot{y} = 0$, $y(0) = 0$. For (i), the reduced problem $1 = 0$ is inconsistent, and while $y = \tau = t/\varepsilon$ is a solution of the form (1.3) we see that Π becomes unbounded as $\tau \to \infty$. For (ii), the reduced problem $0 = 0$ is satisfied for all Y_0, but only the trivial solution $Y_0 = 0$ is a limit of the unique solution $y = 0$.

Campbell and Rose (1978) studied constant coefficient singular singularly-perturbed problems of the form

$$\varepsilon\dot{y} = F(\varepsilon)y \tag{1.5}$$

and showed that the "semistability" condition of Hypothesis (H) is a necessary and sufficient condition for a limiting solution to exist for all $t > 0$ and all initial vectors y^0. O'Malley (1978) obtained asymptotic solutions of (1.1) in the almost-linear case when $f(y,t,0) = F(t)y + g(t)$, assuming that the linear reduced system $F(t)Y_0 + g(t) = 0$ is consistent. A preliminary study of nonlinear systems was reported in O'Malley and Flaherty (1976). Additional work on singular singular-perturbed problems was done by Vasil'eva and others (cf. Vasil'eva (1976) and the references contained therein).

Asymptotic solutions with a different structure than (1.3) might result if initial values were restricted. For example, consider (1.5) with

$$F(\varepsilon) = \begin{bmatrix} 0 & -\varepsilon \\ -1 & 0 \end{bmatrix}$$

and suppose that the initial components satisfy $y_1^0 = \sqrt{\varepsilon} y_2^0$, then we have the trivial solution for $t > 0$, but the boundary layer behaviour is determined by the stretched variable $\sigma = t/\sqrt{\varepsilon}$. More complicated limiting solutions would occur if we allowed "turning points" where the rank of $f_y(y,t,0)$ changes at particular y and t values. Studies of these interesting and difficult problems are contained in the work of Howes (1978) and Kreiss (1978). The latter also contains numerical methods. Two simple scalar examples of such problems are $\varepsilon\dot{y} = -y^3 + \varepsilon y$ and $\varepsilon\dot{y} = (t - 1/2)y$, where the ranks of $f_y(y_0,t,0)$ change at $y = 0$ and $t = 1/2$, respectively.

In Section 3 of this paper we develop asymptotic expansions for the outer solutions $Y(t,\varepsilon)$ of a special class of singular singularly-perturbed problems and in Section 4 we consider more general problems. Some preliminary linear algebra is presented in Section 2. Expansions for the boundary layer correction $\Pi(\tau,\varepsilon)$ and a proof of the asymptotic validity of our solutions have also been obtained and will be reported in O'Malley and Flaherty (1978). In Section 5 we develop a numerical procedure for calculating the limiting solution $Y_0(t)$ which is based on the expansion of Section 4 and in Section 6 we apply this procedure to some examples and discuss the results.

Our primary motivation for this work is the need to develop numerical procedures for singularly-perturbed (or stiff) two-point boundary value problems. However, our results should be applicable to initial value problems in power generation and distribution systems, biological and chemical reactions, and electrical networks. A new application is ill-conditioned nonlinear optimization problems (cf. Boggs and Tolle (1977)).

2. ALGEBRAIC PRELIMINARIES

We shall attempt to find an asymptotic solution of (1.1) in the form given by (1.3). The decay of Π as $\tau \to \infty$ implies that the outer solution $Y(t,\varepsilon)$ should satisfy

$$\varepsilon\dot{y} = f(y,t,\varepsilon) \tag{2.1}$$

as a power series in ε, i.e.,

$$Y(t,\varepsilon) \sim \sum_{j=0}^{\infty} Y_j(t)\varepsilon^j. \tag{2.2}$$

Under Hypothesis (H) we are guaranteed that

$$f_{0y}(y,t) \equiv f_y(y,t,0) \equiv \frac{\partial f}{\partial y}(y,t,0) \tag{2.3}$$

can be put into its reduced echelon form by an orthogonal matrix $E(y,t)$. Golub (1965) discussed a numerical procedure for obtaining E by performing a sequence of k Householder transformations. The differentiability of E follows that of f_0 under the constancy of rank condition (cf. Golub and Pereyra (1976)). We partition E as

$$E = \begin{bmatrix} E_1 \\ E_2 \end{bmatrix} \tag{2.4}$$

where E_1 is $k \times m$, E_2 is $(m-k)\times m$, and

$$E_2 f_{0y} = 0 \tag{2.5}$$

In addition,

$$Ef_{0y} E^T = \begin{bmatrix} S & X \\ 0 & 0 \end{bmatrix} \tag{2.6}$$

where

$$S = E_1 f_{0y} E_1^T \quad , \quad X = E_1 f_{0y} E_2^T \quad , \tag{2.7}$$

Hypothesis (H) guarantees that S has k stable eigenvalues. We note that Clasen *et al* (1978) used such constant orthogonal matrices E to integrate stiff problems, while O'Malley (1978) used time dependent matrices for almost linear problems.

The orthogonality of E further implies that

$$E_1 E_2^T = 0 \ , \ E_1 E_1^T = I_k \ , \ E_2 E_2^T = I_{m-k} \ , \text{ and} \tag{2.8}$$

$$E_1^T E_1 + E_2^T E_2 = I_m \ ,$$

where I_m is the m × m identity matrix. Using the last relation we introduce the complementary orthogonal projections

$$P = E_1^T E_1 \quad , \quad Q = E_2^T E_2 \tag{2.9}$$

which provide a direct sum decomposition of m-space with Q projecting onto $N(f_{0y}^T)$, the null space of f_{0y}^T, and P projecting onto $R(f_{0y})$, the range of f_{0y}.

3. A SPECIAL PROBLEM: E(y,t) = E(t)

In this section we examine special problems (1.1) when the orthogonal matrix E(y,t) introduced in Section 2 is independent of y. This, of course, includes the nearly linear problems where

$$f(y,t,0) = F(t)y + g(t)$$

and "classical" singular perturbation problems having the form

$$\varepsilon \dot{y}_1 = f_1(y_1,y_2,t) + \varepsilon g_1(y_1,y_2,t,\varepsilon)$$

$$\varepsilon \dot{y}_2 = \varepsilon g_2(y_1,y_2,t,\varepsilon) \quad .$$

Here, y_1 is a k-vector, y_2 is an (m-k)-vector, and $\partial f_1/\partial y_1$ is of rank k. In this case $E = I_m$.

We define

$$z = E(t)y, \tag{3.1}$$

and further partition z like E, i.e.,

$$\begin{bmatrix} z_1 \\ z_2 \end{bmatrix} = \begin{bmatrix} E_1 y \\ E_2 y \end{bmatrix}. \tag{3.2}$$

Introducing (3.2) into (1.1) gives the following system for z:

$$\varepsilon \dot{z}_1 = h_1(z_1, z_2, t, \varepsilon) \quad , \quad z_1(0) = E_1(0)y^0 \tag{3.3a}$$

$$\dot{z}_2 = h_2(z_1, z_2, t, \varepsilon)/\varepsilon \quad , \quad z_2(0) = E_2(0)y^0 \tag{3.3b}$$

where,

$$h_i = E_i f(E^T z, t, \varepsilon) + \varepsilon E_i E^T z \quad , \quad i = 1,2. \tag{3.4}$$

We have divided (3.3b) through by ε since

$$h_2(z_1, z_2, t, 0) = 0 \tag{3.5}$$

necessarily follows if the reduced system (1.2) for (1.1) is consistent. This is because

$$\frac{\partial h_2}{\partial z_i}(z_1, z_2, t, 0) = E_2 f_{0y}(E^T z, t) E_i^T = 0 \quad , \quad i = 1,2$$

upon use of (3.4) and (2.5). Thus, $h_2(z_1, z_2, t, 0)$ is a function of t only. However, the reduced system (1.2) implies the corresponding reduced system

$$h_i(z_1, z_2, t, 0) = 0 \quad , \quad i = 1,2$$

for (3.3). Hence $h_2(z_1, z_2, t, 0)$ must be the trivial function of t on $0 \le t \le T$ for any z_1 and z_2, otherwise (1.2) would have no solutions. Tikhonov's results apply to (3.3) because his stability condition that

$$\frac{\partial h_1}{\partial z_1}(z_1,z_2,t,0) = E_1 f_{0y}(E^T z,t)E_1^T \equiv S \tag{3.6}$$

have stable eigenvalues holds for all z_1, z_2 and for all t on $0 \le t \le T$. Thus, (3.3) has an asymptotic solution of the form

$$\begin{aligned} z_1(t,\varepsilon) &= Z_1(t,\varepsilon) + \Lambda_1(\tau,\varepsilon) \\ z_2(t,\varepsilon) &= Z_2(t,\varepsilon) + \varepsilon\Lambda_2(\tau,\varepsilon) \end{aligned} \tag{3.7}$$

(cf. O'Malley (1974)), where the outer solution (Z_1,Z_2) and the boundary layer correction (Λ_1,Λ_2) each have power series expansions in ε with the boundary layer correction decaying to zero as $\tau = t/\varepsilon \to \infty$.

Since the outer solution provides the asymptotic solution for $t > 0$, we must have

$$\varepsilon\dot{Z}_1 = h_1(Z_1,Z_2,t,\varepsilon) \quad , \quad \dot{Z}_2 = h_2(Z_1,Z_2,t,\varepsilon)/\varepsilon \tag{3.8}$$

satisfied as power series

$$Z_i(t,\varepsilon) \sim \sum_{j=0}^{\infty} Z_{ij}(t)\varepsilon^j \quad , \quad i = 1,2, \tag{3.9}$$

in ε. The leading term must necessarily satisfy the limiting problem

$$h_1(Z_{10},Z_{20},t,0) = 0 \tag{3.10a}$$

$$\dot{Z}_{20} = h_{2_\varepsilon}(Z_{10},Z_{20},t,0) \quad , \quad Z_{20}(0) = E_2(0)y^0(0). \tag{3.10b}$$

Its unique solution is obtained since (3.6) and the implicit function theorem imply that the algebraic equation $h_1(Z_{10},Z_{20},t,0) = 0$ can be uniquely solved for the k-vector

$$Z_{10}(t) = \phi(Z_{20}(t), t) \tag{3.11}$$

leaving the nonlinear (m-k) th order initial value problem

$$\dot{Z}_{20} = \frac{\partial h_2}{\partial \varepsilon}(\phi(Z_{20},t), Z_{20},t,0) \; , \; Z_{20}(0) = E_2(0)y^0(0) \tag{3.12}$$

for Z_{20}. We shall assume that the unique solution to (3.12) continues to exist throughout $0 \le t \le T$. Note that the reduced system (1.2) implied that both $h_1 = 0$ and $h_2 = 0$ along $(Z_{10}, Z_{20}, t, 0)$, but it did not provide equation (3.12) needed to uniquely determine the limiting outer solution (Z_{10}, Z_{20}).

Higher order terms in (3.9) satisfy linear problems

$$\frac{\partial h_1}{\partial z_1}(Z_{10}, Z_{20}, t, 0) Z_{1j} + \frac{\partial h_2}{\partial z_2}(Z_{10}, Z_{20}, t, 0) Z_{2j} = g_{1,j-1}(t)$$

$$\dot{Z}_{2j} = \frac{\partial^2 h_2}{\partial z_1 \partial \varepsilon}(Z_{10}, Z_{20}, t, 0) Z_{1j} + \frac{\partial^2 h_2}{\partial z_2 \partial \varepsilon}(Z_{10}, Z_{20}, t, 0) Z_{2j} + \qquad (3.13)$$

$$g_{2,j-1}(t), \quad Z_{2j}(0) = -\Lambda_{2,j-1}(0)$$

with the $g_{i,j-1}(t)$'s determined by lower order terms in the outer expansion. One solves the first equation for Z_{1j} as a linear function of Z_{2j}, and then the resulting linear differential equation for Z_{2j}. Thus, the outer expansion (3.9) can be uniquely generated termwise in $0 \le t \le T$ up to a knowledge of the initial value of the boundary layer correction component $\Lambda_2(0, \varepsilon)$.

The boundary layer correction is obtained by noting that both (z_1, z_2) and (Z_1, Z_2) satisfy the differential equations (3.3). Hence, using (3.8) in (3.3) we have

$$\frac{d\Lambda_1}{d\tau} = h_1(Z_1(\varepsilon\tau, \varepsilon) + \Lambda_1(\tau, \varepsilon), Z_2(\varepsilon\tau, \varepsilon) + \varepsilon\Lambda_2(\tau, \varepsilon), \varepsilon\tau, \varepsilon)$$
$$-h_1(Z_1(\varepsilon\tau, \varepsilon), Z_2(\varepsilon\tau, \varepsilon), \varepsilon\tau, \varepsilon) \qquad (3.14)$$
$$\frac{d\Lambda_2}{d\tau} = [h_2(Z_1(\varepsilon\tau, \varepsilon) + \Lambda_1(\tau, \varepsilon), Z_2(\varepsilon\tau, \varepsilon) + \varepsilon\Lambda_2(\tau, \varepsilon), \varepsilon\tau, \varepsilon)$$
$$-h_2(Z_1(\varepsilon\tau, \varepsilon), Z_2(\varepsilon\tau, \varepsilon), \varepsilon\tau, \varepsilon)].$$

We require Λ_1 and Λ_2 to decay as $\tau \to \infty$ and satisfy the initial condition

$$\Lambda_1(0, \varepsilon) = E_1(0) y^0(0) - Z_1(0, \varepsilon). \qquad (3.15)$$

Taking

$$\Lambda_i(\tau,\varepsilon) \sim \sum_{j=0}^{\infty} \Lambda_{ij}(\tau)\varepsilon^j \qquad i = 1,2 \tag{3.16}$$

we find that the leading term Λ_{10} must satisfy the nonlinear initial value problem

$$\frac{d\Lambda_{10}}{d\tau} = h_1(Z_{10}(0) + \Lambda_{10}(\tau), Z_{20}(0),0,0)$$
$$h_1(Z_{10}(0), Z_{10}(0),0,0) \ , \tag{3.17}$$
$$\Lambda_{10}(0) = E_1(0)y^0(0) - Z_{10}(0) .$$

This problem has a unique exponentially decaying solution $\Lambda_{10}(\tau)$ since (3.6) implies that $\frac{\partial h_1}{\partial z_1}(z_1,z_2,t,0)$ has stable eigenvalues for all arguments. Knowing Λ_{10} we can calculate Λ_{20} and successive terms in (3.14). The details of this calculation are omitted here as they will be reported elsewhere (cf. O'Malley and Flaherty (1978)).

The asymptotic validity of the expansion (3.9) follows from Tikhonov's theorem (cf. Wasow (1976) or Vasil'eva and Butuzov (1973)). Returning to the original variables, we have found a unique asymptotic solution of the form (1.3) with the outer solution given by

$$Y(t,\varepsilon) = E^T(t)Z_1(t,\varepsilon) + E_2^T(t)Z_2(t,\varepsilon)$$

and with the exponentially decaying boundary layer correction given by

$$\Pi(\tau,\varepsilon) = E_1^T(\varepsilon\tau)\Lambda_1(\tau,\varepsilon) + \varepsilon E_2^T(\varepsilon\tau)\Lambda_2(\tau,\varepsilon) .$$

The result will even be valid for all $t \geq 0$ provided that Z_{20} decays exponentially as $t \to \infty$ (cf. Hoppensteadt (1966)).

4. THE ORIGINAL PROBLEM

We now return to the original problem where the orthogonal matrix E can depend on y as well as t. As noted in Section 2, the outer solution (2.2) should satisfy the system (2.1) as a power series in ε for $t \geq 0$. The leading term in the expansion will satisfy (1.2) and, for $j \geq 1$, $f_y(Y_0,t,0)Y_j$ will be successively determined by the preceding Y_ℓ, $\ell = 0,1,\ldots, j-1$. This fails to uniquely determine the Y_j's since $f_y(Y_0,t,0)$ has rank $k < m$. We shall instead find them as solutions of differential equations. To this end, we differentiate (2.1) with respect to t to obtain.

$$f_y(Y,t,\varepsilon)\dot{Y} + f_t(Y,t,\varepsilon) = \varepsilon\ddot{Y} \tag{4.1}$$

and use (1.2) and (4.1) together.

We define $E_{10}(t) = E_1(Y_0(t),t)$ and let $E_{20}(t), P_0(t)$, and $Q_0(t)$ be analogously defined. From (2.8) and (2.9) we see that $P_0 + Q_0 = I_m$; thus, we can write

$$\dot{Y} = P_0\dot{Y} + Q_0\dot{Y} \tag{4.2}$$

and seek to obtain equations for $P_0\dot{Y}$ and $Q_0\dot{Y}$. In particular, (4.1) and (4.2) imply

$$E_{10}f_y(Y,t,\varepsilon)(P_0\dot{Y} + Q_0\dot{Y}) = E_{10}(-f_t + \varepsilon\ddot{Y}).$$

Using the stable matrix

$$S_0 = E_{10}f_y(Y_0,t,0)E_{10}^T \tag{4.3}$$

and (2.9) we have

$$P_0\dot{Y} = -A_0\{[f_y(Y,t,\varepsilon) - f_y(Y_0,t,o)]P_0\dot{Y} + f_y(Y,t,\varepsilon)Q_0\dot{Y} + f_t(Y,t,\varepsilon) - \varepsilon\ddot{Y}\} \tag{4.4}$$

where,

$$A_0 = E_{10}^T S_0^{-1} E_{10}. \tag{4.5}$$

From (2.1) we have

$$Q_0\dot{Y} = Q_0 f(Y,t,\varepsilon)/\varepsilon. \tag{4.6}$$

Using (4.3) and (4.6) in (4.2) we find

$$\dot{Y} = -A_0\{[f_y(Y,t,\varepsilon) - f_y(Y_0,t,0)]\dot{Y} + f_t(Y,t,\varepsilon) - \varepsilon\ddot{Y}\} + B_0Q_0f(Y,t,\varepsilon)/\varepsilon$$

where

$$B_0 = I_m - A_0f_{0y}. \tag{4.8}$$

It may be useful to note that B_0 is a projection with

$$B_0P_0 = 0,\ B_0Q_0 = B_0,\ \text{and } B_0A_0 = 0.$$

Setting $\varepsilon = 0$ in (4.7) yields the limiting nonlinear equation

$$\dot{Y}_0 = -A_0f_t(Y_0,t,0) + B_0Q_0f_\varepsilon(Y_0,t,0) \tag{4.9}$$

We note that the term $Q_0f_{0y}(Y_0,t)Y_1$ is missing since $Q_0f_{0y} = 0$ upon use of (2.5) and (2.9).

In order to obtain further coefficients Y_j it is necessary to consider the coefficients of higher powers of ε in (4.7). Thus, setting

$$f(Y,t,\varepsilon) \sim \sum_{j=0}^{\infty} f_j(Y,t)\varepsilon^j \tag{4.10}$$

and using the expansion (2.2) for Y implies that

$$f(Y,t,\varepsilon) = f_0(Y_0,t) + \varepsilon[f_1(Y_0,t) + f_{0y}(Y_0,t)Y_1] + \varepsilon^2[f_2(Y_0,t) + f_{0y}(Y_0,t)Y_2 + f_{1y}(Y_0,t)Y_1 + \frac{1}{2}(f_{0yy}(Y_0,t)Y_1)Y_1] + O(\varepsilon^3)$$

together with corresponding expansions for $f_t(Y,t,\varepsilon)$ and $f_y(Y,t,\varepsilon)$. The coefficient of ε in (4.7) then provides the following nonlinear equation for Y_1

$$\dot{Y}_1 = -A_0\{f_{1t}(Y_0,t) + f_{0ty}(Y_0,t)Y_1 + [f_{1y}(Y_0,t) + f_{0yy}(Y_0,t)Y_1][P_0\dot{Y}_0 + Q_0f_1(Y_0,t)] - \ddot{Y}_0\} + B_0Q_0\{f_2(Y_0,t) + f_{1y}(Y_0,t)Y_1 + \frac{1}{2}[f_{0yy}(Y_0,t)Y_1]Y_1\}$$

Except for the final quadratic term, this is a nonhomogeneous linearization of the equation for Y_0. Higher order terms Y_j, $j\geq 2$, satisfy linear differential equations with successively known nonhomogeneous terms.

We note that it may be advantageous to obtain differential equations for the successive terms Y_j of the outer expansion even in the special case (see section 3) where $E(y,t)$ is independent of y. In that case, we solved the nonlinear algebraic equation (3.10a) for Z_{10} as a function of Z_{20}, followed by a nonlinear initial value problem (3.10b) for Z_{20}. It may often be numerically simpler to solve the initial value problem (4.9) for $Y_0(t)$ and those for later terms successively. We have not, however, fully explored both possibilities.

We will have to assume, of course, that the nonlinear initial value problems, (4.9) and (4.11), for Y_0 and Y_1 have solutions on $0 \leq t \leq T$. Moreover, since (1.2) and its time derivative (4.1) are built into (4.7), consistency with (1.2) at $t = 0$ implies consistency of the outer expansion for $t > 0$. If consistency failed, the form (1.3) of the solution would be inappropriate. Thus, using (1.2), (2.1), (2.2), and (4.10), we must have

$$
\begin{aligned}
&f_0(Y_0(0),0) = 0, \\
&f_{0y}(Y_0(0),0)Y_1(0) = \dot{Y}_0(0) - f_1(Y_0(0),0) \, , \qquad (4.12) \\
&\cdot \; \cdot \; \cdot \; \cdot \; \cdot
\end{aligned}
$$

These equations always have a solution under our assumptions. For example, in the second equation we must have $\dot{Y}_0(0) - f_1(Y_0(0),0)$ in the range of $f_{0y}(Y_0(0),0)$. Recall, however, that $I_m = P_0 + Q_0$ provides a direct sum decomposition of m space with

$$R(Q_0) = N(f_{0y}^T(Y_0(0),0)) \quad \text{and} \quad R(P_0) = R(f_{0y}(Y_0(0),0)).$$

Thus, the second of (4.12) will be automatically satisfied since $Q_0 f_{0y} = 0$ implies that $Q_0[\dot{Y}_0(0) - f_1(Y_0(0),0)] = 0$.

Because f_{0y} has rank k, k components of $Y_0(0)$ are determined as a function of the remaining m-k components. Indeed, we could attempt to solve (1.2) for $E_{10}(0)Y_0(0)$ in terms of $E_{20}(0)Y_0(0)$ since $S_0(0)$ (cf. (4.3)) is nonsingular. Likewise, for $j > 0$, termwise determination of $f_{0y}(Y_0(0),0)Y_j(0)$ implies that of $E_{10}(0)Y_j(0)$ (by an argument similar to the one preceding (4.3)). Thus, $E_{10}(0)Y_j(0)$, or $P_0(0)Y_j(0)$, is determined termwise while $E_{20}(0)Y_j(0)$, or $Q_0(0)Y_j(0)$, may be specified. The purpose of the boundary layer correction is to compensate for the jump in $P_0(0)(Y_j(0) - y_j^0)$ and to specify the values of $Q_0(0)Y_j(0)$, $j \geq 0$.

Once again, the representation (1.3) and the fact that the differential equation (1.1) is satisfied by both y and Y imply that the boundary layer correction $\Pi(\tau,\varepsilon)$ must satisfy the nonlinear equation

$$\frac{d\Pi}{d\tau} = f(Y(\varepsilon\tau,\varepsilon) + \Pi(\tau,\varepsilon),\varepsilon\tau,\varepsilon) - f(Y(\varepsilon\tau,\varepsilon),\varepsilon\tau,\varepsilon) \tag{4.13}$$

as a power series

$$\Pi(\tau,\varepsilon) \sim \sum_{j=0}^{\infty} \Pi_j(\tau)\varepsilon^j \tag{4.14}$$

in ε and decay to zero as $\tau \to \infty$. Moreover,

$$\Pi(0,\varepsilon) \sim y^0(\varepsilon) - Y(0,\varepsilon). \tag{4.15}$$

The details of the calculation of the boundary layer correction and a proof of the asymptotic validity of the solution are omitted here and will be presented in O'Malley and Flaherty (1978). We summarize our findings, however, in the following theorem.

Theorem: Consider the initial value problem

$$\varepsilon\dot{y} = f(y,t,\varepsilon) \; , \; y(0) = y^0(\varepsilon)$$

for an m-vector y as $\varepsilon \to 0^+$. Assume that:

(i) f is infinitely differentiable in y and t and f and $y^0(\varepsilon)$ have asymptotic series expansions in powers of ε.

(ii) There exists an infinitely differentiable orthogonal matrix $E(y,t)$ for all y and for t in the interval $0 \le t \le T$ such that $E(y,t)f_y(y,t,0)$ is row-reduced and of rank k, $0 \le k < m$. Moreover, partitioning E after its first k rows as in (2.4), we have

$$Ef_y(y,t,0)E^T = \begin{bmatrix} S & X \\ 0 & 0 \end{bmatrix}$$

where $S = E_1 f_y(y,t,0)E_1^T$ is a stable matrix for all values of y and t.

(iii) The nonlinear system

$$f(Y_0(0),0,0) = 0 \tag{4.16a}$$

$$Q(Y_0(0),0)[y^0(0)-Y_0(0) + \int_0^\infty f(Y_0(0) + \Pi_0(\tau),0,0)d\tau] = 0 \tag{4.16b}$$

can be uniquely solved for $Y_0(0)$. Here, $\Pi_0(\tau)$ is the decaying solution of

$$\frac{d\Pi_0}{d\tau} = f(Y_0(0) + \Pi_0(\tau),0,0) - f(Y_0(0),0,0),$$
$$\Pi_0(0) = y^0(0) - Y_0(0).$$

(iv) The matrix

$$I - C_0E_{10}(0)B_0(Y_0(0),0)$$

is invertible for a particular matrix C_0. (This insures that $Y_1(0)$ may be uniquely determined.)

(v) The initial value problems (4.9) and (4.11) have

solutions on the interval $0 \le t \le T$.

Then, the initial value problem (1.1) has a unique solution of the form

$$y(t,\varepsilon) = Y(t,\varepsilon) + \Pi(\tau,\varepsilon).$$

Some comments on this theorem are listed below.

(i) Hypothesis (ii) is guaranteed by out earlier Hypothesis (H).

(ii) The theorem is easily obtained from Tikhonov's result if $E(y,t)$ is independent of y. It is considerably simplified if only $E_2(Y_0(0),0)$, and thereby $Q(Y_0(0),0)$, is independent of $Y_0(0)$. In this case (4.16b) reduces to the linear equation

$$Q_0(0)[y^0(0) - Y_0(0)] = 0 \tag{4.17}$$

and (4.16a) becomes a nonlinear equation for $P_0(0)Y_0(0)$. It can be further shown that the invertibility condition of Hypothesis (iv) then will be automatically satisfied.

(iii) Higher order terms follow without complication under these hypotheses.

(iv) Vasil'eva (1976) considers such problems under a list of ten hypotheses, generally paralleling, but more restrictive than ours. Her most critical assumption involves the existence of a k-dimensional manifold of decaying solutions for $\Pi_0(\tau)$ which can, more or less, be stated in the form

$$E_{20}(0)\Pi_0(\tau) = \Phi(E_{10}(0)\Pi_0(\tau))$$

for a particular function Φ and for all τ and $Y_0(0)$. At $\tau = 0$, we would have

$$E_{20}(0)[y^0(0) - Y_0(0)] = \Phi(E_{10}(0)[y^0(0) - Y_0(0)])$$

where $Y_0(0)$ must also satisfy the reduced equation at $\tau = 0$. This analog of Hypothesis (iii) should uniquely determine $Y_0(0)$ so that the resulting $\Pi_0(0)$ lies on the manifold of initial values corresponding to decaying solutions of $\Pi_0(\tau)$.

5. NUMERICAL ALGORITHM

We have developed an algorithm based on the asymptotic analysis of Section 4 to calculate the leading term $Y_0(t)$ in the outer solution. For most problems it is possible to calculate numerical solutions without explicitly identifying a small parameter ε; thus, we consider initial value problems in the form

$$\dot{y} = \hat{f}(y,t,\varepsilon) \equiv f(y,t,\varepsilon)/\varepsilon \ , \ y(0) = y^{\bullet}(\varepsilon). \tag{5.1}$$

The ε, although shown in (5.1), is to be regarded as unidentifiable. However, if the actual limiting solution is desired, the evaluation of $\dot{y}$ in (5.1) causes overflow, or the order ε terms in f are close to the unit roundoff of the computer relative to the order unity terms in f, then a value of ε can be identified and the initial value problem can be written in the form of equation (1.1). The actual computer code is capable of handling both cases, and all that the user need do is define $\hat{f}$ as in (5.1) or f as in (1.1).

The algorithm consists of two parts: (i) calculating the initial conditions $Y_0(0)$ for the outer problem and (ii) integrating the differential equation (cf., (4.9)) for $Y_0(t)$. We first describe the integration procedure.

The differential equation (4.9) for Y_0 is not stiff; hence, any good code for integrating non-stiff initial value problems may be used. We use both the Adams' methods that are incorporated into the Hindmarsh (1974) version of Gear's code and the IMSL version of the Bulirsch and Stoer (1966) extrapolation procedure. Both of these codes require the evaluation of $\dot{Y}_0$ as a function of Y_0 and t, and we accomplish this as follows:

(i) Calculate $E(Y_0,t)$ by decomposing $\hat{f}_y(Y_0,t,\varepsilon)$. It is not necessary to set ε to 0 unless ε has been explicitly recognized and the actual limiting solution is desired. Golub's (1965) procedure, which uses a sequence of k Householder transformations with column pivoting, is used to obtain E. At the ν *th* step, $1 \le \nu \le k$, of this procedure we have

$$E^{(\nu)}(Y_0,t)\hat{f}_y(Y_0,t,\varepsilon) = \begin{bmatrix} U & V \\ 0 & W \end{bmatrix} K^T$$

where, U is $\nu \times \nu$ and upper triangular, V is $\nu \times (m-\nu)$, and K is a permutation matrix due to the column pivoting. The procedure terminates, and the rank k of $\hat{f}_y$ is determined, when the maximum available pivot element in W is small relative to the diagonal elements of U. We then have $E = E^{(k)}$. The decomposition is not performed at every time step, but, rather a test is used to determine if E has changed by too much. Thus, the same matrix E may be used for several time steps or, when E is constant, for the entire integration. If at any stage of the computation the rank k of $\hat{f}_y$ changes, a turning point has probably been encountered, and the integration is terminated.

(ii) Partition E into E_1 and E_2 as in (2.4). Calculate

$$Q = E_2^T E_2 \text{ and } S = E_1 \hat{f}_y(Y_0,t,\varepsilon) E_1^T.$$

(iii) Calculate $Q\dot{Y}_0 = Q\hat{f}(Y_0,t,\varepsilon)$ and $b = -E_1[\hat{f}_t(Y_0,t,\varepsilon) + \hat{f}_y(Y_0,t,\varepsilon)(Q\dot{Y}_0)]$. When ε is explicitly recognized $Q\dot{Y}_0$ is calculated as $Qf_\varepsilon(Y_0,t,0)$.

(iv) Solve $S(E_1\dot{Y}_0) = b$ for $E_1\dot{Y}_0$ by Gaussian elimination and calculate

$$P\dot{Y}_0 = E_1^T(E_1\dot{Y}_0).$$

(v) Calculate $\dot{Y}_0 = P\dot{Y}_0 + Q\dot{Y}_0$.

We now turn to the calculation of the initial conditions $Y_0(0)$ for the outer problem. This is a difficult task when E_2 depends on y. It requires the solution of the nonlinear system (4.16) and the computation of the boundary layer solution $\Pi_0(\tau)$, which itself depends on $Y_0(0)$. It is possible that the integral in (4.16b) may be adequately approximated by a very crude quadrature rule, which would greatly simplify the computation. Miranker (1973) has successfully used such a technique on stiff problems, but we have not as yet explored this possibility. Our code has only been implemented for problems where E_2 is independent of y; thus, when ε is not explicitly recognized $Y_0(0)$ is determined as the solution of

$$\begin{aligned} &\hat{f}(Y_0(0),0,\varepsilon) = 0 \\ &E_2[Y_0(0) - y^0(\varepsilon)] = 0. \end{aligned} \tag{5.2}$$

A Newton like iteration scheme, which closely parallels the computation of $\dot{Y}_0(t)$ is used to solve this nonlinear system. The procedure is outlined below.

(i) Select an initial guess $X^{(0)}$ for $Y_0(0)$, e.g., $X^{(0)} = y^0$ and set $\mu = 0$.

(ii) Calculate $E^{(\mu)}$ by decomposing $\hat{f}_y(X^{(\mu)},0,\varepsilon)$. This is performed as in step (i) of the procedure for calculating $\dot{Y}_0$. If E_1 is independent of y then this step need only be performed once.

(iii) Calculate $Q^{(\mu)}$ and $S^{(\mu)}$ as in step (ii) of the previous procedure.

(iv) Calculate $q^{(\mu+1)} = Q^{(\mu)}(y^0 - X^{(\mu)})$ and

$$b^{(\mu)} = -E_1^{(\mu)}[\hat{f}(X^{(\mu)},0,\varepsilon) + \hat{f}_y(X^{(\mu)},0,\varepsilon)q^{(\mu+1)}]$$

(v) Solve $S^{(\mu)}[E_1^{(\mu)}(X^{(\mu+1)} - X^{(\mu)})] = b^{(\mu)}$ for $E_1^{(\mu)}(X^{(\mu+1)} - X^{(\mu)})$ and calculate $p^{(\mu+1)} = (E_1^{(\mu)})^T[E_1^{(\mu)}(X^{(\mu+1)} - X^{(\mu)})]$

(vi) Set $X^{(\mu+1)} = X^{(\mu)} + p^{(\mu+1)} + q^{(\mu+1)}$.

If $\| X^{(\mu+1)} - X^{(\mu)} \|$ is less than some prescribed tolerance set $Y_0(0) = X^{(\mu+1)}$, otherwise increment μ by 1 and repeat steps (ii) through (vi).

Of course, if the problem is almost linear then only one iteration need be performed.

The entire procedure was successfully applied to several examples, some of which are discussed in the next section.

6. NUMERICAL EXAMPLES AND DISCUSSIONS OF RESULTS

In this section we present and discuss the results of three examples comparing our method of Section 5 with Hindmarsh's (1974) version of Gear's code for stiff differential equations. Both the Adams' methods that we use to integrate the reduced differential equation and Gear's stiffly stable methods are contained in this code, and the user sets a parameter to select the appropriate method. Hindmarsh's code and the IMSL Bulirsch and Stoer code also require the user to select an estimate for the relative local discretization error and an initial step size for the integration. In all cases we selected the relative error tolerance as 10^{-6}. This is perhaps a bit too severe for our methods, because if the problem is not very stiff the reduced solution will be calculated with more accuracy than necessary. The initial step size was selected as 10^{-4} for Adams' methods, $1/10\varepsilon$ for Gear's methods, and 1 for Bulirsch and Stoer's method. We found that the IMSL code was extremely sensitive to the choice of the initial step size and the times for the integration varied quite dramatically depending on this choice. Our choice of unity seemed near optimal for the problems that we considered.

In the tables that follow five numerical solutions are compared. The solutions labeled "asymptotic" were calculated by our method without explicitly recognizing ε and using either the Hindmarsh (Adams) or the IMSL codes; those labeled "Gear" were solved by Hindmarsh's (Gear) code; and those labeled "reduced" were calculated by our method with ε explicitly set to zero. Additional headings in the tables are as follows:

e is the maximum difference in any component, times 10^6, between a numerical solution and the exact solution at terminal time T. For our asymptotic or limiting solutions

$$e = \max_{1 \le i \le m} |Y_{0i}(T) - y_i(T)| \times 10^6.$$

In general, for the examples considered, the error was fairly constant outside of the initial boundary layer.

d when the exact solution is not known, we have tabulated the maximum difference in any component, times 10^6, between solutions obtained by our method and those by Gear's code at terminal time T.

NFE For our asymptotic or limiting solutions this denotes the number of times that $\dot{Y}_0$ was evaluated during the course of the integration. For Gear's solutions it denotes the number of times that $\dot{y}$ was evaluated.

NJE For Gear's solutions this denotes the number of times that $\hat{f}_y$ was evaluated during the course of the integration. Our methods evaluate $\hat{f}_y$ each time $\dot{Y}_0$ is evaluated.

CP Time in milli-seconds to integrate the problem, excluding input/output and supervisor state time. Except where noted it includes the time necessary to calculate the initial conditions $Y_0(0)$ by our method. In most cases the times are averaged over several runs. All calculations were performed on an IBM 360/67 at the Rensselaer Polytechnic Institute.

CP_{rel} CP time relative to the fastest execution time.

The individual examples are discussed below.

Example 1:

$$\dot{y} = \begin{bmatrix} (1/\varepsilon-1) & 2(1/\varepsilon-1) \\ -(1/\varepsilon-1) & -(2/\varepsilon-1) \end{bmatrix} y, \quad y(0) = \begin{bmatrix} 1 \\ 1 \end{bmatrix}, \quad 0 \le t \le T = 1.$$

This constant coefficient example is an adaptation of one considered by Gear (1971). The exact solution is

$$y(t) = \begin{bmatrix} 4e^{-t} & -3e^{-t/\varepsilon} \\ -2e^{-t} & +3e^{-t/\varepsilon} \end{bmatrix}$$

The results are compared in Table 1 for $\varepsilon = 10^{-i}$, $i=2,4,6,8$. They are typical of the results of subsequent examples in that they show that the accuracy of our method increases as the stiffness increases without an increase in computational effort. On the average, our asymptotic and reduced solutions required 5 milli-seconds to calculate the initial conditions for the outer problem.

TABLE I

Summary of Results for Example 1

METHOD		ε = 10^{-2}	10^{-4}	10^{-6}	10^{-8}
Asymptotic (Adams)	e	11100.	111.	1.81	.0130
	NFE	34	34	34	34
	CP	135	137	138	129
	CP_{rel}	1.57	1.60	1.62	1.51
Asymptotic (IMSL)	e	11100.	110.	1.10	.00171
	NFE	33	33	33	33
	CP	89.1	89.8	91.0	89.8
	CP_{rel}	1.04	1.05	1.05	1.05
Reduced (Adams)	e	.126	.701	.701	.701
	NFE	34			
	CP	135			
	CP_{rel}	1.58			
Reduced (IMSL)	e	.000309	.000309	.000309	.000309
	NFE	33			
	CP	85.5			
	CP_{rel}	1.00			
Gear	e	3.51	7.61	8.85	4.40
	NFE	158	183	188	195
	NJE	17	24	25	27
	CP	332	396	407	422
	CP_{rel}	3.89	4.63	4.75	4.94

TABLE II

Summary of Results for Example 2

METHOD		ε 10^{-2}	10^{-4}	10^{-6}	10^{-8}
Asymptotic (Adams)	e	300.	1.98	.640	.666
	NFE	46	46	46	46
	CP	196	192	188	196
	CP_{rel}	1.36	1.33	1.31	1.37
Asymptotic (IMSL)	e	301.	2.65	.0284	.00223
	NFE	49	49	49	49
	CP	154	150	152	148
	CP_{rel}	1.07	1.05	1.06	1.03
Reduced (Adams)	e	214.	2.81	.688	.667
	NFE	46			
	CP	187			
	CP_{rel}	1.30			
Reduced (IMSL)	e	214.	2.14	.0201	.00175
	NFE	49			
	CP	144			
	CP_{rel}	1.00			
Gear	e	.541	1.26	3.93	3.97
	NFE	167	191	196	203
	NJE	20	25	26	28
	CP	341	399	406	421
	CP_{rel}	2.38	2.78	2.83	2.93

TABLE III

Summary of Results for Example 3

METHOD		ε 10^{-2}	10^{-4}	10^{-6}	10^{-8}
Asymptotic (Adams)	d	317.	3.39	.352	.358
	NFE	30	30	30	30
	CP	170	170	170	172
	CP_{rel}	1.66	1.66	1.66	1.68
Asymptotic (IMSL)	d	317.	3.56	.440	.446
	NFE	21	21	21	21
	CP	108	107	107	107
	CP_{rel}	1.05	1.05	1.05	1.05
Reduced (Adams)	d	1217.	12.0	.269	.358
	NFE	30			
	CP	165			
	CP_{rel}	1.61			
Reduced (IMSL)	d	1217	12.1	.411	.445
	NFE	21			
	CP	102			
	CP_{rel}	1.00			
Gear	NFE	143	144	150	158
	NJE	17	19	21	23
	CP	343	365	380	400
	CP_{rel}	3.35	3.57	3.72	3.91

TABLE IV

Time to integrate from t=0 to t = T = 10ε for Example 3

ε	Asymptotic (Adams)		Gear				
	NFE	CP	NFE	NJE	CP	d	CP_{ratio}
10^{-2}	12	82.9	119	14	316	451.	3.81
10^{-4}	5	52.1	121	15	341	4.81	6.54
10^{-6}	2	37.1	121	15	342	.0361	9.22
10^{-8}	2	37.4	121	15	341	.0231	9.12

TABLE V

Time to integrate from t=0 to t = T = 1 for Example 3 using initial conditions for the outer problem (results for ε = 0 are the reduced solution)

	Asymptotic/ Reduced (Adams)			Asymptotic/ Reduced (IMSL)			Gear			
ε	NFE	CP	CP_{rel}	NFE	CP	CP_{rel}	NFE	NJE	CP	CP_{rel}
10^{-2}	30	142	1.87	21	79	1.04	54	6	103	1.36
10^{-4}	30	141	1.87	21	79	1.04	56	5	102	1.35
10^{-6}	30	141	1.87	21	79	1.04	43	8	100	1.32
10^{-8}	30	143	1.89	21	78	1.04	51	10	118	1.55
0	30	138	1.83	21	76	1.00				

Example 2:

$$\dot{y} = \frac{1}{\varepsilon}\begin{bmatrix} (y_1+y_2)[1 - \frac{1}{2}(y_1+y_2)^2] - \frac{\varepsilon}{\sqrt{2}}(y_2^2-y_1^2) \\ (y_1+y_2)[1 - \frac{1}{2}(y_1+y_2)^2] + \frac{\varepsilon}{\sqrt{2}}(y_2^2-y_1^2) \end{bmatrix}, \quad y(0) = \begin{bmatrix} -2 \\ 0 \end{bmatrix}$$

$$0 \le t \le T = 2$$

This nonlinear problem was contrived so that the orthogonal matrix E is constant and the exact solution is known as

$$y_1(t) = (\xi - \eta)/\sqrt{2} \qquad y_2(t) = (\xi + \eta)/\sqrt{2}$$

with

$$\xi = -(1 - \tfrac{1}{2}e^{-2t/\varepsilon})^{-1/2} \qquad \eta = \sqrt{2}\; e^{-t}\left(\frac{1-1/\xi}{1+1/\sqrt{2}}\right)^{-\varepsilon}$$

The results are presented in Table 2 and generally parallel those for Example 1. The average time required to calculate the initial conditions for the asymptotic and reduced solutions was 24 milli-seconds

Example 3:

$$\dot{y} = \hat{f}(y,\varepsilon) = \frac{1}{\varepsilon}\begin{bmatrix}(y_2^2 - y_1y_3) - \varepsilon y \\ 2(y_1y_3 - y_2^2) + \varepsilon y \\ (y_2^2 - y_1y_3)\end{bmatrix}, y(0) = \begin{bmatrix}1\\0\\1\end{bmatrix}, \quad 0 \le t \le T = 1$$

This example arises in chemical reactions and was studied by Vasil'eva (1976). She did not specify the εy_1 terms in $\hat{f}$ nor the initial conditions and they were selected by us rather arbitrarily. The Jacobian $\hat{f}_y(y,0)$ of this system has rank 1 for all $y \neq 0$ and it may be row-reduced by a constant orthogonal matrix E. The results of this example are presented in Table 3. The average time required to calculate the initial conditions for the asymptotic and reduced solutions was 28 milli-seconds.

Our method is to be used on problems where the boundary layer solution is not of interest; hence, we should be able to calculate the initial conditions for the outer problem faster than a stiff differential equation solver could integrate through the boundary layer. In order to provide some evidence that this is the case we solved Example 3 in the interval $0 \le t \le 10\varepsilon$ (the approximate boundary layer region) using Gear's methods and our asymptotic method with the Adams' integrators. The results are presented in Table 4 for $\varepsilon = 10^{-i}$, $i = 2,4,6,8$. The CP times for our method includes both the times to calculate the initial conditions and to integrate the outer problem from $t = 0$ to 10ε. To make the comparison somewhat more fair we re-evaluated E after each iteration, even though it is constant for this example. For $\varepsilon \leq 10^{-6}$ we see that our method can calculate the solution at the edge of the boundary layer region approximately 9 times faster than Gear's methods.

A comparison of the results in Tables 3 and 4 shows that about 90% of the time required to integrate Example 3 from $t = 0$ to 1 by Gear's code is devoted to the boundary layer region for $\varepsilon \leq 10^{-4}$. This suggests the possibility of using our method to calculate the initial conditions for the outer problem and then using a stiff method to integrate the original differential equation. This test was performed on Example 3, and the results are reported in Table 5. All methods use the same initial conditions, i.e., those generated by our method. The CP times required to calculate these conditions are not included in Table 5. The difference between any two computed solutions is less than 3×10^{-4}. While the results are far from conclusive, they do show the extra computational effort that is required by Gear's method for very stiff problems.

The state of the art of numerical methods for stiff initial value problems for ordinary differential equations is very well developed (cf. Enright *et al* (1975)) and a variety of good techniques exist. Nevertheless, there are many problems, particularly in chemical reactions, where asymptotic methods should be useful. They may be used to calculate accurate solutions of very stiff problems, to furnish initial conditions for standard stiff integration routines, and/or as an analytical tool to provide qualitative information about the solutions of stiff problems. In future papers we hope to extend our calculations to initial value problems where E depends on y and to consider boundary value problems.

ACKNOWLEDGEMENTS

This work was supported in part by the Air Force Office of Scientific Research, Grant Number AFOSR-75-2818 and by the Office of Naval Research, Contract Number N00014-76-C-0326.

REFERENCES

Boggs, P.T. and Tolle, J.W. (1977). "Asymptotic Analysis of a Saddle Point of a Two Parameter Multiplier Function", *Tech. Rep. No. 6, Operations Research and Systems Analysis*, Univ. of North Carolina, Chapel Hill.

Bulirsch, R. and Stoer, J. (1966). "Numerical Treatment of Ordinary Differential Equations by Extrapolation Methods", *Numer. Math.*, 8, 1-13.

Campbell, S.L. and Rose, N.G. (1978). "Singular Perturbations of Autonomous Linear Systems", to appear in *SIAM J. Math. Anal.*, 9.

Clasen, R.J., Garfinkel, D., Shapiro, N.Z. and Roman, G.-C. (1978). "A Method for Solving Certain Stiff Differential Equations", *SIAM J. Appl. Math.*, 34, 732-742.

Enright, W.H., Hull, T.E. and Lindberg, B. (1975). "Comparing Numerical Methods for Stiff Systems of O.D.E.'s", *BIT*, 15, 10-48.

Gear, C.W. (1971). *Numerical Initial Value Problems in Ordinary Differential Equations*, chapter 11. Prentice Hall, Englewood Cliffs, N.J.

Golub, G.H. (1965). "Numerical Methods for Solving Linear Least Squares Problems", *Numer Math.*, 7, 206-216.

Golub, G.H. and Pereyra, V. (1976). "Differentiation of Pseudoinverses, Separable Nonlinear Least Square Problems, and Other Tales" in *Generalized Inverses and Applications*, (M.Z. Nashed, ed.), 303-324.

Hindmarsh, A.C. (1974). "Gear: Ordinary Differential Equation System Solver", UCID-30001 (Rev. 3), Lawrence Livermore Laboratory, University of California, Livermore, California 94550.

Hoppensteadt, F.C. (1966). "Singular Perturbations on the Infinite Interval", *Trans. Amer. Math. Soc.*, 123, 521-535.

Howes, F.A. (1978). "Singularly Perturbed Nonlinear Boundary Value Problems with Turning Points", *SIAM J. Math. Anal.*, 9, 250-271.

Kreiss, H.O. (1978). "Difference Methods for Stiff Ordinary Differential Equations", *SIAM J. Numer. Anal.*, 15, No. 1, 21-58.

Miranker, W.L. (1973). "Numerical Methods of Boundary Layer Type for Stiff Systems of Differential Equations", *Computing*, 11, 221-234.

O'Malley, R.E., Jr. (1974). *Introduction to Singular Perturbations*, Academic Press, New York.

O'Malley, R.E., Jr. (1978). "On Singular Singularly-Perturbed Initial Value Problems", to appear in *Applic. Anal.*

R.E. O'Malley, Jr. and Flaherty, J.E. (1976). "Singular Singular-Perturbation Problems", in *Lecture Notes in Mathematics, No. 594, Singular Perturbations and Boundary Layer Theory*, Springer Verlag, Berlin, 422-436.

O'Malley, R.E., Jr. and Flaherty, J.E. (1978). "Numerical Methods for Singular Singularly-Perturbed Initial Value Problems", in preparation.

Vasil'eva, A.B. (1976). "Singularly Perturbed Systems with an Indeterminacy in their Degenerate Equations", *Soviet Math. Dokl.*, 12, 1227-1235.

Vasil'eva, A.B. and Butuzov, G.F. (1973). *Asymptotic Expansions of Solutions of Singularly Perturbed Differential Equations*, Nauka, Moscow.

Wasow, W. (1976). *Asymptotic Expansions for Ordinary Differential Equations*, Kreiger, Huntington, N.Y.

COERCIVE SINGULAR PERTURBATIONS: STABILITY AND CONVERGENCE

L.S. Frank
Mathematical Institute, The University of Nijmegen
Nijmegen, The Netherlands

ABSTRACT

Singularly perturbed differential and difference boundary value problems are considered for the linearized model of thin elastic plates. Necessary and sufficient conditions for uniform stability are pointed out. Convergence results are stated in the case, when the ratio $\rho = h/\varepsilon$ (with h the mesh-size and ε the original parameter) is bounded from below by some positive constant: $\rho \geq \rho_0 > 0$.

§1. INTRODUCTION

Singularly perturbed linear boundary value problems for Ordinary Differential Operators on a finite interval $U \subset \mathbb{R}$ are considered. We indicate the ellipticity and coerciveness conditions (on the operator in U and the boundary operators on ∂U) which are necessary and sufficient for a two-sided a priori estimate to hold uniformly with respect to the small parameter for the solutions to the singular perturbation in appropriate Sobolev type spaces.

An elliptic coercive singular perturbation might happen to be approximated by finite differences (or finite elements) in such a way that the original stability of the problem is destroyed. We point out the ellipticity and coerciveness conditions on difference approximations, which are necessary and sufficient for a two-sided a priori estimate to hold uniformly with respect to both parameters, the original one and the meshsize, in Sobolev type spaces of meshfunctions.

Convergence of difference solutions to the differential one is also discussed.

For the sake of simplicity, the results are presented for the linearized one-dimensional model of thin elastic plates. For stability results in the general case we refer to [2-4]. Convergence results with proofs will be published elsewhere.

§2. LINEARIZED ONE-DIMENSIONAL MODEL OF THIN ELASTIC PLATES. (DIFFERENTIAL PROBLEM).

Notation and statement of the problem. 2.1. - Let U be the unit interval in $\mathbb{R}$ and $\partial U = \{0,1\}$ the boundary of U. We denote by $\varepsilon \in (0,\varepsilon_0]$ a positive parameter and assume that $\varepsilon_0 << 1$.

For a function $u : (0,\varepsilon_0] \times \overline{U} \to C$ and a vector $s = (s_1,s_2,s_3) \in \mathbb{R} \times \mathbb{N}_+ \times \mathbb{N}_+$, i.e. s_2,s_3 non-negative integers, define the Sobolev type norm of order s:

$$(2.1.1)\quad ||u||_{(s)}=\varepsilon^{-s_1}(||u||_0+||\partial^{s_2}u||_0+\varepsilon^{s_3}||\partial^{s_2+s_3}u||_0),$$
$$|u|_{(s)} = \sup_{\varepsilon} ||u||_{(s)},$$

where $\partial = -id/dx$ and $||.||_0$ is the norm in $L_2(U)$. Denote by $H_{(s)}(U)$ the space of all functions u with the norm (2.1.1) finite. $H_{(s)}(U)$ is a Banach space and for any fixed ε it can be provided with the natural inner product. In fact, using the Fourier transform and the standard partition of unity argument, one defines the norms $||\cdot||_{(s)}$, $|\cdot|_{(s)}$ and the corresponding spaces $H_{(s)}(U)$ for any $s \in \mathbb{R}^3$.

For a function $\phi : (0,\varepsilon_0] \to C$ and $l \in \mathbb{R}$ define the norm of order l:

$$(2.1.2)\quad [\phi]_l = \varepsilon^{-l}|\phi|,\ |[\phi]|_l = \sup_{\varepsilon} [\phi]_l,$$

and denote by C_l the Banach space of all functions ϕ with the norm (2.1.2) finite.

Let

$$(2.1.3)\quad A(x,\varepsilon,\partial) = \varepsilon^2\partial^4+\varepsilon a_1(x)\partial^3+a_2(x)\partial^2+R(x,\varepsilon,\partial)$$

where the remainder $R(x,\varepsilon,\partial)$ is given by the formula

$$(2.1.4)\quad R(x,\varepsilon,\partial)=\varepsilon^{\gamma}(\varepsilon b_1(x,\varepsilon)\partial^3+b_2(x,\varepsilon)\partial^2)+b_3(x,\varepsilon)\partial+b_4(x,\varepsilon)$$

with $\gamma > 0$, $a_j \in C^{\infty}(\overline{U})$, $b_j \in C^{\infty}(\overline{U})$, b_j being continuous on $[0,\varepsilon_0] \times \overline{U}$.

The function

$$(2.1.5)\quad A_0(x,\varepsilon,\xi) = \varepsilon^2\xi^4+\varepsilon a_1(x)\xi^3+a_2(x)\xi^2,\qquad (x,\varepsilon,\xi) \in \overline{U} \times (0,\varepsilon_0] \times \mathbb{R},$$

is the principle symbol of the operator A.

Definition 2.1.1. - The operator A defined by (2.1.3), (2.1.4) is called an elliptic singular perturbation of order $\nu = (0,2,2)$ if its principal symbol A_0 defined by (2.1.5) satisfies the condition:

$$(2.1.6)\quad M_0(x,\eta) \neq 0,\ \ \forall\ (x,\eta) \in \overline{U} \times \mathbb{R}$$

where

$$(2.1.7)\quad M_0(x,\eta) = A_0(x,1,\eta)/A_0(x,0,\eta).$$

One can consider a more general class of singularly perturbed Ordinary Differential operators A whose principal symbol is given by the formula

(2.1.8) $A_0(x,\varepsilon,\xi) = \varepsilon^{-\nu_1}\xi^{\nu_2} \sum_{0\le p\le \nu_3} a_p(x)(\varepsilon\xi)^p$

where $\nu = (\nu_1,\nu_2,\nu_3) \times \mathbb{R} \times \mathbb{N}_+ \times \mathbb{N}_+$, $a_p \in C^\infty(\overline{U})$ and $a_{\nu_3}(x) \equiv 1$, ν being the order of A and the class of all singularly perturbed operators of order ν being denoted by L_ν.

An operator $A \in L_\nu$ is called elliptic of order ν if its principal symbol satisfies (2.1.6) with $M_0(x,\eta)$ given by (2.1.7), (2.1.8).

Let $B_j(x',\varepsilon,\partial) \in L_{\mu_j(x')}$ with $\mu_j(x')=(\alpha_j(x'),m_j(x'),p_j(x'))$, $x' \in \partial U$, $j = 1,2$, and denote by $B_{j0}(x',\varepsilon,\xi)$ the principal symbol of B_j. Assume

(2.1.9) $m_1(x') < m_2(x')$, $\forall\, x' \in \partial U$.

Let $s \in \mathbb{R}^3$ be such that

(2.1.10) $s_2 < \max_{x'\in\partial U} m_2(x') + 1/2$,

$$s_2+s_3 > \max_{x'\in\partial U}\ \max_{j=1,2} \{m_j(x')+p_j(x')\} + 1/2.$$

Consider the singularly perturbed boundary value problem:

(2.1.11) $A(x,\varepsilon,\partial)u(x) = f(x)$, $x \in U$,

(2.1.12) $\pi_0 B_j(x',\varepsilon,\partial)u(x') = \phi_j(x')$, $j = 1,2$, $x' \in \partial U$,

where $f \in H_{(s-\nu)}(U)$, $\phi_1(x') \in C_{\sigma_1(x')}$ with $\sigma_1(x') = s_1-\alpha_1(x')$, $\phi_2(x') \in C_{\tau_2(x')}$ with $\tau_2(x') = s_1+s_2-\alpha_2(x')-m_2(x')-1/2$ and solutions $u(x)$ are sought in $H_{(s)}(U)$ with $s \in \mathbb{R}^3$ satisfying the condition (2.1.10). Besides, the restriction operator $\pi_0 v(x')$ in (2.1.12) is interpreted as the trace of the (continuous on $\overline{U}$) function $v(x)$ on the boundary ∂U.

Denote by $\eta(x')$ the zero of the quadratic equation

$M_0(x',\eta) = 0$, which is contained in the upper complex half plane when $x' = 0$ and in the lower one when $x' = 1$, so that $\operatorname{Im}\eta(0) > 0$ and $\operatorname{Im}\eta(1) < 0$.

Definition 2.1.2. - The problem (2.1.11), (2.1.12) is called the coercive singular perturbation if A in (2.1.11) is elliptic of order $\nu = (0,2,2)$ and the principal symbols of $B_2(x',\varepsilon,\partial)$, $x' \in \partial U$, satisfy the condition:

(2.1.13) $B_{20}(x',1,\eta(x')) \neq 0, \quad \forall\, x' \in \partial U.$

Remark 2.1.3. - For the linearized one-dimensional model of thin elastic plates with $A_0(x,\varepsilon,\partial) = \varepsilon^2\partial^4+\partial^2$ and the Dirichlet boundary conditions: $B_j(x',\varepsilon,\partial) = \partial^{j-1}$, $j = 1,2$, $x' \in \partial U$, one finds: $M_0(x,\eta) = \eta^2+1$, $\eta(0) = i$, $\eta(1) = -i$, so that $B_{20}(x',1,\eta(x')) = \eta(x') = (-1)^{x'}i \neq 0$, $\forall\, x' \in \partial U$.

The reduced problem related with (2.1.11), (2.1.12) is stated as follows:

(2.1.14) $A^0(x,\partial)u^0(x) = f^0(x), \quad x \in U$

(2.1.15) $\pi_0 B_1^0(x',\partial)u^0(x') = \phi_1^0, \quad x' \in \partial U,$

where $f^0(x) = \lim\limits_{\varepsilon\to 0} \varepsilon^{-s_1}f(x)$, $f^0 \in H_{s_2-2}(U)$,

$\phi_1^0(x') = \lim\limits_{\varepsilon\to 0} \varepsilon^{-s_1+\alpha_1(x')}\phi_1(x')$, $\phi_1^0(x') \in C_0$, $\forall\, x' \in \partial U$,

$u^0(x) = \lim\limits_{\varepsilon\to 0} \varepsilon^{-s_1}u(x)$, $u^0 \in H_{s_2}(U)$,

$A^0(x,\partial) = A(x,0,\partial)$, $B_1^0(x',\partial) = \lim\limits_{\varepsilon\to 0} \varepsilon^{\alpha_1(x')} B_1(x',\varepsilon,\partial)$,

provided that the limits exist.

We denote by $\mathfrak{A}^\varepsilon$ the operator, which corresponds to the perturbed boundary value problem (2.1.11), (2.1.12),

(2.1.16) $\mathfrak{A}^\varepsilon \in \operatorname{Hom}(H_{(s)}(U);H_{(s-\nu)}(U) \times \prod\limits_{x'\in\partial U} C_{\sigma_1(x')} \times C_{\tau_2(x')})$

with $\sigma_1(x') = s_1-\alpha_1(x')$, $\tau_2(x') = s_1+s_2-\alpha_2(x')-m_2(x')-1/2$, while $\mathfrak{A}^0$ is the corresponding operator for the reduced problem (2.1.14), (2.1.15):

$$(2.1.17)\ \mathfrak{A}^0 \in \mathrm{Hom}(H_{s_2}(U);\ H_{s_2-\nu_2}(U) \times \prod_{x'\in\partial U} C_0),$$

where, as usual, $\mathrm{Hom}(X_1,X_2)$ is the Banach space of all continuous linear mappings from X_1 into the Banach space X_2.

Statement of the results. 2.2.

Theorem 2.2.1. - Assume that the conditions (2.1.9), (2.1.10) are satisfied. Then the coerciveness condition is necessary and sufficient for the following two-sided a priori estimate to hold uniformly with respect to $\varepsilon \in (0,\varepsilon_0]$ for the solutions of (2.1.11), (2.1.12):

$$(2.2.1)\ ||u||_{(s)} \sim ||f||_{(s-\nu)} + \sum_{x'\in\partial U} [\phi_2(x')]_{\tau_2(x')} + ||u||_{(s')},$$

where again $\tau_2(x') = s_1+s_2-\alpha_2(x')-m_2(x')-1/2$, $||.||_{(s')}$ is any norm weaker than $||.||_{(s)}$ and the constants in the equivalency relation $\sim$ depend only on s,s' and ε_0.

Assume that there exists $s_2 \in \mathbb{R}$ such that

$$(2.2.2)\ \max_{x'\in\partial U} m_1(x')+1/2 < s_2 < \min_{x'\in\partial U} m_2(x')+1/2.$$

Assume also that the reduced problem (2.1.14), (2.1.15) establishes the isomorphism:

$$(2.2.3)\ \mathfrak{A}^0 \in \mathrm{ISO}(H_{s_2}(U);\ H_{s_2-\nu_2}(U) \times \prod_{x'\in\partial U} C_0).$$

Theorem 2.2.2. - Under the conditions (2.1.10), (2.2.2), (2.2.3) and for ε_0 sufficiently small, the coercive singular perturbation (2.1.11), (2.1.12) establishes the isomorphism:

$$(2.2.4)\ \mathfrak{A}^\varepsilon \in \mathrm{ISO}(H_{(s)}(U);\ H_{(s-\nu)}(U) \times \prod_{x'\in\partial U} C_{\sigma_1(x')} \times C_{\tau_2(x')})$$

where $\sigma_1(x')$, $\tau_2(x')$ are the same, as previously in (2.1.16), that is $\sigma_1(x') = s_1-\alpha_1(x')$, $\tau_2(x') = s_1+s_2-\alpha_2(x')-m_2(x')-1/2$, $x' \in \partial U$.

Remark 2.2.3. - For the one-dimensional model of thin elastic plates with the Dirichlet boundary operators $B_j(x',\varepsilon,\partial) = \partial^{j-1}$, $x' \in \partial U$, $j = 1,2$, the conditions of Theorem 2.2.2 are satisfied, since $0 = \max_{x'\in\partial U} m_1(x') < \min_{x'\in\partial U} m_2(x') = 1$, so that one can take, for instance, $s_2 = 1$; besides, $\mathfrak{A}^0 : H_1^0(U) \to H_{-1}(U)$ is a positive operator, so that (2.2.3) is also satisfied. In the latter case the estimate (2.2.1) can be sharpened and rewritten as follows:

$$(2.2.5)\ ||u||_{(s)} \sim ||f||_{(s-\nu)} + \sum_{x'\in\partial U} \{[\phi_1(x')]_{s_1} + [\phi_2(x')]_{s_1+s_2-3/2}$$

where $s \in \mathbb{R}^3$ is supposed to satisfy the inequalities:

$$(2.2.6)\ 1/2 < s_2 < 3/2,\quad s_2+s_3 > 3/2,$$

and the equivalency relation $\sim$ holds uniformly with respect to $\varepsilon \in (0,\varepsilon_0]$ with some constants which depend only on s and ε_0. In particular, for $s = (0,1,1)$ the estimate (2.2.5) says:

$$(2.2.7)\ ||u||_{(0,1,1)} \sim ||f||_{(0,-1,-1)} + \\ + \sum_{x'\in\partial U} \{|\phi_1(x')| + \varepsilon^{\frac{1}{2}}|\phi_2(x')|\}.$$

For the sake of simplicity, we state the convergence results for $B_j(x',\varepsilon,\partial) = \partial^{j-1}$, $j = 1,2$, $x' \in \partial U$ and $A(x,\varepsilon,\partial)$ the operator, arising in the linearized model of thin elastic plates:

$$(2.2.8)\ A(x,\varepsilon,\partial) = \varepsilon^2\partial^4 + \partial a(x)\partial + b(x),$$

where $a(x) \geq a_0 > 0$, $b(x) \geq 0$, $\forall x \in \overline{U}$.

In this case the reduced problem is stated as follows

$$(2.2.9)\ (\partial a(x)\partial + b(x))u^0(x) = f^0(x),\ x \in U,$$

$$(2.2.10)\ \pi_0 u^0(x') = \phi_1^0(x') \qquad ,\ x' \in \partial U.$$

Assume that $f^0 = f$, $\phi_1^0 = \phi$ with f and ϕ_1 the same as for the perturbed problem.

Theorem 2.2.4. - The following error estimate holds uniformly with respect to $\varepsilon \in (0,\varepsilon_0]$:

$$(2.2.11)\quad ||u-u^0||_{(s)} \leq C\{\varepsilon^2||f||_{(s-\nu+2e_2)} + \varepsilon^2 \sum_{x'\in\partial U} [\phi_1(x')]_{s_1} + \\ + \sum_{x'\in\partial U} [\phi_2(x') - \pi_0\partial u^0(x')]_{s_1+s_2-3/2}\},$$

where $1/2 < s_2 < 3/2$, $s_2+s_3 > 3/2$, $e_2 = (0,1,0)$ and the constant $C > 0$ depends only on s and ε_0.

In particular, if $f \in L_2(U)$, $\forall\varepsilon \in [0,\varepsilon_0]$, $\phi_j(x') \in C_0$, $j = 1,2$, $x' \in \partial U$, then by taking in (2.2.11) $s = (0,1,1)$ one gets:

$$(2.2.12)\quad ||u-u^0||_1 \leq C_1\varepsilon^{\frac{1}{2}},$$

where C_1 does not depend on $\varepsilon \in (0,\varepsilon_0]$.

The solution u^0 of the reduced problem, in general, does not comply with the second boundary condition: $\pi_0\partial u(x') = \phi_2(x')$, and that is the reason for a lack of convergence when $s_2 > 3/2$. For $s_2 < \frac{1}{2}$ the restriction operator $\pi_0 u^0(x')$ is not well defined, and, as a consequence of this fact, the solution of the perturbed problem, in general, fails to converge to the solution of the reduced problem in this case as well, the latter being not well defined. Consider the following simplified singular perturbation:

$$(2.2.13)\quad \left(\frac{i\varepsilon\partial}{\sqrt{a(0)}} + 1\right)\left(\frac{-i\varepsilon\partial}{\sqrt{a(1)}} + 1\right)(\partial a(x)\partial + b(x))v^0(x) = f(x),\ x\in U$$

$$(2.2.14)\quad \pi_0\partial^{j-1}v^0(x') = \phi_j(x'),\ j = 1,2,\ x' \in \partial U.$$

Theorem 2.2.5. - The following error estimate holds uniformly with respect to $\varepsilon \in (0,\varepsilon_0]$:

$$(2.2.15)\quad ||u-v^0||_{(s)} \leq C\{||f||_{(s-\nu+e)} + \sum_{x'\in\partial U} ([\phi_1(x')]_{s_1-1} + \sum_{1\leq j\leq 2} [\phi_j(x')]_{s_1+s_2-5/2})\},$$

provided that s satisfies the inequalities:

$$(2.2.16)\quad \tfrac{1}{2} < s_2 < 5/2,\ s_2+s_3 > 3/2,$$

where e = (-1,1,1) and the constant C depends only on s and ε_0.

For s = (0,2,1) the estimate (2.2.15) becomes:

$$(2.2.17)\quad ||u-v^0||_{(0,2,1)} \leq C\varepsilon\{||f||_1 + \sum_{x'\in\partial U} \sum_{1\leq j\leq 2} [\phi_j(x')]_{\frac{1}{2}}\}.$$

In particular, for $f \in H_1(U)$, $\forall\varepsilon \in [0,\varepsilon_0]$, $\phi_j(x') \in C_0$, $j = 1,2$, $x' \in \partial U$, the approximate solution v^0 converges to the exact solution u in $H_2(U)$, the rate of convergence being $O(\varepsilon^{\frac{1}{2}})$.

Remark 2.2.6. - Using an iterative process, consisting of successively solving the boundary value problem (2.2.13), (2.2.14) with different (known) data $\{f,\phi_j\}$, one constructs an approximation $v^{(p-1)}$ of order p to the solution u of (2.1.11), (2.1.12) such that with e = (-1,1,1) uniformly with respect to $\varepsilon \in (0,\varepsilon_0]$ holds:

$$(2.2.18)\quad ||u-v^{(p-1)}||_{(s)} \leq C\{||f||_{(s-\nu+pe)} + \sum_{x'\in\partial U} ([\phi_1(x')]_{s_1-p} + \sum_{1\leq j\leq 2} [\phi_j(x')]_{s_1+s_2-p-3/2})\},$$

the last error estimate being valid for s satisfying the conditions:

$$(2.2.19)\quad \tfrac{1}{2} < s_2 < p+3/2,\ s_2+s_3 > 3/2.$$

The above-mentioned procedure is equivalent to the V-L method to construct the asymptotic formulas for the exact solution u (see [7]).

Remark 2.2.7. - If $f \in H_2(U)$ one can further simplify the problem (2.2.13), (2.2.14) replacing the equation (2.2.13) by the following one

$$(2.2.20)\quad \left(\frac{i\varepsilon\partial}{\sqrt{a(0)}}+1\right)\left(\frac{-i\varepsilon\partial}{\sqrt{a(1)}}+1\right)\{(\partial(x)\partial+b(x))w^0(x)-f(x)\}=0,\quad x \in U.$$

In the last case one has for $||u-w^0||_{(0,2,1)}$ the same estimate (2.2.17) where $||f||_1$ in the right-hand side is replaced by $||f||_2$.

§3. LINEARIZED ONE-DIMENSIONAL MODEL OF THIN ELASTIC PLATES. (DIFFERENCE PROBLEM).

Notation, statement of the problem, stability results. 3.1.

Let U_k be the uniform mesh in $\overline{U}$ with the meshsize h. We define the discrete analogues of the spaces $H_{(s)}(U)$ in the natural way using sums in (2.1.1) instead of integrals and replacing ε by $\varepsilon+h$. For any $s \in \mathbb{R}^3$ the corresponding difference norm of order s is defined via discrete Fourier transform and partition of unity argument for meshfunctions. We denote the discrete analogue of the norm (2.1.1) by $||\cdot||_{(s);h}$ and put: $|u|_{(s)} = \sup_{\varepsilon,h} ||u||_{(s);h}$. The notation $[\phi]_1$ is again used for the boundary norm (2.1.2) where ε is replaced by $\varepsilon+h$, while $|[\phi]|_1 = \sup_{\varepsilon,h} [\phi]_1$. Any approximation by finite differences of the principal symbol $A_0(x,\varepsilon,\partial)$ in (2.1.3) can be written as follows:

$$(3.1.1)\quad a_0(x,\varepsilon,h,\Theta) = Q_0(x,\rho,\Theta)D\overline{D}$$

where $\rho = h/\varepsilon$, Θ is the shift operator on U_h in the positive x-direction, $D = -ih(\Theta-1)$, $\overline{D} = -ih(1-\Theta^{-1})$ and $Q_0(x,\rho,\theta)$ is a smooth function of $x \in \overline{U}$, $\rho > 0$ and $\theta \in S = \{\theta \in \mathbb{C} \mid |\theta| = 1\}$. Usually, when the approximation is of local type, $Q_0(x,\rho,\theta)$ is a polynomial of some fixed order in variables θ, θ^{-1}. In any case, Q_0 with $\varepsilon = 1$ is supposed to be an approximation of the second order operator $\partial^2 + a_1(x)\partial + a_2(x)$, that is to say that

$$(3.1.2)\quad Q_0(x,h,\exp(ih\xi)) = (\xi^2 + a_1(x)\xi + a_2(x))(1+O((h\xi)^q)),\quad h \to 0$$

uniformly with respect to $x \in \overline{U}$; ξ belonging to any compact set in $\mathbb{R}$, the integer $q > 0$ is the accuracy of the approximation.

Definition 3.1.1. - A difference approximation of (2.1.3) is called elliptic of order $\nu = (0,2,2)$ if its principal symbol a_0 satisfies the inequalities

$$(3.1.3)\quad C^{-1}(1+\rho^2|\theta-1|^2) \leq |Q_0(x,\rho,\theta)| \leq C(1+\rho^{-2}|\theta-1|^2)$$

where the constant $C > 0$ does not depend on $x \in \overline{U}$, $\rho > 0$, $\theta \in S$.

A difference approximation of (2.1.3) is called weakly elliptic for $\rho \in I \subset (0,+\infty)$ if (3.1.3) holds with some constant $C > 0$ which depends on the interval I.

Remark 3.1.2. - The extension of the ellipticity notion to the higher order operators is obvious.

It follows from (3.1.3) that the function $\theta \to Q_0(x,\rho,\theta)$ does not have zeros $\theta = \theta(x,\rho) \in S$ for any $x \in \overline{U}$, any $\rho \in (0,+\infty)$ (or $\rho \in I$ in the case of the weak ellipticity).

Condition 3.1.3. - Assume that the function $\theta \to Q_0(x',\rho,\theta)$, $x' \in \partial U$, has only one zero $\theta(x',\rho)$ such that

$$(3.1.4)\quad |\theta(0,\rho)| < 1,\ |\theta(1,\rho)| > 1,\ \forall\rho > 0.$$

Let $b_{20}(x',\varepsilon,h,\theta)$ be a difference approximation of the principal symbol $B_{20}(x',\varepsilon,\partial)$ of the boundary operator B_2 in the boundary condition (2.1.12).

Definition 3.1.4. - A difference approximation of (2.1.11), (2.1.12) is called coercive if the approximation of $A(x,\varepsilon,\partial)$ is elliptic of order $\nu = (0,2,2)$, *Condition 3.1.3* is satisfied and the following inequalities hold:

$$(3.1.5)\quad C_2 \geq (1+\rho)^{\alpha_2(x')+m_2(x')} |b_{20}(x',1,\theta(x',\rho))| \geq C_1,$$
$$\forall\rho > 0,\ \forall x' \in \partial U$$

where $\theta = \theta(x',\rho)$ is the zero of Q_0 satisfying (3.1.4) and the constants C_j, $j = 1,2$ do not depend on $\rho > 0$, $x' \in \partial U$.

If $B_2(x',\varepsilon,\partial) = \partial$, $\forall x' \in \partial U$, and $b_2(0,\varepsilon,h,\Theta) = D$, $b_2(1,\varepsilon,h,\Theta) = \bar{D}$, then for the approximation

$$(3.1.6)\quad \varepsilon^2(D\bar{D})^2 + a(x)D\bar{D}$$

of the principal symbol of the operator (2.2.8) the coercivity condition is satisfied. Indeed, (3.1.6) is an elliptic difference singular perturbation of order $\nu = (0,2,2)$, the function $\theta \to Q_0(x,\rho,\theta)$, $Q_0(x,\rho,\theta) = \rho^{-2}(2-\theta-\theta^{-1})+a(x)$, has precisely one zero satisfying (3.1.4) and the inequalities (3.1.5) hold with

$$C_1 = \min\{1,\min_{x'\in\partial U} \sqrt{a(x')}\},\quad C_2 = 3\max\{2,\max_{x'\in\partial U} \sqrt{a(x')}\}.$$

An analoguous definition of weak coerciveness does not assume Condition 3.1.3 and envolves the requirement that a certain inequality (similar to (3.1.5)) holds uniformly with respect to $\rho \in I$, $x' \in \partial U$. In the latter case the number of boundary conditions for the difference problem might be greater than in (2.1.12) (in fact, it must be greater or equal than the number of zeros of the function $\theta \to Q_0(x',\rho,\theta)$ which satisfy the inequalities (3.1.4)).

For the linearized model of thin elastic plates the corresponding coercive difference problem can be stated as follows:

(3.1.7) $(\varepsilon^2(D\bar{D})^2+tDa(x)\bar{D}+(1-t)\bar{D}a(x)D+b(x))v(x) = \pi_h f(x),$
$x \in U_h \cap (h,1-h),$

(3.1.8) $\pi_0 v(x') = \phi_1(x'), \quad x' \in \partial U,$

(3.1.9) $\pi_0 Dv(0) = \phi_2(0), \quad \pi_0 \bar{D}v(1) = \phi_2(1),$

where t is a parameter, $t \in [0,1]$, and π_h is a continuous projector from $H_{(s-\nu)}(U)$ onto the corresponding discrete analogue of $H_{(s-\nu)}(U)$. If f is smooth enough than the simple restriction of f(x) onto U_h will yield to the latter condition. For $t = \frac{1}{2}$ the accuracy of the difference scheme (3.1.7) is: $q = 2$ uniformly with respect to $\varepsilon \in [0,\varepsilon_0]$.

Instead of the approximation (3.1.9) of the boundary operator ∂ one can consider the approximation: $-(2ih)^{-1}(3-4\Theta+\Theta^2)$ at the left endpoint $x' = 0$ and the one: $(2ih)^{-1}(3-4\Theta^{-1}+\Theta^{-2})$ at the right endpoint $x' = 1$. The accuracy of this approximation is: $q = 2$. The corresponding difference boundary value problem is again coercive for $\rho \in (0,+\infty)$, $x' \in \partial U$, the constants C_j, $j = 1,2$, in (3.2.5) being the same as previously.

Under the condition (2.2.6), i.e. $\frac{1}{2} < s_2 < 3/2$, $s_2+s_3 > 3/2$, one establishes for the solution v of (3.1.7)-(3.1.9) (or of the difference problem, where the boundary condition (3.1.9) is replaced by the second order approximation of ∂, mentioned above), a two-sided a priori estimate similar to (2.2.5) which holds uniformly with respect to $\varepsilon \in (0,\varepsilon_0]$, $h \in (0,h_0]$ in corresponding spaces of mesh-functions, provided that ε_0,h_0 are sufficiently small. Also for ε_0,h_0 sufficiently small, one shows the existence of the unique solution to the difference problem (3.1.7)-(3.1.9), the existence and uniqueness result in the difference case being similar to the claim (2.2.4). Of course, one uses the fact that the condition (2.2.3) is satisfied in the case of thin elastic plates with the Dirichlet boundary conditions.

Convergence results. 3.2.

In this section we assume that $\rho \geq \rho_0 > 0$, as this case is the most interesting for applications. We also assume that $f \in C^\infty(\overline{U})$, $\forall\varepsilon \in [0,\varepsilon_0]$, $\phi_j(x') \in C_0$, $j = 1,2$, $x' \in \partial U$, where f,ϕ_j are the data for the differential singular perturbation (2.2.8) with the Dirichlet boundary conditions. We stress that even in this case the solution of the differential problem is not a smooth function (uniformly with respect to $\varepsilon \in (0,\varepsilon_0])$ and the most one can expect in the considered case is that the solution of the differential singular perturbation belongs to a bounded set in $H_{s_2}(U)$ with any $s_2 < 3/2$ for all $\varepsilon \in (0,\varepsilon_0]$. We begin with a new definition of the approximation.

Definition 3.2.1. - A difference singular perturbation with symbol $a(x,\varepsilon,h,\theta)$ approximates on the curve $\varepsilon = bh^\gamma$, $\gamma \geq 1$, the differential one with symbol $A(x,\varepsilon,\xi)$ if the following holds:

$$(3.2.1) \quad |A(x,bh^\gamma,\xi) - a(x,bh^\gamma,h,\exp(ih\xi))| \leq Ch^q$$

where $x \in \overline{U}$, $h \in (0,h_0]$, $|\xi| \leq L$ and the constant C depends only on L and h_0; the last $q > 0$ such that (3.2.1) holds is the accuracy of the approximation on the curve $\varepsilon = bh^\gamma$.

Since $f \in C^\infty(\overline{U})$, $\forall\varepsilon \in [0,\varepsilon_0]$, one can take as a projection operator π_h in the right hand side of (3.1.7) the trace of $f(x)$ on $U_h \cap (h,1-h)$.

Remark 3.2.2. - For all $t \in [0,1]$ the approximation (3.1.7)-(3.1.9) has the accuracy $q = 1$ on any curve $\varepsilon = bh^\gamma$, $\gamma \geq 1$. If $t = \frac{1}{2}$ and the boundary operator (3.1.9) is replaced by the three point-approximation of the operator ∂ mentioned above, then the accuracy for the corresponding coercive difference singular perturbation is: $q = 2$ on any curve $\varepsilon = bh^{\gamma}$, $\gamma \geq 1$.

Theorem 3.2.3. - Let q be the accuracy for the coercive difference approximation of the Dirichlet problem for the singular perturbation (2.2.8). Then the following error estimate holds uniformly with respect to $\varepsilon \in (0,\varepsilon_0]$, $\rho \geq \rho_0$:

$$(3.2.2) \quad ||u-v||_{(s);h} \leq C(h^{3/2-s_1-s_2}+h^{q-s_1}),$$

where v is the solution to the difference problem, $s_2 < 3/2$, and the constant C depends only on $s,\varepsilon_0,h_0,\rho_0,f$ and $\{\phi_j\}$.

In particular, for $s = (0,1,0)$ one gets

$$(3.2.3) \quad ||u-v||_{1;h} \leq Ch^{\frac{1}{2}}$$

while with $s = (0,\frac{1}{2}+\delta,0)$, $0 < \delta < \frac{1}{2}$ via the difference imbedding theorem, the estimate (3.2.2) yields:

$$(3.2.4) \quad ||u-v||_{C^0(U_h)} \leq C_\delta h^{1-\delta}$$

uniformly with respect to ε,h, $\rho \geq \rho_0$ for all $\delta > 0$.

The solution v to the difference problem (3.1.7)-(3.1.9) does not approximate well enough in the boundary layer the solution u to the differential singular perturbation, since the scheme (3.1.7)-(3.1.9) does not reflect in a satisfactory way the asymptotic behaviour of the exact solution. For second order Ordinary Differential Singular Perturbations, Difference schemes adjusted to the asymptotic behaviour of the solutions were considered previously by N.S. Bachvalov and A.M. Il'yn. We refer also with respect to this matter to W.L. Miranker, The Computational Theorie of Stiff Differential Equations, Rome (1975).

Consider the following difference approximation of (2.2.13), (2.2.14):

$$(3.2.5)\ (1-\exp(-\rho\sqrt{a(0)})(1-\exp(-\rho\sqrt{a(1)})f(x) =$$
$$= (1-\Theta^{-1}\exp(-\rho\sqrt{a(0)})(1-\Theta\exp(-\rho\sqrt{a(1)})(tDa(x)\bar{D} +$$
$$+ (1-t)\bar{D}a(x)D + b(x))w(x), \quad x \in U_h \cap (h,1-h),$$

$$(3.2.6)\ \pi_0 w(x') = \phi_1(x'), \quad x' \in \partial U,$$

$$(3.2.7)\ \pi_0\sqrt{a(0)}(\Theta-1)w(0) = i\varepsilon(1-\exp(-\rho\sqrt{a(0)})\phi_2(0),$$
$$\pi_0\sqrt{a(1)}(1-\Theta^{-1})w(1) = i\varepsilon(1-\exp(-\rho\sqrt{a(1)})\phi_2(1),$$

where again $t \in [0,1]$ is a parameter.

Remark 3.2.4. - On any curve $\varepsilon = bh^\gamma$, $\gamma \geq 1$, the accuracy for the approximation (3.2.5)-(3.2.7) of the Dirichlet problem for the singular perturbation (2.2.8) is: $q = 1$. If $\varepsilon = bh^\gamma$, $\gamma \geq 2$, $a(0) = a(1)$ and $t = \frac{1}{2}$, then the difference operator (3.2.5) approximates (2.2.8) with the accuracy $q=2$.

One checks that (3.2.5)-(3.2.7) is a coercive difference singular perturbation.

Theorem 3.2.5. - The following error estimate holds uniformly with respect to $\varepsilon \in (0,\varepsilon_0]$, $h \in (0,h_0]$, $\rho \geq \rho_0$:

$$(3.2.8)\ ||u-w||_{(s);h} \leq C(h^{5/2-s_1-s_2}+h^{1-s_1})$$

where $s_2 < 5/2$ and the constant C depends only on $\varepsilon_0, h_0, \rho_0$, f and $\{\phi_j\}$, $j = 1,2$.

In particular, with $s = (0,2,0)$ the estimate (3.2.8) yields:

$$(3.2.9)\ ||u-w||_{2;h} \leq Ch^{\frac{1}{2}}.$$

Taking in (3.2.8) $s = (0,1,0)$ and using the difference imbedding theorem, one gets

$$(3.2.10)\ ||u-w||_{C^0(U_h)} \leq Ch$$

with $s = (0,3/2+\delta,0)$, $\delta > 0$, via the difference imbedding theorem, the error estimate (3.2.8) yields:

(3.2.11) $||u-w||_{C^1(U_h)} \leq C_\delta h^{1-\delta}$

with any $\delta > 0$, where $C^1(U_h)$ is the difference analogue of the space $C^1(\overline{U})$.

Remark 3.2.6. - For coercive difference approximations of coercive singular differential perturbations we indicate algebraic conditions, which are necessary and sufficient for the improvement of the convergence with respect to the smoothness. In other words, these conditions are necessary and sufficient in order to have the convergence for s_2 satisfying less restrictive conditions, so that, in fact, difference solutions have more derivatives converging uniformly with respect to $\varepsilon \in (0,\varepsilon_0]$, $\rho \geq \rho_0 > 0$ to the corresponding derivatives of the exact solution.

The proofs of the results stated above are based upon an accurate investigation of the asymptotic solutions (asymptotic Poisson's kernels) for the differential and difference singular perturbations, respectively. Of course, the situation in the difference case is complicated by the presence of the second parameter h. Nevertheless, the difficulties can be overcome by separating the case when $\varepsilon \leq Ch$ and the one, when $h << \varepsilon$, the latter being to a certain extent similar to the numerical treatment of usual non-stiff differential problems.

REFERENCES

1. Bachvalov, N.S., On the optimisation of methods for the solution of Ordinary Differential Equations with highly oscillating solutions, *J. of Computational Mathematics and Math. Physics*, 11, no.5 (1971), p.1318-1322.
2. Frank, L.S., General Boundary Value Problems for Ordinary Differential Equations with Small Parameter, *Annali di Mat. Pura ed Applicata*, (IV), vol.CXIV, p.27-67, (1977).

3. Frank, L.S., Perturbations singulières et Différences Finies *C.R. Acad. sc.*, t.283, Série A (1976).
4. Frank, L.S., Difference Singular Perturbations, I. A priori estimates (to appear).
5. Il'yn, A.M., Difference scheme for the differential equation whose highest derivative is affected by the presence of a small paramter, *Math. Notices*, 6, no.2, p.237-248, (1969).
6. Miranker, W.L., *The Computational Theory of Stiff Differential Equations*, Rome (1975).
7. Vishik, M.I., Lynsternik, L.A., Regular degeneracy and boundary layer for linear differential equations with a small parameter, *Uspekhi Mat. Nauk* 12:5, p.3-122 (1957).

TOWARDS TIME-STEPPING ALGORITHMS FOR CONVECTIVE-DIFFUSION

D. F. Griffiths
Mathematics Department, University of Dundee, Scotland

ABSTRACT

In this paper we discuss the evolution of convective-diffusion phenomena governed by equations of the type

$$\frac{\partial u}{\partial t} = \varepsilon\frac{\partial^2 u}{\partial x^2} - k\frac{\partial u}{\partial x}$$

on the strip $(x,t) \in (0,1) \times (0,\infty)$ and subject to given initial data and a variety of homogeneous boundary conditions.

Spacial discretizations of this equation by either finite elements or finite differences leads to a system of ordinary differential equations which we analyse with particular emphasis on the effect of the different boundary conditions when $k/\varepsilon \gg 1$.

1. INTRODUCTION

Many idealizations of physical situations involving both convective and diffusive effects can be modelled by an equation of the form

$$\frac{\partial u}{\partial t} = \varepsilon\frac{\partial^2 u}{\partial x^2} - k\frac{\partial u}{\partial x}, \quad (\varepsilon > 0 , k > 0) \tag{1.1}$$

on the semi-infinite strip $(x,t) \in (0,1) \times (0,\infty)$.

We shall consider three problems associated with this

equation corresponding to the initial condition

$$u(x,0) = u_0(x) \ , \ x \in (0,1) \tag{1.2}$$

and a variety of boundary conditions:

Problem (1) $u(x,t)$ periodic in x with unit period,

Problem (2) $u(0,t) = u(1,t) = 0 \ , \ t > 0 \ ,$ (1.3)

Problem (3) $u(0,t) = u_x(1,t) = 0 \ , \ t > 0 \ .$ (1.4)

Throughout this paper we shall assume that convection dominates diffusion ($k/\varepsilon \gg 1$) so that, in Problem (1) the initial data drifts to the right with speed k with only a small amount of dissipation. On the other hand the boundary condition (1.3) impedes the drift of the initial data and consequently a boundary layer of width $O(\varepsilon/k)$ develops at the right hand boundary.

We shall concern ourselves with the analysis of the systems of ordinary differential equations generated by the approximation of Problems (1) and (2) by semi-discrete generalised Galerkin methods. This analysis will be undertaken from the point of view of phase errors, decay rates and the presence or otherwise of unphysical oscillations in the solutions. We shall also accumulate information relating to the stiffness of the systems of ordinary differential equations which will enable us at some future date to choose optimal time-stepping algorithms.

Among many recent papers devoted to a numerical treatment of these problems we cite the work of Price, Cavendish and Varga (1963), Price and Varga (1970), Siemieniuch and Gladwell (1976), Starius (1978), Mitchell and Griffiths (1978).

Note that further difficulties in approximating the steady state solution are encountered when equation (1.1) is subject to inhomogeneous Dirichlet data. This aspect has also received considerable attention recently; see for instance Miller (1975), Christie, Griffiths, Mitchell and Zienkiewicz (1976), Griffiths and Lorenz (1978), Hemker (1977).

2. SEMI-DISCRETE GENERALISED GALERKIN METHODS

Only the semi-discrete approximation to Problem (1) will be described; the other formulations follow in an obvious manner.

Let Π_h denote a uniform partition of $[0,1]$ into elements of length $h\ (=(N+1)^{-1})$ with knots $x_i = ih$, $i = 0, 1, \ldots, N+1$. Associated with Π_h we have two finite dimensional subspaces Φ_h, Ψ_h of $\mathring{H}^1[0,1]$ with the properties that

1. Φ_h and Ψ_h have equal dimension, say D;
2. the restriction of any function out of either subspace to a subinterval of Π_h is a polynomial.

Φ_h and Ψ_h are commonly referred to as spaces of trial and test functions respectively.

If $\{\phi_1, \ldots, \phi_D\}$ and $\{\psi_1, \ldots, \psi_D\}$ form bases for the spaces Φ_h and Ψ_h respectively, the semi-discrete generalised Galerkin method for solving Problem (1) is defined as: Find

$$U(x,t) = \sum_{j=1}^{D} U_j(t)\phi_j(x) \tag{2.1}$$

so that

$$\left(\frac{\partial U}{\partial t},\psi_i\right) + (U',\varepsilon\psi_i'+k\psi_i) = 0\ ,\ i = 1, 2, \ldots, D \tag{2.2}$$

where a dash denotes differentiation with respect to x. Substituting (2.1) into (2.2) leads to a system of D ordinary differential equations of the form

$$h^2 \underset{\sim}{M}\dot{\underset{\sim}{U}} = \varepsilon \underset{\sim}{S}\underset{\sim}{U} \ , \ \underset{\sim}{U}(0) = \underset{\sim}{U}_0 \tag{2.3}$$

where the dot denotes differentiation with respect to t and the 'mass' and 'stiffness' matrices M and S have elements defined by

$$(h^2M)_{ij} = (\phi_j,\psi_i) \ , \ (\varepsilon S)_{ij} = (\phi_j',\varepsilon\psi_i'+k\psi_i) \ . \tag{2.4}$$

We now turn our attention to specific finite element spaces and the forms of the resulting matrices M and S.

(i) Low Order Methods

We begin by choosing a piecewise linear trial space with the usual basis functions given by $\phi_j(x) = \phi(x/h - j)$, $i = 1, 2, \ldots, N$, where

$$\phi(S) = \begin{cases} 0 & |S| \geq 1 \\ 1-S & |S| < 1 \ . \end{cases}$$

Basis functions for Ψ_h are constructed from these by defining $\psi_j(x) = \psi(x/h - j)$, $j = 1, 2, \ldots, N$, where

$$\psi(S) = \phi(S) + \alpha\sigma(S) \ ,$$

$$\sigma(S) = \begin{cases} 0 & |S| > 1 \\ -3S(1-S) & 0 \leq S \leq 1 \\ -\sigma(-S) & -1 \leq S \leq 0 \end{cases}$$

and α is an arbitrary parameter. These test spaces were first introduced by Christie et al. (1976) as a means of simulating upstream differences in finite element methods. In fact, $\alpha = 1$ produces the upstream (backward) difference for the first order space derivative whereas $\alpha = 0$ produces the central difference approximation. An analysis of this combination of trial and test functions in the steady convective-diffusion problem can be found in Griffiths and

Lorenz (1978).

Evaluation of the expression (2.4) with these trial and test functions leads to the $N \times N$ system (2.3) with

$$M = I + \frac{1}{6}A - \frac{1}{4}\alpha B \;,\; S = (1+\alpha L)A - LB \tag{2.5}$$

where A and B are the familiar matrices

$$A = \begin{bmatrix} -2 & 1 & & \\ 1 & -2 & 1 & \\ & \cdot & \cdot & \cdot \\ & & 1 & -2 \end{bmatrix} \qquad B = \begin{bmatrix} 0 & 1 & & \\ -1 & 0 & 1 & \\ & \cdot & \cdot & \cdot \\ & & -1 & 0 \end{bmatrix} \tag{2.6}$$

and L is the <u>cell Peclet number</u>

$$L = kh/2\varepsilon \;. \tag{2.7}$$

For fixed values of k and ε, a straightforward truncation error analysis shows that the consistency error is $O(h^2+\alpha h)$.

(ii) Piecewise Quadratic Galerkin Method

Choosing identical trial and test spaces spanned by piecewise quadratic functions with a Lagrangian basis leads to a system of the form (2.3), of dimension $D = 2N + 1$, in which

$$\underset{\sim}{U} = \{U_{\frac{1}{2}}, U_1, \ldots, U_{N+\frac{1}{2}}\}^T \;, \tag{2.8}$$

$$M = \frac{1}{10}\begin{bmatrix} 8 & 1 & & & & & \\ 2 & 8 & 2 & -1 & & & \\ & 1 & 8 & 1 & & & \\ & -1 & 2 & 8 & \cdot & \cdot & \\ & & & \cdot & \cdot & \cdot & \\ & & & \cdot & \cdot & \cdot & \cdot \quad \cdot \\ & & & & & 1 & 8 \end{bmatrix} , \tag{2.9}$$

$$S = \begin{bmatrix} -8 & 2(2-L) & & & & & \\ 4(2+L) & -14 & 4(2-L) & -(1-L) & & & \\ & 2(2+L) & -8 & 2(2-L) & & & \\ & -(1+L) & 4(2+L) & -14 & \cdot & \cdot & \\ & & & \cdot & \cdot & \cdot & \\ & & & \cdot & \cdot & \cdot & \cdot \\ & & & & & -(1+L) & -8 \end{bmatrix} \tag{2.10}$$

3. ANALYSIS OF SEMI-DISCRETE SYSTEMS: PROBLEM 1

For convenience we begin by analysing the system (2.3) arising from the semi-discretization of Problem 1. In this case equation (1.1) has a general solution of the form

$$u(x,t) = \Sigma a_p u_p(x,t) \tag{3.1}$$

where

$$u_p(x,t) = \exp(-\lambda t + i\omega(x-kt)) , \tag{3.2}$$

$\lambda = \varepsilon\omega^2$ and $\omega = p\pi$, $p = 1, 2, \ldots$. Our analysis investigates the degree to which the semi-discrete approximations reproduce the decay rates λ and drift speed k at large values of the cell Peclet number L .

(i) Low Order Method

From section 2(i) we find that the nodal variables $\{U_j(t)\}$ satisfy

$$\begin{aligned} &h^2\{(1+\tfrac{3}{2}\alpha)\dot{U}_{j-1} + 4\dot{U}_j + (1-\tfrac{3}{2}\alpha)\dot{U}_{j+1}\}/6 = \\ &\varepsilon\{(1+\alpha L+L)U_{j-1} - 2(1+\alpha L)U_j + (1+\alpha L-L)U_{j+1}\} , j = 0, 1, \ldots \end{aligned} \tag{3.3}$$

with $U_{N+1+j}(t) = U_j(t)$ for all j. By seeking a solution of (3.3) in the form

$$U_j(t) = \exp(-\lambda^h t + i\omega(x - k^h t)) \tag{3.4}$$

we find that

$$\frac{k^h}{k} = \frac{a + (\frac{1}{L} + \alpha)\sin^2 \frac{1}{2}\theta}{a^2 + \frac{1}{4}\alpha^2 \sin^2\theta} P(\theta) \tag{3.5}$$

and $$\frac{\lambda^h}{\lambda} = \frac{a + \frac{1}{3}\alpha L \sin^2 \frac{1}{2}\theta}{a^2 + \frac{1}{4}\alpha^2 \sin^2\theta} P^2(\tfrac{1}{2}\theta) \tag{3.6}$$

where $a = (2+\cos\theta)/3$, (3.7)

$$P(\theta) = \sin\theta/\theta \tag{3.8}$$

and $\theta = \omega h \equiv p\pi h$, $p = 1, 2, \ldots, N+1$. An indication of the behaviour of k^h/k and λ^h/λ can be obtained from Figures 1 and 2 where they are plotted for $\alpha = 0, 1$ and $L = 0.1$, 1 and 10 . Note that there is a persistent phase error when $k = 0$ $(L = 0)$ and persistent dissipation when $\varepsilon = 0$ and $\alpha \neq 0$. Both these effects are significant only at high frequencies. We note also that there is a phase lead in all but the higher frequencies but this is compensated, especially when $\alpha = 1$, by overdamping.

We can compare the performance of this family of low-order finite element methods with corresponding low-order finite difference methods by replacing the mass matrix M of (2.5) by the identity matrix ('mass-lumping'). Repeating the earlier analysis we find that, corresponding to (3.5) and (3.6),

$$k^h/k = P(\theta) \ , \tag{3.9}$$

$$\lambda^h/\lambda = (1+\alpha L)P^2(\tfrac{1}{2}\theta) \ . \tag{3.10}$$

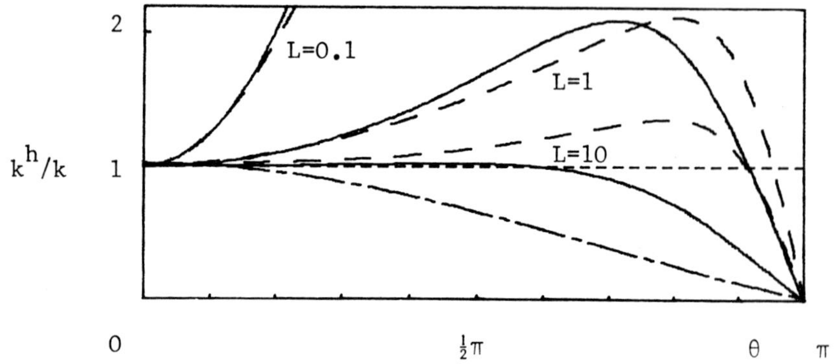

Fig. 1. k^h/k vs θ for Galerkin methods with $\alpha = 0$ (——), $\alpha = 1$ (- - -) and also the lumped mass systems (—-—).

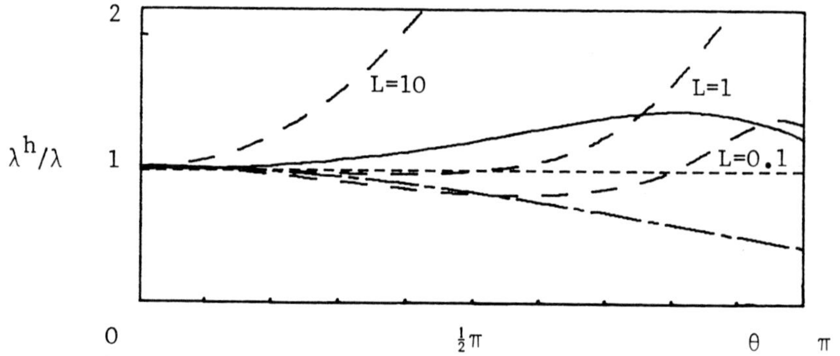

Fig. 2. λ^h/λ vs θ for Galerkin methods with $\alpha = 0$ (——), $\alpha = 1$ (- - -) and also the lumped mass system with $\alpha = 0$ (—-—).

Referring to graphs of these functions for $\alpha = 0$ in Figures 1 and 2 it is seen that there is a considerable underestimate of the speed of propagation of the higher frequencies. This is accompanied by an underestimate of the damping of these components when $\alpha = 0$ which suggests that the initial profile will break up with time. However when $\alpha = 1$, the additional factor on the right of (3.10) means that every component is strongly overdamped leading to excessive artificial dissipation.

(ii) Quadratic Galerkin Method

The 2N+2 nodal variables $U_0(t)$, $U_{\frac{1}{2}}(t)$, ..., $U_{N+\frac{1}{2}}(t)$ satisfy a system of three and five-point difference-differential equations whose precise terms are easily deduced from (2.9) and (2.10). Static condensation is employed to eliminate the half-integer node variables to give

$$d_1U_{j-1} + d_2U_j + d_3U_{j+1} = 0 \ , \ j = 0, 1, \ldots, N \qquad (3.11)$$

with

$$d_1 = -5D^2 + 4D + 4LD - 3 - 3L - L^2$$

$$d_2 = 2(15D^2+26D+3+L^2)$$

$$d_3 = -5D^2 + 4D - 4LD - 3 + 3L - L^2$$

and

$$D \equiv (h^2/20\varepsilon)d/dt \ .$$

Defining $z = (\lambda^h+i\omega k^h)h^2/20\varepsilon$ and seeking a solution of (3.11) in the form (3.4) requires that z satisfy

$$5z^2(3-\cos\theta) - 2z(13+2\cos\theta-2iL\sin\theta) + (3+L^2)(1-\cos\theta) + 3iL\sin\theta = 0 \qquad (3.12)$$

where we have again used $\theta = \omega h$. Since we are primarily interested in the behaviour of solutions at large values of L it is convenient to solve (3.12) in an asymptotic expansion of the form

$$z = iLz_0 + z_1 + iz_2/L + \ldots \ . \qquad (3.13)$$

We then find that

$$k^h/k = \sigma_0 + \sigma_1/L^2 + \ldots \qquad (3.14)$$

and

$$\lambda^h/\lambda = \rho_0 + \rho_1/L^2 + \dots$$

where

$$\begin{aligned} \sigma_o &= \{-4\cos\tfrac{1}{2}\theta \pm (38-2\cos\theta)^{\frac{1}{2}}\}P(\tfrac{1}{2}\theta)/(3-\cos\theta) \\ \rho_0 &= \{60(\sigma_0 - P(\theta))/\theta^2 - 4\sigma_0 P(\tfrac{1}{2}\theta)^2\}/(4P(\theta)+(3-\cos\theta)\sigma_o) \; . \end{aligned} \qquad (3.15)$$

It can be shown that $\max|\sigma_1(\theta)| = O(10^{-2})$ for $\theta \in (0,\pi)$ and so we shall ignore terms of $O(1/L^2)$ in (3.14).

For each $\theta \in (0,\pi)$ there are two values of σ_0 and ρ_0. From Table 1 it is clearly seen that both drift and dissipation are accurately reproduced for all frequencies when the positive sign is taken in (3.15a). Adopting the negative sign in (3.15a) generates fast components which move in the wrong direction. Fortunately these are heavily damped and will not therefore contaminate the solution to any noticeable extent.

TABLE 1

	Primary root		Secondary root	
θ/π	σ_0	ρ_0	σ_0	ρ_0
0.0	1.0000	1.000	-5.00	∞
0.2	1.0000	1.001	-4.42	134
0.4	1.0005	1.006	-3.25	24.6
0.6	1.0022	1.022	-2.22	7.40
0.8	1.0050	1.054	-1.50	2.73
1.0	1.0066	1.115	-1.01	1.11

4. ANALYSIS OF SEMI-DISCRETE SYSTEMS: PROBLEM 2

The solution of (1.1) subject to homogeneous Dirichlet boundary conditions may be written in the form

$$u(x,t) = \Sigma a_p u_p(x,t) \tag{4.1}$$

with

$$u_p(x,t) = e^{-\lambda t}\exp(kx/2\varepsilon)\sin p\pi x , \tag{4.2}$$

$$\lambda = \varepsilon\omega^2 + k^2/4\varepsilon \quad \text{and} \quad \omega = p\pi .$$

By seeking a solution of the semi-discrete system (2.3) in the form

$$\underset{\sim}{U}(t) = \underset{\sim}{V} \exp(-\lambda^h t) \tag{4.3}$$

where $\underset{\sim}{V}$ is a constant D-vector, we are lead to a generalised eigenvalue problem

$$-\lambda^h h^2 M\underset{\sim}{V} = \varepsilon S\underset{\sim}{V} . \tag{4.4}$$

A technique for determining closed expressions for the eigenvalues λ^h of such systems, for particular forms of the matrices M and S, has been developed by Fletcher and Griffiths (1978). The application of this technique to the matrices M and S encountered in this paper has been detailed in Mitchell and Griffiths (1978) and so we shall restrict our attention to the behaviour of λ^h for large values of L.

For notational convenience we shall define $[x]$ to be the largest integer less than x, $\gamma \equiv \gamma_p \equiv \cos p\pi h$ and $\bar{\lambda} = \lambda h^2/\varepsilon$.

(i) Low Order Method

With M and S given by (2.5), it can be shown that $\bar{\lambda}$

satisfies the family of quadratic equations

$$[4-(1-\tfrac{9}{4}\alpha^2)\gamma^2]\bar{\lambda}^2 - 6[4(1+\alpha L)+(2-\alpha L)\gamma^2]\bar{\lambda} + 36[(1+\alpha L)^2-\{(1+\alpha L)^2-L^2\}\gamma^2] = 0 \quad (4.5)$$

for $p = 1, 2, \ldots [\frac{1}{2}(N+1)]$. Note that $\bar{\lambda}_{N+1-p} = \bar{\lambda}_p$ and, when N is odd, there is a single real root given by

$$\bar{\lambda} = 3(1+\alpha L) \quad \text{when} \quad p = [\tfrac{1}{2}(N+1)] . \quad (4.6)$$

For $\gamma \neq 0$, all roots of (4.5) are complex when

$$L^2 - 3\alpha L - 3 > 0 \quad (4.7)$$

and their asymptotic behaviour can be described by

$$\bar{\lambda}_p \simeq \frac{6(2+\gamma^2)}{4-\gamma^2} \pm \frac{6iL\gamma}{4-\gamma^2} \qquad (\alpha = 0) \quad (4.8)$$

$$p = 1, 2, \ldots, [\tfrac{1}{2}(N+1)]$$

$$\bar{\lambda}_p \simeq \frac{12(4-\gamma^2)L}{16+5\gamma^2} \pm \frac{12i\gamma(13-\gamma^2)^{\frac{1}{2}}L}{16+5\gamma^2} \qquad (\alpha = 1) \quad (4.9)$$

Both real and imaginary parts of $\bar{\lambda}$ are proportional to L when $\alpha = 1$ whereas for $\alpha = 0$, the real part of $\bar{\lambda}$ is constant, independent of L.

For large times, the dominant components in the solution will be those with smallest absolute real parts. These are easily shown to be

$$\begin{aligned} \bar{\lambda} &= 3 && (p = [\tfrac{1}{2}(N+1)]) && \text{when} \quad \alpha = 0 \\ \bar{\lambda} &= (12 \pm i\sqrt{12})L/7 && (p = 1) && \text{when} \quad \alpha = 1 . \end{aligned} \quad (4.10)$$

The corresponding fundamental solutions may be written

$$U_p(x_j, t) = \exp(-\bar{\lambda}\varepsilon t/h^2)E^j \sin p\pi x_j \quad (4.11)$$

where $x_j = jh$ and E^j in some way attempts to approximate the term $\exp(kx/2\varepsilon) \equiv (\exp(L))^j$ in (4.2). From (4.10) and (4.11) it is seen that the dominant component for large t

is of high frequency when $\alpha = 0$ (wavelength 2h) compared to a wavelength 2 when $\alpha = 1$. It is therefore to be expected that the low order method with $\alpha = 0$ will develop oscillatory solutions.

For low-order finite difference methods for this problem we again replace the mass matrix M of (2.5) by the identity. A solution of the resulting system is then given in the form (4.3) with

$$\bar{\lambda} = 2(1+\alpha L) + 2[(1+\alpha L)^2 - L^2]^{\frac{1}{2}} \cos p\pi h \qquad (4.12)$$

for $p = 1, 2, \ldots, N$. All eigenvalues are real and positive provided $L \le 1/(1-\alpha)$, $|\alpha| \le 1$. When L exceeds this bound all eigenvalues have the same real parts and this leads to a complete break-up of the initial data profile. Note that each component may be written in the form (4.11) with

$$E = \{(1+\alpha L+L)/(1+\alpha L-L)\}^{\frac{1}{2}} .$$

(ii) Quadratic Galerkin Method

Using the techniques described in Fletcher et al. (1978) it can be shown that the eigenvalues $\bar{\lambda}_p$ ($p = 1, 2, \ldots, 2N+1$) satisfy a family of quartic polynomial equations analogous to (4.5). An analysis of this system reveals that the dominant component for large times has the form (4.3) with $\lambda^h = 10h^2/\varepsilon$ and

$$\underset{\sim}{V} = (1, 0, \mu, 0, \mu^2, \ldots, 0, \mu^N)^T \qquad (4.13)$$

where $\mu = (4L+9)/(4L-9)$. The corresponding solution $U(x,t)$ for large t vanishes at all knots of the mesh Π^h and is positive at the mid-points of each element. For moderate values of L, say $L = 5$, we find that $\mu = 29/11$ and so the effect of this component will be noticeable in only a few elements near $x = 1$. Its presence will, however, tend to contaminate the entire interval for larger values of L .

5. NUMERICAL EXAMPLES

In order to confirm the preceding analysis, numerical results are presented for Problems (1), (2) and (3) when the initial data is a 'spike' of unit height and width 2h :

$$u_0(x) = \begin{cases} 0 & , \ |x-\xi| > h \\ |(x-\xi)|/h , & |x-\xi| \le h . \end{cases}$$

In all experiments we have taken a mesh size $h = 1/10$ and centered the spike at $\xi = 3h$. When the remaining parameters are set to $k = 1$, $\varepsilon = 0.005$ we obtain a cell Peclet number of $L = 10$.

Integration of the system (2.3) was accomplished by first-order backward difference method using a time increment which was small enough to ensure that the additional error introduced was negligible.

The numerical solutions are graphed in Figures 3-5 and, for comparison, the exact solutions are also plotted for the case when periodic boundary conditions are applied. Because of difficulties experienced in computing the exact solutions of the differential equation for Problems (2) and (3), we use instead the solution of the pure initial value problem $(-\infty < x < \infty)$. This differs from the required solutions only within a narrow boundary layer near $x = 1$.

The results in Figures 3 and 5 are seen to agree well with the findings of our analysis. The low order method with $\alpha = 0$ performs particularly well on Problem (1) in strong contrast to the corresponding lumped mass system (Figure 3a).

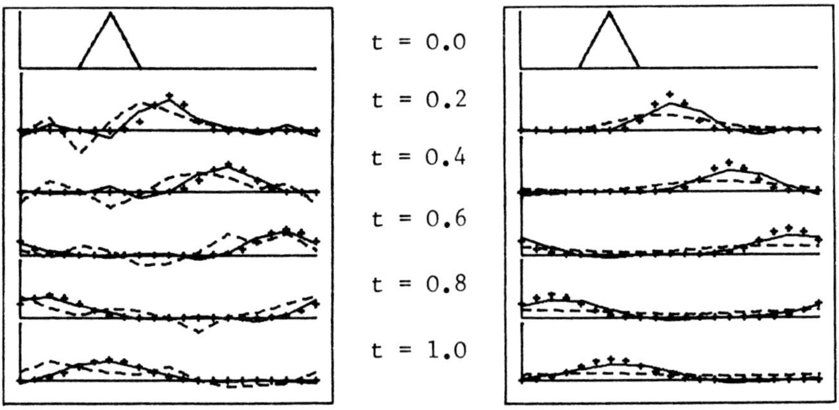

(a) $\alpha = 0$ (b) $\alpha = 1$

Periodic boundary conditions

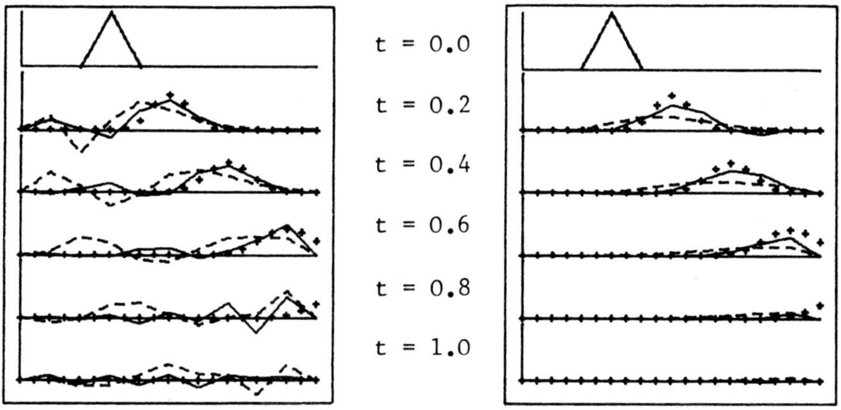

(c) $\alpha = 0$ (d) $\alpha = 1$

Dirichlet boundary conditions

Fig. 3. Low order Galerkin methods (full lines) and lumped mass systems (broken lines). The symbol + indicates the exact solution for (a) and (b) and exact solution of the pure initial value problem for (c) and (d).

The excessive dissipation of the $\alpha = 1$ methods is also evident for all three problems but is again far superior to the lumped mass system with $\alpha = 1$.

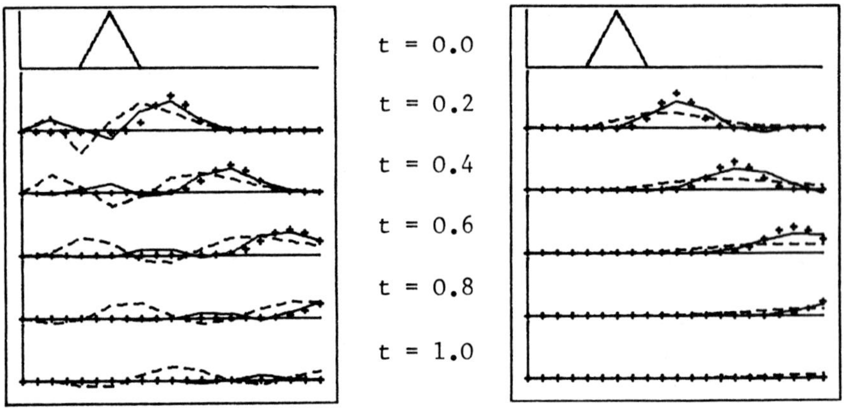

(a) $\alpha = 0$ (b) $\alpha = 1$

Neumann boundary condition at $x = 1$

Fig. 4. Low order Galerkin methods (full lines) and lumped mass systems (broken lines). The symbol + depicts the exact solution of the pure initial value problem.

A feature common to all methods is that their performance on all three problems is virtually identical up until the time that the spike touches the boundary ($t = 0.4$). Thereafter there is a marked deterioration shown in Figures 3(c) and 5(b). For this latter case, compare the solution at $t = 0.8$ with that discussed in Section 4(ii).

Finally we note the rather disappointing performance of the low-order method with $\alpha = 0$ on Problem (3) which has only a very mild boundary layer at $x = 1$.

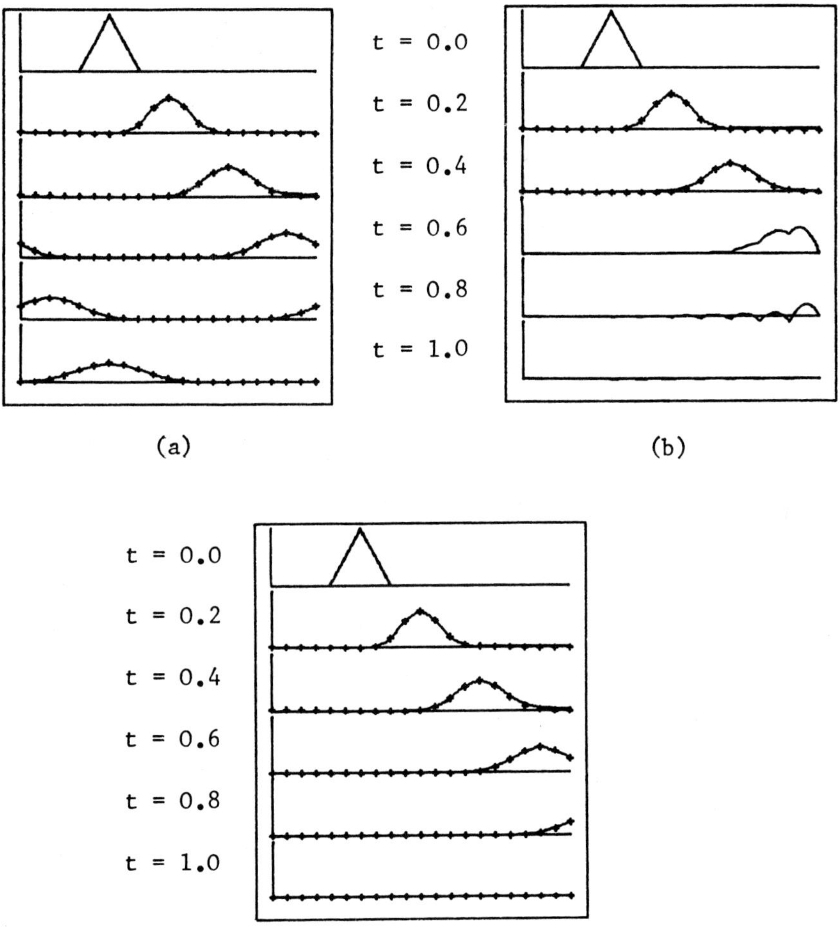

Fig. 5. Quadratic Galerkin methods. (a) Periodic boundary condition, (b) Dirichlet boundary condition and (c) Neumann boundary condition at $x = 1$. The exact solutions are depicted by + .

REFERENCES

Christie, I., Griffiths, D. F., Mitchell, A. R. and Zienkiewicz, O. C. (1976). Finite Element Methods for Second Order Differential Equations with Significant First Derivatives. International Journal for Numerical Methods in Engineering. 10, 1389-1396.

Fletcher, R. and Griffiths, D. F. (1978). The Generalized Eigenvalue Problem for Certain Unsymmetric Band Matrices. Dundee University Numerical Analysis Report NA/26.

Griffiths, D. F. and Lorenz, J. (1978). An Analysis of the Petrov-Galerkin Finite Element Method. To appear in Computing Methods in Applied Mechanics and Engineering.

Hemker, P. W. (1977). A Numerical Study of Stiff Two-Point Boundary Problems. Mathematical Centre Tracts 80, Amsterdam.

Miller, J. J. (1975). A Finite Element Method for a Two Point Boundary Value Problem with a Small Parameter Affecting the Highest Derivative. Trinity College Dublin Mathematics School Report TCD-1975-11.

Mitchell, A. R. and Griffiths, D. F. (1978). Semi-Discrete Generalised Galerkin Methods for Time-Dependent Conduction-Convection Problems. To appear in: The Mathematics of Finite Elements and Applications III (J. Whiteman ed.).

Price, H. S., Cavendish, J. C. and Varga, R. S. (1963). Numerical Methods of Higher-Order Accuracy for Diffusion-Convection Equations. Society of Petroleum Engineering Journal.

Price, H. S. and Varga, R. S. (1970). Error Bounds for Semi-discrete Galerkin Approximations to Parabolic Problems with Applications to Petroleum Reservoir Mechanics. SIAM-AMS Proceedings, 2.

Siemieniuch, J. L. and Gladwell, I. (1976). Some Explicit Finite Difference Methods for the Solution of the Model Diffusion-Convection Equation. University of Manchester Numerical Analysis Report No. 16.

Starius, G. (1978). Numerical Treatment of Boundary Layers for Perturbed Hyperbolic Equations. University of Uppsala Computer Science Report No. 69.

ERROR BOUNDS FOR EXPONENTIALLY FITTED GALERKIN METHODS APPLIED TO STIFF TWO-POINT BOUNDARY VALUE PROBLEMS

P.P.N. de Groen
University of Technology, Eindhoven, The Netherlands

P.W. Hemker
Mathematical Centre, Amsterdam, The Netherlands

ABSTRACT

A linear second order singularly perturbed two-point boundary-value problem is considered. Discretisation by means of Petrov-Galerkin methods of finite element type, where the trial spaces contain piecewise exponentials, is studied. Error bounds, both pointwise and in the energy norm, are derived. The relation with other special difference schemes is shown and the error bounds obtained are compared with numerical results.

1. INTRODUCTION

We study special Galerkin methods for computation of numerical approximations to the singularly perturbed boundary-value problem on the interval [a,b]

$$\begin{aligned} L_\varepsilon u &:= -\varepsilon u'' + pu' + qu = f, \qquad (' = d/dx) \\ u(a) &= u(b) = 0 \end{aligned} \tag{1.1}$$

where ε is a small positive parameter and where p, q and f are sufficiently smooth functions which satisfy

$$\left.\begin{array}{l} p(x) \geq p_0 > 0 \\ q(x) - \tfrac{1}{2}p'(x) \geq 1 \end{array}\right\} \quad \forall x \in [a,b]. \tag{1.2}$$

It is well known that $y_\varepsilon \in H_0^1(a,b)$ is a solution of problem (1.1) if and only if it is a solution of the Galerkin (or weak) form

$$\left\{\begin{array}{l} u \in H_0^1(a,b) \quad \text{and} \\ B_\varepsilon(u,v) := \varepsilon(u',v') + (pu'+qu,v) = (f,v) \qquad \forall v \in H_0^1(a,b), \end{array}\right. \tag{1.3}$$

where $(\cdot,\cdot)$ denotes the usual innerproduct in $L^2(a,b)$. Moreover, both problems have a unique solution, which we shall denote by y_ε in the sequel.

By choosing in $H_0^1(a,b)$ subspaces S^h and V^h of equal finite dimension we obtain the Petrov-Galerkin discretisation of problem (1.1) : find $y_\varepsilon^h \in S^h$ such that

$$B_\varepsilon(y_\varepsilon^h,v) = (f,v) \quad \forall v \in V^h. \tag{1.4}$$

The space S^h is called the *solution space* and V^h the *test space*, whereas both spaces are called *trial spaces*.
For non-stiff two-point boundary value problems both the solution and the test space are usually chosen to be equal to the space P_k^h of piecewise polynomials of degree $\leq k$ on a quasi-uniform mesh Δ,

$$\Delta := \{x_i \mid i=0,1,\ldots,n\}, \quad a = x_0 < x_1 < x_2 < \ldots < x_n = b, \tag{1.5}$$

$$h_i := x_i - x_{i-1}, \quad h := \max_i h_i, \quad \min_i h_i/h \geq \mu > 0.$$

$$P_k^h := \{u \in H_0^1(a,b) \mid D^{k+1}u\,\big|_{(x_{i-1},x_i)} = 0\}, \tag{1.6}$$

where D stands for differentiation and $u_{|I}$ denotes the

restriction of the function u to the open interval I.
When such trial spaces are used for non-stiff problems, the Galerkin discretisation yields an approximation to the solution which is almost as good as the best approximation of the solution in the solution space. Moreover, the Galerkin approximation shows "superconvergence" at the mesh-points, since the test space contains good approximations of Green's function at the mesh-points (cf. Douglas and Dupont (1974)).

In our stiff problems, where ε is a small parameter (i.e. the ratio $hp(x)/\varepsilon$ is large), piecewise polynomial spaces (in general) do not contain satisfactory approximations to the solution and to Green's function. The reason is that the solution of (1.1) and Green's function have narrow boundary layers in which their slope is very large. In order to improve the approximation properties of the solution space we add to P_k^h in each subinterval a piecewise exponential that is a local approximation to the singular (i.e. the rapidly varying) solution of the equation $L_\varepsilon u = 0$. On the subinterval $[x_{i-1},x_i]$ the principal (singular) part of L_ε is $-\varepsilon D^2 + p(x_i)D$ whose singular solution is an increasing exponential. Therefore, with a non-negative "fitting function" $\alpha(x)$, we define a finite dimensional space E_k^h by

$$E_k^h := \{u \in H_0^1(a,b) \mid D^{k+1}(D-\alpha(x_i))u \,|_{(x_{i-1},x_i)} = 0, \quad i = 1,\ldots,n\} \tag{1.7}$$

With $\alpha(x) \equiv p(x)/\varepsilon$, this space is fitted exponentially to the singular part of L_ε and it indeed contains a good approximation of the solution y_ε of (1.1).
Likewise we improve the approximation properties of the testspace by adding local approximations to the singular

solution of the adjoint equation $L_\varepsilon^* u = 0$. The principal singular part of L_ε^* on (x_{i-1}, x_i) is $-\varepsilon D^2 - p(x_{i-1})D$, whose singular solution is an exponential decaying to the right. Therefore we define the finite dimensional space F_k^h by

$$F_k^h := \{u \in H_0^1(a,b) \mid D^{k+1}(D+\alpha(x_{i-1}))u \,|_{(x_{i-1},x_i)} = 0, \; i = 1,\dots,n\} \qquad (1.8)$$

With $\alpha(x) \equiv p(x)/\varepsilon$, this space is fitted exponentially to the singular part of L_ε^* and it contains good approximations of $G_\varepsilon(x_i,\cdot)$, $i = 1,\dots, n-1$, Green's function of (1.1) at the nodes.

The dimension of E_k^h and F_k^h is given by $\dim(E_k^h) = \dim(F_k^h) = nk + n - 1$. We see that $P_k^h \subset E_k^h \cap F_k^h$ for any fitting function α and we notice that both spaces E_k^h and F_k^h coincide if $\alpha(x_i) = 0$, $i = 0,1,2,\dots,n$, in which case $E_k^h = F_k^h = P_{k+1}^h$. If $\alpha(x_i) \neq 0$ the space E_k^h contains the exponential $\exp(+\alpha(x_i)x)$ on (x_{i-1},x_i) and F_k^h contains the exponential $\exp(-\alpha(x_i)x)$ on (x_i,x_{i+1}).

In this paper we shall consider only exponentially fitted spaces with fitting function $\alpha(x) \equiv p(x)/\varepsilon$, which is the natural choice for a problem of type (1.4). With the aid of these spaces we obtain several different Petrov-Galerkin discretisations for problem (1.1). For each of these discretisations existence of a unique solution is guaranteed by an a priori estimate of the following type

$$\exists d > 0 \quad \forall u \in S^h \quad \exists v \in V^h : B_\varepsilon(u,v) \geq d \, \|u\|_\varepsilon \|v\|_\varepsilon , \qquad (1.9)$$

where $\|\cdot\|_\varepsilon$ denotes the energy-norm related to B_ε,

$$\|u\|_\varepsilon^2 := \varepsilon \|u'\|^2 + \|u\|^2. \qquad (1.10)$$

Error estimates for the solutions of the discretised problems can be derived both pointwise at the nodes and in the energy norm (see also De Groen (1978)). The orders of the error estimates are given in table 1.

TABLE 1

The order of the error estimates obtained for exponentially fitted Galerkin methods. The dimension of all trial spaces is nk + n - 1. *For comparison with the numerical experiments see table* 3.

	S_h	V_h	Order of the error		restrictions in the proof
			in the $\Vert\cdot\Vert_\varepsilon$ norm	at mesh points	
1	P_{k+1}	P_{k+1}	1	1	none
2	E_k	E_k	$\varepsilon+h^k$	$\varepsilon+h^k$	none
3	F_k	F_k	1	$\varepsilon+h^k$	none
4	$E_{k-1}+F_{k-1}$	$E_{k-1}+F_{k-1}$	$\varepsilon+h^{k-1}$	ε^2+h^{2k-2}	none
5	E_k	P_{k+1}	$\varepsilon+h^k$	$\varepsilon+h^k$	$h+\frac{\varepsilon}{h}<\gamma$
6	E_k	F_k	$\varepsilon+h^k$	ε^2+h^{2k}	$h+\frac{\varepsilon}{h}<\gamma$
7	P_{k+1}	F_k	1	$\varepsilon^2/h+h^{2k+1}$	$h+\frac{\varepsilon^2}{h}<\gamma$

The most remarkable of these results is 7, in which the solution space has no special virtues for approximation of the singular solution and in which nevertheless a high accuracy is obtained at the points of the mesh.

In section 2 of this paper we describe the construction

of exponentially fitted finite element schemes and we show the relation to other difference schemes. In section 3 we give the proof of the error bounds for the cases $S^h = E_k^h$ and $S^h = P_{k+1}^h$, $V^h = F_k^h$. In section 4 we report results from numerical experiments and we compare them with the error bounds derived.

2. EXPONENTIALLY FITTED FINITE DIFFERENCE SCHEMES

In this section, first we describe sets of basis functions for exponentially fitted trial spaces which are suitable for computational purposes. Thereafter, using these basis functions, we give some examples of exponentially fitted finite element methods and, for some special cases, we compute the resulting difference schemes. Finally we show their relation to difference schemes as proposed by Il'in (1969) and Abrahamsson, Keller and Kreiss (1974).

(2a) *Basis functions in* E_k^h *and* F_k^h

Let $\{\phi_i \mid i = 1,\dots,m\}$ and $\{\psi_i \mid i = 1,\dots,m\}$ be bases in the solution space S^h and the testspace V^h respectively. Applying Petrov-Galerkin methods, we seek an approximation y_ε^h of the form

$$y_\varepsilon^h = \sum_{j=1}^{m} a_j \phi_j \ , \tag{2.1}$$

which satisfies the m equations

$$B_\varepsilon(y_\varepsilon^h \ , \psi_i) = (f,\psi_i), \ i = 1,\dots,m. \tag{2.2}$$

Hence, for actual construction of a Petrov-Galerkin discretisation, the selection of a proper set of basis functions is a major issue.

The following two practical considerations give an indication how to find suitable sets of functions $\{\phi_i\}$ and $\{\psi_i\}$.

1. If n-1 basis functions have the support (x_{i-1}, x_{i+1}) for $i = 1,2,\ldots, n-1$ and the nk remaining basis functions have their support in a single subinterval only, then the resulting linear system is block-tridiagonal and can be reduced to a tridiagonal system by static condensation.
2. In order to obtain discretisations in which a subset $\{a_{m_i} \mid i = 1,\ldots, n-1\}$ of the coefficients $\{a_j\}$ yields the values of the approximation y_ε^h at the nodes, one has to select the basis functions $\{\phi_j\}$ such that

$$\phi_j(x_i) = \delta_{j,m_i}, \quad 1 \le i \le n-1, \ 1 \le j \le nk+n-1.$$

For k = 0 these considerations determine the basis functions in E_k^h and F_k^h uniquely because $\dim(E_0^h) = \dim(F_0^h) = n-1$ and there are n-1 values $y(x_i)$ to compute. The requirements

$$\begin{aligned} &\phi_j \in E_0^h \\ &\text{support } (\phi_j) \subset (x_{i-1}, x_{i+1}) \\ &\phi_i(x_i) = \delta_{ij} \end{aligned}$$

yield the set of basis functions $\{\phi_i\}_{i=1}^{n-1}$ in E_0^h;

$$\phi_i(x) = \begin{cases} 1 - \Psi((x - x_{i-1})/h_i, \alpha_i h_i), & x \in (x_{i-1}, x_i), \\ \Psi((x - x_i)/h_{i+1}, \alpha_{i+1} h_{i+1}), & x \in (x_i, x_{i+1}), \\ 0, & x \notin (x_{i-1}, x_{i+1}), \end{cases} \tag{2.3}$$

where we use the notations

$$\Psi(\xi,\alpha) := \frac{e^{\alpha\xi} - e^{\alpha}}{1 - e^{\alpha}} \quad \text{and} \quad \alpha_i := \alpha(x_i) \ . \tag{2.4}$$

Analogously the basis functions in F_0^h are given by

$$\psi_i(x) = \begin{cases} 1 - \Psi((x - x_{i-1})/h_i, -\alpha_{i-1} h_i), & x \in (x_{i-1}, x_i), \\ \Psi((x - x_i)/h_{i+1}, -\alpha_i h_{i+1}), & x \in (x_i, x_{i+1}), \\ 0, & x \notin (x_{i-1}, x_{i+1}). \end{cases} \tag{2.5}$$

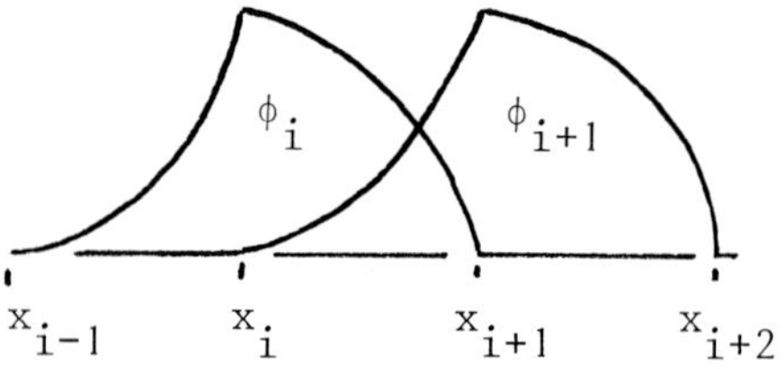

Fig. 1a.
Basis function ϕ_i *in* E_0^h.

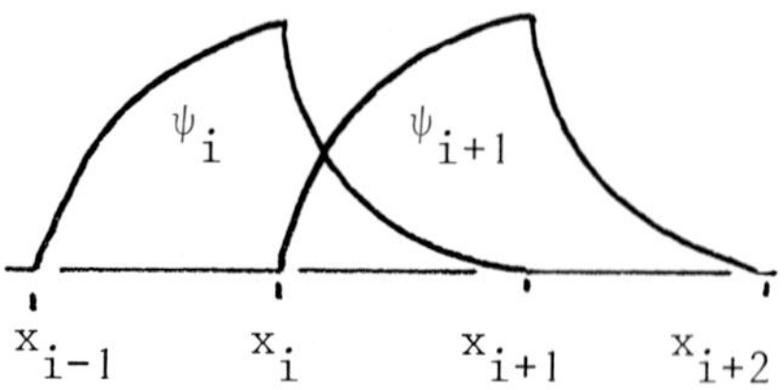

Fig. 1b.
Basis function ψ_i *in* F_0^h.

We notice that for $h/\varepsilon \to 0$ the exponentially fitted basis functions tend to the usual piecewise linear hat-functions and that for $h/\varepsilon \to \infty$, ϕ_i tends to the characteristic function of (x_i, x_{i+1}) and ψ_i tends to the characteristic function of (x_{i-1}, x_i).

For $k > 0$ there are several possibilities to form bases in E_k^h or F_k^h which satisfy the above mentioned two considerations.

(1.) We can extend the usual set of k-th degree C^0 - piecewise polynomials which form a Lagrange type finite element basis in P_k^h to a basis for E_k^h or F_k^h. To complete the basis it should be supplemented by the exponential. For $k > 0$ we can find this exponential basis function with a support in a single interval by taking in (x_{i-1}, x_i) a linear combination of the exponential and a polynomial from P_k^h such that the resulting function vanishes at x_{i-1} and x_i

(1 A.) If this Lagrange type finite element basis in P_k^h on (x_{i-1}, x_i) is based on a subdivision

$$x_{i-1} = \xi_0 < \xi_1 < \ldots < \xi_k = x_i,$$

this polynomial can be taken such that the exponential basis function vasishes at $\xi_0, \xi_1, \ldots, \xi_k$.

(1 B.) This polymonial can also be taken linear such that the exponential basis function on (x_{i-1}, x_i) for E_k^h becomes

$$(\exp(\alpha_i x) - \exp(\alpha_i x_{i-1}))h_i +$$
$$- (\exp(\alpha_i x_i) - \exp(\alpha_i x_{i-1}))(x - x_{i-1}) \qquad (2.6)$$

and the exponential basis function for F_k^h on (x_{i-1}, x_i) is

$$(\exp(-\alpha_{i-1} x) - \exp(-\alpha_{i-1} x_{i-1}))h_i +$$
$$- (\exp(-\alpha_{i-1} x_i) - \exp(-\alpha_{i-1} x_{i-1}))(x - x_{i-1}). \qquad (2.7)$$

Only in the case where $\alpha_i = 0$ or $\alpha_{i-1} = 0$ the functions (2.6) and (2.7) vanish on (x_{i-1}, x_i), and have to be replaced by a (k+1)-th degree polynomial which vanishes at x_{i-1} and x_i.

(2.) Given a subdivision $x_{i-1} = \xi_0 < \xi_1 < \ldots < \xi_{k+1} = x_i$, another basis can be found in E_k^h by taking on (x_{i-1}, x_i) a Lagrange-type polynomial base on $\xi_0, \xi_1, \ldots, \xi_k$ (polynomials that do not vanish at $\xi_{k+1} = x_i$), by adding the exponential function $1 - \Psi((x - x_{i-1})/h_i, \alpha_i h_i)$ and by correcting the k+1 polynomials by this exponential such that the resulting basis functions vanish at x_i (cf. Hemker (1977)).

Bases in F_k^h can be formed analogously.

(2b) *Exponentially fitted finite element / finite difference schemes*

With the above basis functions in the equations (2.1) and (2.2), the discretisation of the problem (1.1) leads to a block-tridiagonal linear system which, by static condensation, can be reduced to a tridiagonal system. The result is that a three-term difference scheme is obtained. For the general case the explicit description of such schemes is rather laborious. A full description of some of these schemes is given in Hemker (1977). In this paper we shall restrict ourselves to some simple examples which already show the main features of the more general and higher order methods.

$$k = 0, \; \alpha \gg 1 \qquad\qquad k = 0, \; \alpha \approx 1 \qquad\qquad k = 0, \; \alpha = 0$$

$$k = 1, \; \alpha \neq 0 \qquad\qquad k = 1, \qquad \alpha \neq 0 \qquad\qquad k = 1, \; \alpha = 0$$
$$\text{type 1a, 1b} \qquad\qquad \text{type 2} \qquad\qquad\qquad \text{type 1a, 1b, 2}$$

$$k = 2, \; \alpha \neq 0 \qquad\qquad k = 2, \; \alpha \neq 0 \qquad\qquad k = 2, \; \alpha = 0$$
$$\text{type 1b} \qquad\qquad\qquad \text{type 1a} \qquad\qquad\qquad \text{type 1a}$$

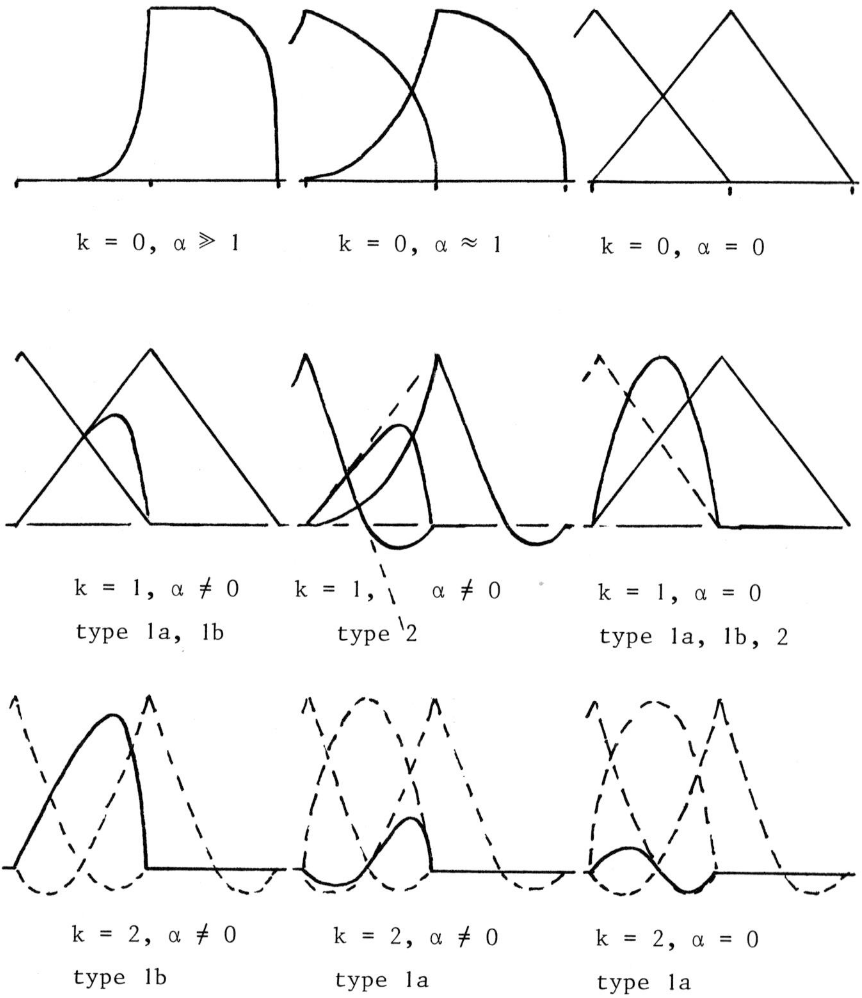

Fig. 2. *Several basis functions in* E_k^h, $k = 0,1,2$.

Example 1

If, for the discretisation of the model equation

$$-\varepsilon y'' + y' = 0, \tag{2.8}$$

with inhomogeneous boundary conditions and on a uniform mesh, we apply the Petrov-Galerkin method with the solution space $S^h = P_1^h$ and the testspace $V^h = F_0^h$ with $\alpha(x) - p(x)/\varepsilon = 1/\varepsilon$, then we obtain the difference scheme

$$\{ -\frac{\varepsilon}{h} - \frac{1}{2}(1+m)\}y_{i-1} + \{ \frac{2\varepsilon}{h} + m \}y_i + \\ + \{ -\frac{\varepsilon}{h} + \frac{1}{2}(1-m)\}y_{i+1} = 0, \tag{2.9}$$

where $m = \coth(\frac{h}{2\varepsilon}) - \frac{2\varepsilon}{h}$. This difference scheme is equivalent with Il'in's scheme, cf. Il'in (1969). In the limit for $h/\varepsilon \to 0$ it is equal to central differences and in the limit for $\varepsilon/h \to 0$ it is backward differences.

We remark that in this example the solution of the discretized problem is exact at the nodes, due to the fact that Green's function $G_\varepsilon(x_i,\cdot)$ of this problem is an element of the test space, cf. (3.44).

Example 2

If we apply the same Galerkin method as in the previous example to the constant coefficient equation

$$- \varepsilon y'' + py' + qy = f, \tag{2.10}$$

(i.e. we take $S^h = P_1^h$ and $V^h = F_0^h$ with $\alpha(x) \equiv p/\varepsilon$), then we obtain the following element stiffness matrix A and element loading vector b;

$$A := \frac{\varepsilon}{h}\begin{pmatrix}+1 & -1\\ -1 & +1\end{pmatrix} + \frac{p}{2}\begin{pmatrix}-1+m & 1-m\\ -1-m & 1+m\end{pmatrix} + \frac{qh}{4}\begin{pmatrix}2-s-m & s-m\\ s+m & 2-s-m\end{pmatrix},$$

$$B := \frac{fh}{2}\begin{pmatrix}1-m\\ 1+m\end{pmatrix}, \tag{2.11}$$

where $m := \coth\left(\frac{ph}{2\varepsilon}\right) - \left(\frac{2\varepsilon}{ph}\right)$ and

$$s := 1 - \frac{2\varepsilon}{ph} m = 1 - \frac{2\varepsilon}{ph}\left\{\coth\left(\frac{ph}{2\varepsilon}\right) - \left(\frac{2\varepsilon}{ph}\right)\right\}.$$

Note that

$$\lim_{\varepsilon/h \to 0} m = 1 \quad ; \quad \lim_{\varepsilon/h \to \infty} m = 0 \quad ;$$

$$\lim_{\varepsilon/h \to 0} s = 1 \quad ; \quad \lim_{\varepsilon/h \to \infty} s = 2/3.$$

Hence we obtain

$$\lim_{\varepsilon/h \to 0} A = p\begin{pmatrix} 0 & 0 \\ -1 & -1 \end{pmatrix} + \frac{qh}{2}\begin{pmatrix} 0 & 0 \\ 1 & 1 \end{pmatrix},$$

$$\lim_{\varepsilon/h \to 0} \mathbf{b} = fh\begin{pmatrix} 0 \\ 1 \end{pmatrix}.$$

Clearly, the reduced scheme reads

$$(-p + \frac{qh}{2})y_{i-1} + (p + \frac{qh}{2})y_i = fh\,,$$

$$i = 1,2,\ldots, n-1. \qquad (2.12)$$

The same scheme is obtained by applying the trapezoïdal rule to the reduced equation $pu' + qu = f$. For the constant coefficient equation the scheme (2.12) is equivalent with the box-scheme to which the method of Abrahamsson, Keller and Kreiss (1974) reduces for $\varepsilon \to 0$.

In the limit for $h/\varepsilon \to 0$ we obtain the scheme

$$(-\frac{\varepsilon}{h^2} - \frac{p}{2h} + \frac{q}{6})y_{i-1} + (\frac{2\varepsilon}{h^2} + \frac{2}{3}q)y_i +$$

$$+ (-\frac{\varepsilon}{h^2} + \frac{p}{2h} + \frac{q}{6})y_{i+1} = f\,,$$

which has also 2nd order accuracy.

For the non-constant coefficient equation the difference

schemes contain integrals in which the coefficient functions p, q and f form part of the integrands. If the integrals are approximated by quadrature, the difference schemes obtained depend on the particular quadrature rule used.

Example 3

Now we discretize the model problem of example 1 on a uniform mesh by the Petrov-Galerkin method with $S^h = P_2^h$ and $V_h = F_1^h$ without prescribing the fitting function $\alpha(\mathbf{x})$ in advance. We obtain the element stiffness matrix

$$\begin{pmatrix} \frac{\varepsilon}{h} - \frac{1}{2} & \frac{1}{6} & -\frac{\varepsilon}{h} + \frac{1}{2} \\ -1 & R & 1 \\ -\frac{\varepsilon}{h} + \frac{1}{2} & -\frac{1}{6} & \frac{\varepsilon}{h} + \frac{1}{2} \end{pmatrix} \tag{2.13}$$

where

$$R := \frac{2\varepsilon}{h} + \frac{2}{\beta} + \left(\frac{6}{\beta} - 3\coth(\beta/2)\right)^{-1},$$

$$\beta := -\alpha(x_i)h.$$

If we apply exponential fitting (i.e. if we take $\alpha(x) \equiv 1/\varepsilon$), then $R = -1/3m$, where m is defined as in example 1. After static condensation this leads to the same difference scheme as in example 1, which yields the exact solution at the meshpoints.

If we consider the method in the limit for $h/\varepsilon \to 0$ (i.e. if we set $h\alpha(x) \equiv 0$), we obtain $R = \frac{2\varepsilon}{h}$ and after static condensation this leads to the 4-th order scheme:

$$\left[-\left(\frac{\varepsilon}{h} + \frac{h}{12\varepsilon}\right) - \frac{1}{2}\right]y_{i-1} + 2\left(\frac{\varepsilon}{h} + \frac{h}{12\varepsilon}\right)y_i + \\ + \left[-\left(\frac{\varepsilon}{h} + \frac{h}{12\varepsilon}\right) + \frac{1}{2}\right]y_{i-1} = 0.$$

The latter scheme corresponds to the (2,2) - Padé approximation of $e^{h/\varepsilon}$ and hence shows no oscillations for $\varepsilon \to 0$, (cf. Van Veldhuizen, these proceedings).

3. RIGOROUS ERROR BOUNDS

In this section we shall derive rigourous error bounds for the Galerkin approximations $y_\varepsilon^h \in E_k^h$ and $\tilde{y}_\varepsilon^h \in P_{k+1}^h$ which satisfy the equations, cf. (1.4),

$$B_\varepsilon(y, v) = (f, v) \qquad \forall v \in F_k^h. \tag{3.1}$$

Error bounds for other combinations of solution and test spaces can be derived in the same way, cf. De Groen (1978). We have chosen these combinations, since they yield the best approximations. Moreover, the error bound for $\tilde{y}_\varepsilon^h$ is very remarkable; although the piecewise polynomial trial space P_{k+1}^h has no special virtues for approximation of the singular solution and although the error of $\tilde{y}_\varepsilon^h$ in the energy norm is of order unity, the error at the mesh-points is quite small.

We shall first sketch how a priory estimates and how error estimates for the best approximations in the trial spaces are obtained. Thereafter we shall give full proofs of the error estimates for y_ε^h and $\tilde{y}_\varepsilon^h$.

NOTE: C denotes a generic (positive) constant, which may differ on each occurrence; C may depend on the data a, b, f, p, q of the problem, the uniformity μ of the mesh, cf. (1.5) and on the degree k of the polynomials in the trial spaces. It certainly does not depend on ε and h.

(3a) *A priori estimates*

A priori estimates are used for comparison of the error of the Galerkin approximation with the error of the best approximation.

LEMMA 1:

$$\|u\|_\varepsilon^2 \le \|L_\varepsilon u\|^2 + |u(a)|^2 + |u(b)|^2, \quad \forall u \in H^2(a,b), \tag{3.2}$$

$$\|u\|_\varepsilon^2 \le B_\varepsilon(u,u), \qquad \forall u \in H_0^1(a,b), \tag{3.3}$$

$$B_\varepsilon(u,v) \le \left\{ \begin{matrix} C\ \|u\|_\varepsilon \|v\|_1 \\ C\ \|u\|_1 \|v\|_\varepsilon \end{matrix} \right\} \le C\varepsilon^{-\frac{1}{2}} \|u\|_1 \|v\|_1 , \qquad \forall u,v \in H_0^1(a,b) \tag{3.4}$$

PROOF: cf. De Groen (1978), lemmas 1,2. The inequalities are derived easily by integration by parts. □

In order to derive lower bounds for B_ε on $E_k^h \times F_k^h$ and on $P_{k+1}^h \times F_k^h$ we define the exponentials $w_i^\pm$ by

$$\begin{aligned} w_i^+(x) &:= \Psi((x-x_i)/h_i,\ \alpha_i h_i), \\ w_{i-1}^-(x) &:= \Psi((x-x_{i-1})/h_i,\ -\alpha_{i-1} h_i), \end{aligned} \tag{3.5}$$

where $\alpha_i := p(x_i)/\varepsilon$, cf. (2.4). They satisfy

$$D(\varepsilon D \mp p(x_i))w_i^\pm = 0,\ w_i^\pm(x_i) = 1,\ w_i^\pm(x_{i\mp 1}) = 0. \tag{3.6}$$

The restriction to (x_{i-1},x_i) of an element $v \in F_k^h$ can be written as the sum of a polynomial π_i of degree $\le k$ plus a multiple of w_{i-1}^-,

$$v(x) = \pi_i(x) + \lambda_i w_{i-1}^-(x), \quad \text{if } x_{i-1} \le x \le x_i. \tag{3.7}$$

For $v \in F_k^h$, decomposed in this way, and $x \in [x_{i-1},x_i]$ we define the maps $M^h : F_k^h \to E_k^h$ and $N_k^h : F_k^h \to P_{k+1}^h$ by

$$M^h v(x) = \pi_i(x) + \lambda_i(-1)^k\{P_k(\xi_i(x)) - w_i^+(x)\} \tag{3.8a}$$

$$N^h v(x) = \pi_i(x) + \tfrac{1}{2}\lambda_i(-1)^k\{P_k(\xi_i(x)) - P_{k+1}(\xi_i(x))\} \tag{3.8b}$$

where $\xi_i(x) := (2x-x_i-x_{i-1})/(x_i-x_{i-1})$ and where P_k stands for the k-th Legendre polynomial. By counting dimensions it is easily seen that the maps M^h and N^h are one-to-one from F_k^h onto E_k^h and P_{k+1}^k respectively. With the aid of these maps we find a priori estimates of type (1.9):

LEMMA 2: *A constant* $\gamma > 0$ *exists, such that*

$$\left.\begin{aligned} B_\varepsilon(M^h v, v) &\geq \tfrac{1}{2}\|v\|_\varepsilon \|M^h v\|_\varepsilon \qquad &(3.9a)\\ B_\varepsilon(N^h v, v) &\geq \tfrac{1}{2}\|v\|_\varepsilon \|N^h v\|_\varepsilon \qquad &(3.9b)\end{aligned}\right\} \forall v \in F_k^h,$$

provided $h + \varepsilon/h \leq \gamma$.

PROOF: Using the coercivity relation (3.3) we find

$$B_\varepsilon(M^h v, v) = B_\varepsilon(v,v) + B_\varepsilon(M^h v - v, v) \geq \|v\|_\varepsilon^2 - |B_\varepsilon(M^h v - v, v)|.$$

Since $M^h v - v$ is zero at the mesh-points by definition, we may integrate the second term by parts,

$$B_\varepsilon(M^h v - v, v) = (M^h v - v, L_\varepsilon^* v).$$

Using the orthogonality properties of $P_k(\xi_i)$ on each subinterval separately we can show

$$|(M^h v - v, L_\varepsilon^* v)| \leq C(h+\varepsilon/h)^{\frac{1}{2}}\|v\|_\varepsilon^2.$$

Moreover, since we have the estimate

$$\|M^h v\|_\varepsilon^2 \leq (1+Ch+C\varepsilon/h)\|v\|_\varepsilon^2, \qquad (3.10)$$

we can find a constant $\gamma > 0$, such that (3.9a) is true for all ε and h satisfying $h+\varepsilon/h < \gamma$. The proof of (3.9b) is analogous. For details we refer to De Groen (1978), lemmas 4 & 5. □

(3b) *Best approximations*

Best approximation of the solution y_ε of problem (1.1) in E_k^h and of Green's function G_ε in F_h^k are derived from asymptotic approximations, which are constructed by the method of "matched asymptotic expansions", cf. Eckhaus (1973) or O'Malley (1974).

The approximation of y_ε consists of a regular part (outer or regular expansion) and a singular part (boundary layer expansion). The lowest order terms $r_0 + \varepsilon r_1$ of the regular expansion are defined by the equations

$$pr_0'+qr_0 = f, \quad pr_1'+qr_1 = r_0'', \quad r_0(a) = r_1(a) = 0. \tag{3.11}$$

The lowest order terms $\tilde{s}_0 + \varepsilon\tilde{s}_1$ of the singular part are defined by

$$\tilde{s}_i(x) := s_i((x-b)/\varepsilon), \ (i=0,1) \text{ and } \zeta := (x-b)/\varepsilon,$$

$$-\ddot{s}_0+p(b)\dot{s}_0 = 0, \ -\ddot{s}_1+p(b)\dot{s}_1 = -\zeta p'(b)\dot{s}_0-q(b)s_0, \tag{3.12}$$

$$s_i(0) = -r_i(b), \ \lim_{\zeta\to-\infty} s_i(\zeta) = 0 \quad (i = 0,1).$$

We note that r' means differentiation with respect to the independent variable x and $\dot{s}$ means differentiation with respect to the boundary layer variable $\zeta := (x-b)/\varepsilon$. The equations (3.11-12) imply

$$\|f - L_\varepsilon(r_0+\varepsilon r_1)\| \le C\varepsilon^2, \ \|L_\varepsilon(\tilde{s}_0+\varepsilon\tilde{s}_1)\| \le C\varepsilon^{3/2} \tag{3.13}$$

and in conjunction with the a priori estimate (3.2) this yields

$$\|y_\varepsilon-r_0-\varepsilon r_1-\tilde{s}_0-\varepsilon\tilde{s}_1\|_\varepsilon \le C\varepsilon^{3/2}; \tag{3.14a}$$

from Sobolev's inequality $|u(x)| \le C\|u\|_1 \le C\varepsilon^{-\frac{1}{2}}\|u\|_\varepsilon$ we infer

$$\max_{a\le x\le b} |y_\varepsilon-r_0-\varepsilon r_1-\tilde{s}_0-\varepsilon\tilde{s}_1| \le C\varepsilon. \tag{3.14b}$$

Approximations of higher order may be derived analogously.

Likewise we construct an asymptotic approximation to Green's function $G_\varepsilon(x,\xi)$ for fixed $x \in (a,b)$. As a function of ξ it satisfies

$$\begin{aligned} &L_\varepsilon^* G_\varepsilon(x,\cdot) = \delta_x \ (= \text{Dirac's delta function}), \\ &G_\varepsilon(x,a) = G_\varepsilon(x,b) = 0, \end{aligned} \tag{3.15}$$

and it has boundary layers at the right-hand sides of the points $\xi = x$ and $\xi = a$. From a regular and a singular (approximate) solution of the equation $L_\varepsilon^* u = 0$ we construct a function whose derivative has the same jump at $\xi = x$ as

$G_\varepsilon(x,\cdot)$ has.

The regular approximate solution $\rho(x,\xi) := \rho_0(x,\xi) + \varepsilon\rho_1(x,\xi)$ is defined by

$$(p\rho_0)' - q\rho_0 = 0, \qquad \rho_0(x,x) = 1, \tag{3.16}$$
$$(p\rho_1)' - q\rho_1 = -\rho_0'', \quad \rho_1(x,x) = 0,$$

where the accent denotes differentiation with respect to ξ. Consequently ρ satisfies the estimate

$$\| L_\varepsilon^* \rho(x,\cdot) \|_{L^2(a,x)} \leq C\varepsilon^2. \tag{3.17}$$

The singular approximate solution $\tilde{\sigma}(x,\xi)$, (the boundary layer term at $\xi = x+0$), is defined by

$$\tilde{\sigma}(x,x+\varepsilon\zeta) := \sigma_0(x,\zeta) + \varepsilon\sigma_1(x,\zeta), \quad \zeta := (\xi-x)/\varepsilon,$$
$$\ddot{\sigma}_0 + p(x)\dot{\sigma}_0 = 0,$$
$$\ddot{\sigma}_1 + p(x)\dot{\sigma}_1 = (q(x) - p'(x))\sigma_0 - \zeta p'(x)\dot{\sigma}_0, \tag{3.18}$$
$$\sigma_0(x,0) = 1,\ \sigma_1(x,0) = 0,\ \lim_{\zeta \to \infty} \sigma_i(x,\zeta) = 0 \ (i = 1,2),$$

where the dot denotes differentiation with respect to the boundary layer variable ζ (at $x + 0$). As a consequence of (3.18) we find the estimate

$$\| L_\varepsilon^* \tilde{\sigma}(x,\cdot) \|_{L^2(x,b)} \leq C\varepsilon^{3/2}. \tag{3.19}$$

From these approximate solutions ρ and $\tilde{\sigma}$ we assemble an approximation of G_ε; its regular and singular parts G_ε^r and G_ε^s are defined by

$$G_\varepsilon^r(x,\xi) := -\beta\tilde{\sigma}(x,b)\rho(b,\xi) + \begin{cases} 0, & \text{if } x < \xi \leq b, \\ \beta\rho(x,\xi), & \text{if } a \leq \xi < x, \end{cases}$$
$$G_\varepsilon^s(x,\xi) := \beta\tilde{\sigma}(a,\xi)\{\rho(b,a)\tilde{\sigma}(x,b)-\rho(x,a)\} + \begin{cases} \beta\tilde{\sigma}(x,\xi), & \text{if } x < \xi \leq b, \\ 0, & \text{if } a \leq \xi < x. \end{cases} \tag{3.20}$$

It is easily seen that the sum $G_\varepsilon^r + G_\varepsilon^s$ is continuous at $\xi = x$. The multiplier β is chosen such that the jump of the ξ - derivative of $G_\varepsilon^r + G_\varepsilon^s$ at $\xi = x$ is equal to $1/\varepsilon$. Simple computation shows

$$\beta(x,\varepsilon) = 1/p(x) + \mathcal{O}(\varepsilon). \tag{3.21}$$

From (3.2) and (3.17-19-21) we find the estimate

$$\| G_\varepsilon(x,\cdot) - G_\varepsilon^r(x,\cdot) - G_\varepsilon^s(x,\cdot)\|_\varepsilon \le C\varepsilon^{3/2} \tag{3.22}$$

uniformly with respect to $x \in [a,b]$.

From these asymptotic approximations we construct approximations which are in the exponentially fitted trial spaces E_k^h and F_k^h . Comparison of these approximations with the Galerkin approximation finally yields the desired error estimate for the latter. In order to obtain the highest possible order with respect to ε we have to deal with the regular and singular parts separately.

The regular approximation $r_0 + \varepsilon r_1$ of y_ε is non-zero at $x = b$, so we look for an approximation of it in the inhomogeneous linear manifold $\phi_\varepsilon + E_k^h$, where ϕ_ε is the linear polynomial

$$\phi_\varepsilon(x) := (r_0(b) + \varepsilon r_1(b))(x-a)/(b-a).$$

Well-known interpolation theorems imply that an approximation r_ε^h exists, such that

$$r_\varepsilon^h \in \phi_\varepsilon + P_k^h \subset \phi_\varepsilon + E_k^h.$$

$$\| r_\varepsilon^h - r_0 - \varepsilon r_1\|_0 \le Ch^{k+1}, \quad \| r_\varepsilon^h - r_0 - \varepsilon r_1\|_1 \le Ch^k. \tag{3.23}$$

Likewise the approximation S_ε^h of the singular part $\tilde{s}_0 + \varepsilon\tilde{s}_1$ has to be in $-\phi_\varepsilon + E_k^h$; for $x \in [x_{i-1},x_i]$ we define it by

$$s_\varepsilon^h(x) := -\phi_\varepsilon(b)\{\exp(\alpha_i(x-x_{i-1}))\prod_{j=i}^{n}\exp(-\alpha_j h_j) + \qquad (3.24)$$
$$- \frac{b-x}{b-a}\prod_{j=1}^{n}\exp(-\alpha_j h_j)\},$$

where $\alpha_j := p(x_j)/\varepsilon$, cf. (1.9). It is easily seen that this approximation satisfies

$$\| s_\varepsilon^h - \tilde{s}_0 - \varepsilon\tilde{s}_1\|_\varepsilon \le C\varepsilon, \qquad (3.25)$$

$$\| (-\varepsilon D+p)(s_\varepsilon^h-\tilde{s}_0-\varepsilon\tilde{s}_1)\|_\varepsilon \le C\varepsilon^{3/2}. \qquad (3.26)$$

By linearity $r_\varepsilon^h + s_\varepsilon^h \in E_k^h$ is a good approximation of y_ε.

Analogously we construct approximations to the regular and singular parts of Green's function. Since the derivative of $G_\varepsilon(x,\cdot)$ has a jump at $\xi = x$, we can find a satisfactory approximation in the space (of piecewise smooth functions) F_k^h only if this jump happens to coincide with a mesh-point. The regular approximation $G_\varepsilon^r(x,\cdot)$ has a jump at $\xi = x$ and is non-zero at $\xi = a$ and $\xi = b$, hence we construct approximations to it in $\psi_i + F_k^h$, where ψ_i is the piecewise linear polynomial $(i = 1,\ldots,n-1)$

$$\psi_i(\xi) := \begin{cases} (G_\varepsilon^r(x_i,a)(\xi-x_i)-G_\varepsilon^r(x_i,x_i-0)(\xi-a))/(a-x_i) & \text{if } a \le \xi < x_i, \\ (G_\varepsilon^r(x_i,b)(\xi-x_i)-G_\varepsilon^r(x_i,x_i+0)(\xi-b))/(b-x_i) & \text{if } x_i < \xi \le b. \end{cases} \qquad (3.27)$$

Analogously to above we find approximations

$$\rho_{\varepsilon,i}^h \in \psi_i + F_h^k \quad \text{and} \quad \sigma_{\varepsilon,i}^h \in -\psi_i + F_k^h,$$

which for $i = 1,\ldots,n-1$ satisfy the estimates

$$\| G_\varepsilon(x_i,\cdot) - \rho^h_{\varepsilon,i} - \sigma^h_{\varepsilon,i} \|_\varepsilon \le C(\varepsilon + h^{k+1}), \tag{3.28}$$

$$\| (Dp-q)(\rho^h_{\varepsilon,i} - G^r_\varepsilon(x_i,\cdot)) \| \le Ch^k, \tag{3.29}$$

$$\| (\varepsilon D+p)(\sigma^h_{\varepsilon,i} - G^s_\varepsilon(x_i,\cdot)) \| \le C\varepsilon^{3/2}. \tag{3.30}$$

(3c) *Error estimates for the Galerkin approximations*

From the approximations constructed above we derive the following theorem:

THEOREM 1: *Let* $y^h_\varepsilon \in E^h_k$ *be the solution of the Galerkin equations*

$$B_\varepsilon(y,v) = (f,v) \qquad \forall v \in F^h_k. \tag{3.31}$$

If $h + \varepsilon/h \le \gamma$, *then* y^h_ε *satisfies the global estimate*

$$\| y_\varepsilon - y^h_\varepsilon \|_\varepsilon \le C(\varepsilon + h^k) \tag{3.32}$$

and it is superconvergent at the nodes,

$$|y^h_\varepsilon(x_i) - y_\varepsilon(x_i)| \le C(\varepsilon^2 + h^{2k}), \; i = 1,\ldots,n-1. \tag{3.33}$$

PROOF: We shall derive error estimates for the regular and the singular part of y^h_ε separately. Let $u_\varepsilon \in \phi_\varepsilon + H^1_0(a,b)$ be the solution of

$$L_\varepsilon u_\varepsilon = f, \; u_\varepsilon(a) = 0, \; u_\varepsilon(b) = r_0(b) + \varepsilon r_1(b) \tag{3.34}$$

and let $u^h_\varepsilon \in \phi_\varepsilon + E^h_k$ satisfy the Galerkin equations (3.31) for this problem. Let $z_\varepsilon \in -\phi_\varepsilon + H^1_0(a,b)$ be the solution of

$$L_\varepsilon z_\varepsilon = 0, \; z_\varepsilon(a) = 0, \; z_\varepsilon(b) = -r_0(b) - \varepsilon r_1(b)$$

and let $z^h_\varepsilon \in -\phi_\varepsilon + E^h_k$ satisfy the Galerkin equations for this problem,

$$B_\varepsilon(z,v) = 0 \qquad \forall v \in F^h_k.$$

Linearity implies $u_\varepsilon + z_\varepsilon = y_\varepsilon$ and $u^h_\varepsilon + z^h_\varepsilon = y^h_\varepsilon$. Formulae (3.11-12-13) imply

$$\|u_\varepsilon - r_0 - \varepsilon r_1\| \le C\varepsilon^2,\ \|z_\varepsilon - \tilde{s}_\varepsilon - \varepsilon\tilde{s}_1\|_\varepsilon \le C\varepsilon^{3/2}; \qquad (3.35)$$

clearly u_ε and z_ε represent (in first order) the regular and singular parts of y_ε.

An error bound for u_ε^h is obtained by comparing it with r_ε^h, cf. (3.23). Since $u_\varepsilon^h - u_\varepsilon$ satisfies

$$B_\varepsilon(u_\varepsilon^h - u_\varepsilon, v) = 0 \qquad \forall v \in F_k^h,$$

we find

$$B_\varepsilon(u_\varepsilon^h - r_\varepsilon^h, v) = B_\varepsilon(u_\varepsilon - r_0 - \varepsilon r_1, v) + B_\varepsilon(r_0 + \varepsilon r_1 - r_\varepsilon^h, v). \qquad (3.36)$$

Using (3.4-35) we estimate the first term,

$$B_\varepsilon(u_\varepsilon - r_0 - \varepsilon r_1, v) \le C\varepsilon^{3/2}\|v\|_\varepsilon$$

and using (3.23) we find for the second term

$$B_\varepsilon(r_0 + \varepsilon r_1 - r_\varepsilon^h, v) \le Ch^k\|v\|_\varepsilon. \qquad (3.37)$$

Hence, lemma 2 and the choice $M^h v := u_\varepsilon^h - r_\varepsilon^h$ yield the estimate

$$\|u_\varepsilon^h - r_\varepsilon^h\|_\varepsilon \le C\left(\varepsilon^{3/2} + h^k\right), \quad (\text{if } h + \varepsilon/h \le \gamma). \qquad (3.38)$$

Likewise an error bound for z_ε^h is obtained by comparing z_ε^h and s_ε^h, cf. (3.24). Since $z_\varepsilon^h - z_\varepsilon$ satisfies

$$B_\varepsilon(z_\varepsilon^h - z_\varepsilon, v) = 0 \qquad \forall v \in F_k^h,$$

we find

$$B_\varepsilon(z_\varepsilon^h - s_\varepsilon^h, v) = B_\varepsilon(z_\varepsilon - \tilde{s}_0 - \varepsilon\tilde{s}_1, v) + B_\varepsilon(\tilde{s}_0 + \varepsilon\tilde{s}_1 - s_\varepsilon^h, v). \qquad (3.39)$$

From (3.4-35) we find

$$B_\varepsilon(z_\varepsilon - \tilde{s}_0 - \varepsilon\tilde{s}_1, v) \le C\varepsilon\|v\|_\varepsilon. \qquad (3.40)$$

For the second term in (3.39) we use the estimate

$$B_\varepsilon(u,v) = (\varepsilon u' - pu, v') + ((q-p')u, v) \le$$

$$\le \|v\|_\varepsilon \{\varepsilon^{-\frac{1}{2}}\|u' - pu\| + C\|u\|\} \qquad \forall u,v \in H_0^1(a,b). \qquad (3.41)$$

In conjunction with (3.26) this implies

$$B_\varepsilon(\tilde{s}_0+\varepsilon\tilde{s}_1-s^h_\varepsilon,v) \le C\varepsilon\|v\|_\varepsilon \quad \forall v \in F^h_k.$$

Hence, lemma 2 and the choice $M^h v := z^h_\varepsilon - s^h_\varepsilon$ yield the estimate

$$\|z^h_\varepsilon - s^h_\varepsilon\|_\varepsilon \le C\varepsilon, \quad (\text{if } h+\varepsilon/h \le \gamma). \tag{3.42}$$

We remark that this estimate does not depend on the degree of the polynomials in E^h_k. Formulae (3.38-42) imply (3.32).

In order to prove the superconvergence we use the identity

$$y(x) = B_\varepsilon(y,G_\varepsilon(x,\cdot)) \quad \forall y \in H^1_0(a,b),\ a < x < b. \tag{3.43}$$

Clearly, the error $e^h_\varepsilon := y^h_\varepsilon - y_\varepsilon$ satisfies the equations

$$e^h_\varepsilon(x) = B_\varepsilon(e^h_\varepsilon,G_\varepsilon(x,\cdot)) = B_\varepsilon(e^h_\varepsilon,G_\varepsilon(x,\cdot)-v) \ \forall v \in F^h_k, \tag{3.44}$$

cf. Douglas & Dupont (1974). If x is a node, F^h_k contains a good approximation of Green's function $G_\varepsilon(x,\cdot)$, namely $\rho^h_\varepsilon + \sigma^h_\varepsilon$. Hence, for i = 1,...,n-1, formula (3.44) implies

$$e^h_\varepsilon(x_i) = B_\varepsilon(e^h_\varepsilon,G_\varepsilon-G^r_\varepsilon-G^s_\varepsilon) + B_\varepsilon(e^h_\varepsilon,G^r_\varepsilon-\rho^h_{\varepsilon,i}) + B_\varepsilon(e^h_\varepsilon,G^s_\varepsilon-\sigma^h_{\varepsilon,i}). \tag{3.45}$$

Formulae (3.4-22) yield an estimate for the first term:

$$|B_\varepsilon(e^h_\varepsilon,G_\varepsilon-G^r_\varepsilon-G^s_\varepsilon)| \le C\varepsilon\|e^h_\varepsilon\|_\varepsilon. \tag{3.46}$$

Since $G^r_\varepsilon - \rho^h_{\varepsilon,i}$ and $G^s_\varepsilon - \sigma^h_{\varepsilon,i}$ both are in $H^1_0(a,b)$ by definition, we can use for the former the estimate

$$|B_\varepsilon(y,v)| \le \varepsilon\|y'\|\|v'\| + \|(Dp-q)y\|\|v\| \tag{3.47}$$

and for the latter the estimate

$$|B_\varepsilon(y,v)| \le \|\varepsilon v' + p\mathbf{v}\|\|y'\| + \|qy\|\|v\|. \tag{3.48}$$

Hence, by (3.29-30) we find

$$\begin{aligned} |B_\varepsilon(e^h_\varepsilon,G^r_\varepsilon(x_i,\cdot) - \rho^h_{\varepsilon,i})| &\le Ch^k\|e^h_\varepsilon\|_\varepsilon, \\ |B_\varepsilon(e^h_\varepsilon,G^s_\varepsilon(x_i,\cdot) - \sigma^h_{\varepsilon,i})| &\le C\varepsilon\,\|e^h_\varepsilon\|_\varepsilon. \end{aligned} \tag{3.49}$$

The formulae (3.32-45-46-49) now imply the superconvergence (3.33). □

In an analogous fashion we derive:

THEOREM 2: *Let* $\tilde{y}_\varepsilon^h \in P_{k+1}^h$ *be the solution of* (3.31). *If* $h+\varepsilon/h < \gamma$, $\tilde{y}_\varepsilon^h$ *satisfies for* $i=1,\ldots,n-1$ *at the mesh points the estimates*

$$|\tilde{y}_\varepsilon^h(x_i) - y_\varepsilon(x_i)| \le \begin{cases} C(\varepsilon+h^k), \\ C(h^{2k+1} + \frac{\varepsilon}{h}^2(1+\frac{\varepsilon^{\frac{1}{2}}}{h})); \end{cases} \tag{3.50}$$

the second estimate is valid only if $\varepsilon|\log\varepsilon| < p_0h$.

PROOF: Although the error of $\tilde{y}_\varepsilon^h$ in energy norm is of order unity, the error at the mesh points is of order $O(\varepsilon+h^k)$, since the test space contains an approximation of Green's function of that order. If ε/h is small enough, the $O(1)$-error in energy norm results from the poor approximation of the singular part of y_ε only. It is committed almost completely in the subinterval (x_{n-1},b), where it is cancelled by the smallness of Green's function. Thus we can improve the estimate at the nodes.

Let u_ε and z_ε be the regular and singular parts of the solution as defined in (3.34) and let $\tilde{u}_\varepsilon^h \in \phi_\varepsilon + P_{k+1}^h$ and $\tilde{z}_\varepsilon^h \in -\phi_\varepsilon + P_{k+1}^h$ be their Galerkin approximations,

$$B_\varepsilon(\tilde{u}_\varepsilon^h,v) = (f,v),\ B_\varepsilon(\tilde{z}_\varepsilon^h,v) = 0, \quad \forall v \in F_k^h.$$

Let $\tilde{r}_\varepsilon^h \in \phi_\varepsilon + P_{k+1}^h$ interpolate $r_0 + \varepsilon r_1$, such that

$$\|\tilde{r}_\varepsilon^h - r_0 - \varepsilon r_1\|_1 \le Ch^{k+1},$$

analogously to (3.23). Analogously to (3.38) we find

$$\|\tilde{u}_\varepsilon^h - \tilde{r}_\varepsilon^h\|_\varepsilon \le C(\varepsilon^{3/2}+h^{k+1}), \text{ if } h+\varepsilon/h \le \gamma.$$

Inserting this estimate in (3.45-46-49) we find for the regular part the error estimate

$$|u_\varepsilon(x_i) - \tilde{u}^h_\varepsilon(x_i)| \le C(\varepsilon^{3/2}+h^{k+1})(\varepsilon+h^k), \qquad (3.51)$$

for $i = 1,\ldots,n-1$, provided $h+\varepsilon/h \le \gamma$.

The error estimate for the singular part $\tilde{z}^h_\varepsilon$ is more involved since the solution space $-\phi_\varepsilon + P^h_k$ does not contain an approximation of z_ε, whose error is better than $\mathcal{O}(1)$ in energy norm. We start with a preliminary error estimate at the knots by the superconvergence trick. Thereafter we improve this estimate by considering the errors on the subintervals $I := (a,x_{n-1})$ and $J := (x_{n-1},b)$ separately. We shall denote the restrictions of $\|\cdot\|_\varepsilon$ and B_ε to I and J by $\|\cdot\|_{\varepsilon,I}$, $\|\cdot\|_{\varepsilon,J}$, $B_{\varepsilon,I}$ and $B_{\varepsilon,J}$ respectively.

Using lemma 2 with $v := (M^h)^{-1}(\tilde{z}^h_\varepsilon+\phi_\varepsilon)$ we find

$$\tfrac{1}{2}\|\tilde{z}^h_\varepsilon + \phi_\varepsilon\|_\varepsilon \|v\|_\varepsilon \le B_\varepsilon(\tilde{z}^h_\varepsilon+\phi_\varepsilon,v) = B_\varepsilon(\phi_\varepsilon,v) \le C\|v\|_\varepsilon \,;$$

hence $\|\tilde{z}^h_\varepsilon\|_\varepsilon = \mathcal{O}(1)$. Analogously we find $\|z_\varepsilon\|_\varepsilon = \mathcal{O}(1)$. In the same way as in (3,44-48-49) we find from these energy norm estimates the pointwise estimate

$$|\tilde{z}^h_\varepsilon(x_i) - z_\varepsilon(x_i)| \le C(\varepsilon+h^k)\|\tilde{z}^h_\varepsilon - z_\varepsilon\|_\varepsilon \le C(\varepsilon+h^k). \qquad (3.52)$$

If $(b-x_{n-1})p_0 > \varepsilon|\log\varepsilon|$, ($p_0$ as in (1.2)), the boundary layer is contained in the subinterval (x_{n-1},b) entirely and we have

$$|z_\varepsilon(x_{n-1})| \le C\varepsilon, \text{ hence } \zeta_\varepsilon := \tilde{z}^h_e(x_{n-1}) \le C(\varepsilon+h^k). \qquad (3.53)$$

In P^h_{k+1} we now define the function w_ε by

$$w_\varepsilon := \begin{cases} \theta\{P_{k+1}(\xi_n)+P_k(\xi_n)+\eta h(P_k(\xi_n)+P_{k-1}(\xi_n))\} + \\ \qquad - (-1)^k\zeta_\varepsilon P_{k+1}(\xi_n), & \text{if } x \in J \\ \\ \zeta_\varepsilon(x-a)/(x_{n-1}-a), & \text{if } x \in I, \end{cases} \qquad (3.54)$$

where $\xi_n := (2x-x_n-x_{n-1})/(x_n-x_{n-1})$, cf. (3.8), and where θ and η are defined by

$$\eta := \frac{2k+1}{2k-1}\,\frac{(k+1)p'(b)+q(b)}{p(b)}\ , \quad \theta := \frac{\phi_\varepsilon(b)+(-1)^k\zeta_\varepsilon}{2+2h\eta}\ .$$

On I we find the estimates

$$\|w_\varepsilon\|_{1,I} \le C(\varepsilon+h^k) \qquad \text{and}$$

$$B_{\varepsilon,I}(w_\varepsilon,v) \le C(\varepsilon+h^k)\|v\|_\varepsilon\,, \ \forall v \in F_k^h.$$

Hence $\|w_\varepsilon - z_\varepsilon\|_{\varepsilon,I}$ and $\|w_\varepsilon - \tilde{z}_\varepsilon^h\|_{\varepsilon,I}$ both are of the order $\mathcal{O}(\varepsilon+h^k)$ and this implies, cf. (3.38-42-49),

$$\|\tilde{z}_\varepsilon^h - z_\varepsilon\|_{\varepsilon,I} \le C(\varepsilon+h^k) \qquad \text{and}$$

$$B_{\varepsilon,I}(\tilde{z}_\varepsilon^h - z_\varepsilon,\ G_\varepsilon(x_i,\cdot) - \rho_{\varepsilon,i}^h - \sigma_{\varepsilon,i}^h) \le C(\varepsilon+h^k)^2, \qquad i = 1,\dots,n-1. \tag{3.55}$$

On J we find the estimates

$$\|w_\varepsilon\|_{\varepsilon,J} \le C(h+\varepsilon/h)^{\frac{1}{2}} \qquad \text{and}$$

$$B_{\varepsilon,J}(w_\varepsilon,v) \le C(h+\varepsilon/h)^{\frac{1}{2}}\|v\|_\varepsilon\,, \ \forall v \in F_k^h;$$

in the proof of the second estimate we use the same trick as in the proof of lemma 2. These estimates imply

$$B_\varepsilon(w_\varepsilon - \tilde{z}_\varepsilon^h, v) \le C(h+\varepsilon/h), \ \forall v \le F_k^h\ .$$

In conjunction with lemma 2 we infer

$$\|\tilde{z}_\varepsilon^h\|_{\varepsilon,J} \le C(h+\varepsilon/h)^{\frac{1}{2}}$$

and since $\tilde{z}_\varepsilon^h$ is a polynomial it satisfies the estimate

$$|\tilde{z}_\varepsilon^h(x)| + h\left|\frac{d}{dx}\tilde{z}_\varepsilon^h(x)\right| \le C(1+\varepsilon^{\frac{1}{2}}/h),$$

$$|\tilde{z}_\varepsilon^h(x)| \le C(\varepsilon+h^k + \left|\frac{d}{dx}\tilde{z}^h(x_{n-1})\right|\,|x-x_{n-!}|).$$

Straightforward computation now yields

$$B_{\varepsilon,J}\left(\tilde{z}_\varepsilon^h - z_\varepsilon,\ G_{\varepsilon,i} - \rho_{\varepsilon,i}^h - \sigma_{\varepsilon,i}^h\right) \le C\left(h^{2k} + \frac{\varepsilon^2}{h}\left(1+\frac{\varepsilon^{\frac{1}{2}}}{h}\right)\right). \tag{3.56}$$

In conjunction with (3.55) this yields the estimate

$$|\tilde{z}_\varepsilon^h(x_i) - z_\varepsilon(x_i)| \le C\left(h^{2k} + \frac{\varepsilon^2}{h}\left(1+\frac{\varepsilon^{\frac{1}{2}}}{h}\right)\right) \quad (3.57)$$

for $i = 1,\ldots,n-1$. If the term h^{2k} is dominant in this error estimate, it can be improved by repeating the process from formula (3.53) on, using the better estimate (3.57) instead of (3.52). So we obtain the desired estimate (3.50). □

4. RESULTS OF NUMERICAL EXPERIMENTS

Several numerical experiments were performed with the exponentially fitted methods described in the previous sections. The accuracy of the computed solution is considered at the mesh-points and this accuracy is compared with the error bounds derived.

(4a) *The experiments*

For the trial spaces S^h and V^h the spaces P^h_{k+1}, E^h_k, F^h_k and $E^h_{k-1} + F^h_{k-1}$ were used with $k = 1,2,3$. With these spaces the seven combinations for the solution and test space were used as they are mentioned in table 1. The partition of the interval of integration was taken quasi-uniformly with $n = 4$ and $n = 8$. For different values of h and ε the accuracy obtained at the mesh-points was compared in order to determine the dependence of the error on these two parameters. For ε the following sequence was used:

$$\varepsilon = 1,\ 0.1,\ 10^{-2},\ 10^{-3},\ 10^{-4},\ 10^{-5},\ 10^{-10}.$$

Mesh selection

In order to eliminate effects possibly due to a uniform partition, the experiments were done with non-uniform partitions, where the mesh-points were selected by

$$x_i = ih + 0.15\,\rho h, \quad i = 1,2\ldots,n-2,$$

where $h = 1/n$ and ρ is a random variable distributed uniformly in $[-1,+1]$.

Quadrature

In the experiments the computation of the integrals was executed by means of an automatic quadrature routine which computed the integrals with an absolute or relative accuracy of 10^{-7} on each subinterval of the grid separately. We are convinced that automatic quadrature is not an efficient procedure. However, since our purpose is to compare the error bounds derived with the actual errors for the methods described, we do not want to consider effects introduced by numerical quadrature. Hence we approximate the exact value of all the integrals involved as good as possible. For efficient quadrature techniques for the exponentially fitted methods we refer to Hemker (1977).

The environment

The experiments were performed in single precision on a CDC-CYBER computer, using the CDC ALGOL 68 compiler (version 1.2.0). The accuracy of a real number is about 14 decimal digits.

The problems

The following three problems were used in the experiments.

PROBLEM 1:

$$-\varepsilon y'' + (2+\cos(\pi x))y' + y = (1+\varepsilon\pi^2)\cos(\pi x) - (2+\cos(\pi x))\pi\sin(\pi x),$$

$$y(0) = 1, \qquad y(1) = -1.$$

The solution is $y(x) = \cos(\pi x)$; the solution has no boundary layer.

PROBLEM 2:

$$-\varepsilon y'' + y' + (1+\varepsilon)y = 0,$$
$$y(0) = 1 + \exp(-(1+\varepsilon)/\varepsilon),$$
$$y(1) = 1 + \exp(-1).$$

The solution is $y(x) = \exp((1+\varepsilon)(x-1)/\varepsilon) + \exp(-x)$; the equation has constant coefficients and the solution has a boundary layer.

PROBLEM 3:

$$-\varepsilon y'' + \cos(\alpha-x)y' + y = \sin(\alpha-x)(1+\varepsilon+\sin(\alpha-x)) - 1 + \\ + \exp((x-1)/\varepsilon)(1-2\sin((\alpha-x)/2)^2/\varepsilon),$$
$$y(0) = \sin(\alpha) + \exp(-1/\varepsilon),$$
$$y(1) = \sin(\alpha-1) + 1.$$

The solution is $y(x) = \sin(\alpha-x) + \exp((x-1)/\varepsilon)$. The equation has non-constant coefficients and the solution has a boundary layer at $x = 1$. In order to prevent results which may be flattered because $p'(b) = 0$, we have experimented both with $\alpha = 1$ and with $\alpha = 5\pi/12$, which imply $p'(1) = 0$ and $p'(1) \neq 0$ respectively.

(4b) *The numerical results*

In order to give an impression of the actual accuracy of the methods we give some examples of the results obtained for problem 3 ($\alpha = \frac{5}{12}\pi$) in table 2.

A summary of a complete series of experimental results is given in table 3 and 4.

TABLE 2

Errors at the meshpoints: $\max\limits_{i=0,\ldots,n} |y_\varepsilon(x_i) - y_\varepsilon^h(x_i)|$.

h \ ε	1	0.1	10^{-2}	10^{-3}	10^{-4}	10^{-5}	10^{-10}
1/4	3.6 (-7)	1.4 (-3)	6.8 (-3)	9.2 (-4)	1.4 (-4)	6.1 (-5)	5.8 (-5)
1/8	2.2 (-8)	9.4 (-5)	4.4 (-3)	7.1 (-4)	7.8 (-5)	1.4 (-5)	7.4 (-6)
	$S_h = V_h = F_1^h$						
1/4	3.7 (-7)	7.5 (-4)	1.4 (-3)	1.8 (-5)	2.0 (-6)	1.4 (-6)	1.4 (-6)
1/8	2.5 (-8)	4.9 (-5)	1.4 (-3)	2.8 (-5)	3.4 (-6)	1.2 (-7)	9.9 (-8)
	$S_h = P_2^h,\ V_h = F_1^h$						
1/4	9.0 (-10)	5.1 (-6)	1.2 (-3)	1.6 (-5)	1.5 (-6)	1.8 (-8)	1.0 (-9)
1/8	1.0 (-11)	3.5 (-7)	6.6 (-4)	3.9 (-5)	3.1 (-6)	3.3 (-8)	3.8 (-11)
	$S_h = P_3^h,\ V_h = F_2^h$						
1/4	3.9 (-13)	1.0 (-6)	7.1 (-4)	1.3 (-5)	1.9 (-6)	1.3 (-8)	3.3 (-11)
1/8	4.1 (-13)	6.4 (-9)	1.8 (-4)	4.5 (-5)	3.9 (-6)	2.9 (-8)	3.4 (-11)
	$S_h = P_4^h,\ V_h = F_3^h$						

TABLE 3

The orders of the error for several Petrov-Galerkin methods

	S_h	V_h	order of the error at mesh points: $h \ll \varepsilon$	order of the error at mesh points: $\varepsilon \ll h$	Remark
1	P_{k+1}^h	P_{k+1}^h	h^{2k+2}	1	
2	E_k^h	E_k^h	h^{2k+2}	1	
3	F_k^h	F_k^h	h^{2k+2}	$\varepsilon + h^{2k}$	*)
4	$E_{k-1}^h + F_{k-1}^h$	$E_{k-1}^h + F_{k-1}^h$	h^{2k+2}	$\varepsilon^2 + h^{2k-2}$	*)
5	E_k^h	P_{k+1}^h	h^{2k+2}	$\varepsilon + h^{k+1}$	
6	E_k^h	F_k^h	h^{2k+2}	$\varepsilon^2 + h^{2k+1}$	
7	P_{k+1}^h	F_k^h	h^{2k+2}	$\frac{\varepsilon^2}{h} + h^{2k+1}$	*)

*) the order of the h-term for $\varepsilon \ll h^2$ might be slightly pessimistic. For details see table 4.

TABLE 4

Experimentally determined orders of convergence for $\varepsilon \ll h$.

	S_h	V_h	k=1	k=2	k=3
3	F_k	F_k	$\varepsilon+h^3$	$\varepsilon+h^{4.5}$	*)
4	$E^h_{k-1}+F^h_{k-1}$	$E^h_{k-1}+F^h_{k-1}$	no experiment	ε^2+h^3	ε^2+h^4
7	P_{k+1}	F_k	$\frac{\varepsilon^2}{h}+h^{3.5}$	$\frac{\varepsilon^2}{h}+h^5$	*)

*) the error was too small to determine the rate of convergence.

In the case $p(x)h \ll \varepsilon$, table 3 shows an error which is much smaller than the theoretical error in table 1. This error of order $O(h^{2k+2})$ is easily understood since, in the case where $p(x)h \ll \varepsilon$, the trial spaces E^h_k, F^h_k and $E^h_{k-1}+F^h_{k-1}$ differ only slightly from the piecewise polynomial spaces P^h_{k+1} and, in fact, have nearly the same approximation properties for smooth functions.

In the more interesting case $\varepsilon \ll p(x)h$, the theoretical error bounds consist of ε-dependent and h-dependent parts. To perceive in an error of the form $\varepsilon^p + h^q$ the orders of both parameters separately, we have performed experiments both with $\varepsilon^p \ll h^q$ and with $\varepsilon^p \gg h^q$.

(4c) *Numerical instability for* $S^h = V^h = E^h_k$

In table 3 we notice that the 2nd method ($S^h = V^h = E^h_k$) does not follow the theoretically derived error bound. This is due to the fact that for small $\alpha_i h_i$ the basis functions on the subinterval (x_{i-1}, x_i) are almost linearly dependent and a jump occurs at the right-hand side of the subinterval. Hence, the 2nd up to the (k+1)st row in the element stiffness matrix are almost linearly dependent.

This causes cancellation of digits when the assembled stiff-ness-matrix is solved. After static condensation we obtain a tridiagonal matrix with elements on the subdiagonal of order ε, as follows

$$\begin{matrix} 1 & 0 & & & \\ \varepsilon & 1 & 1 & & \\ & \varepsilon & 1 & 1 & \\ & & \varepsilon & 1 & 1 \\ & & & 0 & 1 \end{matrix} .$$

This shows that, in first approximation, the reduced equation is solved with the right-hand boundary condition, whereas the original problem is, again in first approximation, the solution of the reduced problem with the left-hand boundary condition. Thus, it is easily seen that for a problem of which the solution contains a singular part, the numerical approximation has an error of order unity.

(4d) *Conclusions*

The theoretical and the experimental results agree as far as the ε-dependence of the error is concerned, except for the case where $S^h = V^h = E_k^h$. For this particular combination of the solution and test space the Petrov-Galerkin method is numerically unstable.

Concerning the order of the error with respect to h, the theoretical bounds given in theorem 1 and 2 seem to be pessi-mistic. In some cases (the methods 3, 5 and 6 in table 3) it seems that the order of the error can be increased by one.

Taking into account that the order of the error of the best approximation in E_k^h is $\mathcal{O}(\varepsilon+h^{k+1})$ in energy norm, one might expect that the Galerkin approximation has the same error in energy norm. This implies that the approximation at the mesh-points would improve by one order in h.

This better estimate would agree with the experiments. Moreover, it would yield a reasonable error estimate for the case k = 0. However, how the better estimate can be proved remains an open question.

REFERENCES

Abrahamsson, L.R., Keller, H.B. and Kreiss, H.O. (1974). "Difference approximations for singular perturbations of systems of ordinary differential equations". *Numer. Math.*, 22 (1974) 367-391.

Douglas, J. and Dupont, T. (1974). "Galerkin approximations for the two-point boundary problem using continuous piecewise polynomial spaces". *Numer. Math.* 22 (1974) 99-109.

Eckhaus, W. (1973). *Matched asymptotic expansions and singular perturbations*. North-Holland Publ. Comp., Amsterdam, 1973.

De Groen, P.P.N. (1978). "A finite element method with a large mesh-width for a stiff two-point boundary value problem", *to appear*.

Hemker, P.W. (1977). *A numerical study of stiff two-point boundary problems*. MC-Tract 80, Mathematical Centre, Amsterdam.

Il'in, A.M. (1969). "Differencing scheme for a differential equation with a small parameter affecting the highest derivative". *Math. Notes Acad. Sc. USSR* 6 (1969) 596-602.

O'Malley, R.E. (1974). *Introduction to singular perturbations*. Academic Press, New York, 1974.

Van Veldhuizen (1979). "Higher order schemes of positive type for Singular Perturbation Problems". *These proceedings*.

SOLUTION OF NONLINEAR SECOND ORDER DIFFERENTIAL EQUATIONS WITH SIGNIFICANT FIRST DERIVATIVES BY A PETROV-GALERKIN FINITE ELEMENT METHOD

J.C. Heinrich and O.C. Zienkiewicz
Department of Civil Engineering,
University College of Swansea, Swansea, Wales

1. INTRODUCTION

In the last years a considerable effort has been made by finite element users to be able to model numerically physical problems dominated by convection phenomena. Typical examples of such problems which arise in a variety of situations in science and engineering are convective and conductive heat (or mass) transport and the flow of viscous incompressible fluids at high Reynolds numbers.

The difficulties inherent in the numerical solution of such problems stem in part from the fact that the differential operators in the governing equations are non-selfadjoint and hence our understanding of them is very limited. Furthermore, for highly convective situations, we are faced with the solution of a second order equation with dominant first derivative terms where the application of boundary conditions causes problems of overconstraint. The discretisations may require the use of double precision words to be used, due to the vast range in the order of magnitude of the elements of the matrices, and even if this is done, ill-conditioned systems often arise after applying a numerical scheme.

From the point of view of finite element method users the need for special ways to treat convective situations was first recognised in Zienkiewicz *et al* (1975) where it was stated that a finite element equivalent to upwind differences was needed to avoid the problems of oscillatory solutions obtained with standard numerical approximations. For the linear, steady state convective-diffusion equation some solutions were given by Miller (1975), Piva and Di Carlo (1975) and Blackburn (1976). For a one-dimensional such equation a major breakthrough was made by Christie *et al* (1976) using piecewise linear basis functions and piecewise quadratic test functions. This work was extended to two dimensional situations using rectangular elements in Heinrich *et al* (1977) and also to piecewise quadratic basis functions in Heinrich and Zienkiewicz (1977) for both one and two dimensions. A further algorithm using piecewise linear triangular elements in two-dimensions was proposed by Tabata (1977), and a further evaluation of the methods as well as a comparison with equivalent finite differences formulations for the convective-diffusion equation has been given in Zienkiewicz and Heinrich (1978).

In the present work we investigate the application of these techniques, of a Petrov-Galerkin kind (Anderssen and Mitchell (1977)), to nonlinear problems where the nonlinearity appears in the convective terms. We look at the solution of the Navier-Stokes equations using primitive variables and problems of coupled convective and conductive heat transfer where an energy equation coupled with the Navier-Stokes equations are to be solved simultaneously (Heinrich *et al*, (1978)).

To solve the nonlinear equations we have used both the method of successive-substitutions and Newton's method; their performance is compared through the solution of the one-dimensional Burger's equation. Our results indicate that although the methods developed through Christie *et al* (1976), Heinrich *et al* (1977) and Heinrich *et al* (1977) can be successfully applied to some nonlinear problems, there is much to be done in order to improve our capabilities for solving many practical situations and achieve a full understanding of the problems.

2. ONE DIMENSIONAL CASE

We will restrict ourselves to the solution of the steady-state (Burger's) equation

$$\mathbf{B}\phi \equiv \frac{d^2\phi}{dx^2} + \lambda\phi\frac{d\phi}{dx} = 0 \qquad (1)$$

defined over the open domain $D = (0,1)$ with boundary conditions

$$\phi(0) = 1, \quad \phi(1) = 0 \qquad (2)$$

where $\lambda>0$ is assumed constant. Extensions to more general situations such as a non-homogeneous righthandside in (1), etc., are immediate, and the essential features of the problem are well represented by equations (1) and (2). Also, equation (1) has the same structure as the equations of conservation of linear momentum in flow of viscous fluids. An understanding of its behaviour will undoubtedly help in the more complex case of flow in two and three dimensions.

Under the above conditions, an analytical solution for equation (1) can be obtained and is of the form

$$\phi(x) = \frac{A}{\lambda}\left(\frac{1+Be^{Ax}}{1-Be^{Ax}}\right) \qquad (3)$$

where A and B are to be determined from the boundary conditions. It is also easily shown that for all $x \varepsilon D$ we have

$$|\phi(x)|\leq 1 \ , \ \frac{d\phi}{dx}\leq 0 \quad \text{and} \quad \left|\frac{d\phi}{dx}\right| \leq \frac{2A}{1-e^{-2A}} \tag{4}$$

Since equation (1) is nonlinear, an iterative procedure is required which will involve a linear operator at each iteration. Here we opt to linearise the differential equation before discretising; we choose this approach since it gives a better insight to the problem. (Had we chosen to discretise the problem first and apply any of the two procedures to be described below to the resulting system of nonlinear algebraic equations, the final system of linear equations to be solved at each iteration would be exactly the same since, at least in these cases, the discretisation and linearisation processes commute.)

The first method we consider is the successive substitutions (SS) iterative procedure, in which the solution of ϕ^n at the nth iteration is given by the solution of the linear differential equation

$$L_s^{n-1}\phi^n \equiv L_s(\phi^{n-1})\phi^n = 0 \tag{5}$$

where the operator L_s is given by

$$L_s(\phi^{n-1}) \equiv -\frac{d^2}{dx^2} + \lambda\phi^{n-1}\frac{d}{dx} \tag{6}$$

We will also use the Newton-Raphson (NR) incremental iterative procedure; in this case ϕ^n at the nth iteration is given by

$$\phi^n = \phi^{n-1} + \Delta\phi^n \tag{7}$$

where $\Delta\phi^n$ is obtained at each step as the solution of

$$L_N^{n-1}\Delta\phi^n \equiv L_N(\phi^{n-1})\Delta\phi^n = -\mathbf{B}\phi^{n-1} \tag{8}$$

and the linear operator L_N is given as

$$L_N(\phi^{n-1}) \equiv -\frac{d^2}{dx^2} + \lambda\phi^{n-1}\frac{d}{dx} + \lambda\frac{d\phi^{n-1}}{dx} \tag{9}$$

2.1 Discretisation

To construct finite element approximations to problem (1), (2), we first need to establish its weak formulation. We will denote by $H^1(\Omega)$ the space of functions ϕ with domain $\Omega = [0,1]$ such that

$$\| \phi \|_1 \equiv \left(\int_0^1 (\phi^2 + (\frac{d\phi}{dx})^2)dx\right)^{\frac{1}{2}} \tag{10}$$

is finite. Important subsets of the space $H^1(\Omega)$ are the space $\overset{o}{H}{}^1(\Omega) \equiv \{\phi\varepsilon H^1(\Omega)/\phi(0) = \phi(1) = 0\}$ and the set $\bar{H}^1(\Omega) = \{\phi\varepsilon H^1(\Omega)/\phi(0) = 1, \phi(1) = 0\}$. The weak form of (1), (2) can then be written.

Find $\phi\varepsilon\bar{H}^1(\Omega)$ such that

$$\int_0^1 (\frac{d\psi}{dx}\frac{d\phi}{dx} + \lambda\psi\phi\frac{d\phi}{dx})dx = 0 \tag{11}$$

for each $\phi\varepsilon\overset{o}{H}{}^1(\Omega)$.

We assume that Ω has been subdivided into N subintervals of the same size h with nodal points $x_i = ih$, $i = 0,1, N$ (regularity of the mesh is assumed only for simplicity of presentation). The function ϕ is then assumed to be approximated piecewise over each element by a function

$$\phi(x) = \sum_{i=0}^{N} N^i(x)\phi_i \tag{12}$$

where $\{N^i\}_{i=0}^N$ are basis functions for a finite dimensional subspace of $H^1(\Omega)$ and $\{\phi_i\}_{i=0}^{N+1}$ nodal values. Piecewise linear and quadratic basis functions are considered although for brevity we will only give details for the linear case. We have

$$N^i(x) = \begin{cases} (x-x_{i-1})/h & x_{i-1}\le x\le x_i \\ 1-(x-x_i)/h & x_i\le x\le x_{i+1} \\ 0 & \text{otherwise} \end{cases} \tag{13}$$

In a Petrov-Galerkin finite element formulation the space of test functions ψ is not necessarily the same as the space of trial functions for which (13) defines a basis (details can be found in Griffiths and Lorenz (1977)). We consider test functions of the form

$$\psi^i(x,\alpha_i,\alpha_{i+1}) = \begin{cases} N^i(x) + \alpha_i F(x) & x_{i-1}\le x\le x_i \\ N^i(x) - \alpha_{i+1} F(x) & x_i\le x\le x_{i+1} \\ 0 & \text{otherwise} \end{cases} \tag{14}$$

and the function F is given over each interval by

$$F(x) = 3(x-x_{i-1})(1-(x-x_{i-1})/h)/h \tag{15}$$

The iterative procedures previously defined reduce to the solution of a linear problem at each step. The linear operators being different in each case, we consider each of them separately.

2.1.1 Successive substitutions iterative method After (5) and (11), at each iteration we have to find $\phi^n \varepsilon \bar{H}^1(\Omega)$ such that

$$\int_0^1 \left(\frac{d\psi}{dx}\frac{d\phi^n}{dx} + \lambda\psi\phi^{n-1}\frac{d\phi^n}{dx}\right)dx = 0 \tag{16}$$

for each $\psi \varepsilon \overset{o}{H}{}^1(\Omega)$. Further replacement into (16) of (13) and (14) leads to the solution of a system of linear equations of the form

$$\underset{\sim\sim}{A\phi} = \underset{\sim}{B} \tag{17}$$

where

$$a_{ij} = \int_0^1 \left(\frac{d\psi^i}{dx}\frac{dN^j}{dx} + \lambda\psi^i\phi^{n-1}\frac{dN^j}{dx}\right)dx \tag{18}$$

If we now assume that ϕ^{n-1} is piecewise constant, that is, constant over each element, we can write a typical i^{th} equation of system (17) as

$$\left(1 + \frac{\gamma_{i+1}}{2}(\alpha_{i+1}-1)\right)\phi_{i+1} - \left(2 + \frac{\gamma_i}{2}(\alpha_i+1) + \frac{\gamma_{i+1}}{2}(\alpha_{i+1}-1)\right)\phi_i + \left(1 + \frac{\gamma_i}{2}(\alpha_i + 1)\right)\phi_{i-1} = 0 \tag{19}$$

with $\gamma_i = \lambda\phi^{n-1}h$ and α_i constant values over the interval (x_{i-1},x_i). The solution of (19) is given by

$$\phi_j = A+B.\prod_{i=1}^{\dot{\gamma}}\left(\frac{1 + \frac{\gamma_i}{2}(\alpha_i+1)}{1 + \frac{\gamma_{i+1}}{2}(\alpha_{i+1}-1)}\right) \tag{20}$$

and will be stable if

$$\alpha_{i+1} \geq 1 - \frac{2}{\gamma_{i+1}}.$$

We know from Christie *et al* (1976) that for constant γ, i.e., $\gamma_i = \gamma$ and $\alpha_i = \alpha$ for each i = 1, ..., N, the exact solution to the linear problem is obtained at each nodal point by choosing

$$\alpha = \alpha opt = (\coth \frac{\gamma}{2}) - \frac{2}{\gamma} \tag{21}$$

Here our experience indicates that the use of this value for each element produces a substantial improvement in accuracy in the case of variable coefficients as well, the optimal value being within the limit of stability is preferable to the more obvious choice $\alpha = 1$ which reproduces an upwind differences scheme but at the same time introduces an excessive amount of numerical diffusion (Heinrich *et al* (1977)). Scheme (19) with α_i computed elementwise using equation (21) is very similar to the "locally exact" finite differences scheme introduced by Barret (1973) (the only difference being that in the latter for a node x_i, γ_i is considered constant over $x_i - h/2 < x < x_i + h/2$ and α_i chosen accordingly), which is exponentially accurate in some cases.

2.1.2 Newton-Raphson's incremental method At each iteration we look for $\Delta\phi^n \epsilon \overset{o}{H}{}^1(\Omega)$ such that

$$\int_0^1 \left(\frac{d\psi}{dx}\frac{d\Delta\phi^n}{dx} + \lambda\psi\phi^{n-1}\frac{d\Delta\phi^n}{dx} + \lambda\psi\frac{d\phi^{n-1}}{dx}\Delta\phi^n\right)dx = -\int_0^1 \psi\mathbf{B}\phi^{n-1}dx \quad (22)$$

for each $\psi \epsilon \overset{o}{H}{}^1(\Omega)$. Assuming once again that the coefficients are piecewise constant, a typical equation in the linear system generated by (22) has the form

$$\left(1 + \frac{\gamma_{i+1}}{2}(\alpha_{i+1} - 1) + \frac{\theta_{i+1}}{12}(3\alpha_{i+1} - 2)\right)\phi_{i+1} - \left(2 + \frac{\gamma_i}{2}(\alpha_i + 1)\right.$$

$$\left. + \frac{\gamma_{i+1}}{2}(\alpha_{i+1} - 1) + \frac{\theta_i}{12}(4 + 3\alpha_i) + \frac{\theta_{i+1}}{12}(4 - 3\alpha_{i+1})\right)\phi_i \quad (23)$$

$$+ \left(1 + \frac{\gamma_i}{2}(\alpha_i + 1) - \frac{\theta_i}{12}(3\alpha_i + 2)\right)\phi_{i-1} = b_i$$

with $\theta_i = \lambda\frac{d\phi^{n-1}}{dx}h^2$ constant over each interval (x_{i-1}, x_i).

The solution of (23) involves fairly complicated expressions for the roots of the characteristic polynomial and it is not possible to give a simple expression for the limit of stability. However, it is clear that the choice $\alpha_i = \pm 1$, $i = 1, \ldots N$, where the sign of α depends on the sign of γ_i

will produce a differences scheme of an upwind type which will be stable for

$$\lambda\frac{d\phi^{n-1}}{dx} \geq - \frac{12}{h^2} .$$

Although this method requires much more work at each iteration than the SS method it is a second order method while the SS method is only of the first order. However, it is not clear from the numerical experiments carried on so far which of them offers more advantages. This will be illustrated later on in the examples.

2.2 Piecewise quadratic basis functions

Although for one dimensional situations linear elements are most of the time sufficient for the solution of a problem, this is not the case in two (and three) dimensional situations, where their extension to linear triangles or bilinear rectangles often produces approximations which are not accurate enough and we have to resort to the more sophisticated quadratic and biquadratic elements. In the one dimensional case the basis functions are

$$N^i(x) = \begin{cases} (x-x_{i-2})(2(x-x_{i-2})/h-1)/h & x_{i-2}\leq x\leq x_i \\ 1 + (x-x_i)(2(x-x_i)/h-3)/h & x_i\leq x\leq x_{i+2} \\ 0 & \text{otherwise} \end{cases} \tag{24}$$

For nodes located at the ends of the elements, which we will refer to as corner nodes. For the nodes interior to the elements, which we call middle nodes, we have

$$N^i(x) = \begin{cases} =(x-x_{i-1})(1-2(x-x_i)/h)/h & x_{i-1}\leq x\leq x_{i+1} \\ 0 & \text{otherwise} \end{cases} \tag{25}$$

Test functions are chosen so that for middle nodes we have

$$\psi^i(x, \alpha_i) = \begin{cases} N^i(x) + 4\alpha_i F(x) & x_{i-1} \le x \le x_{i+1} \\ 0 & \text{otherwise} \end{cases} \tag{26}$$

and for corner nodes

$$\psi^i(x, \beta_i, \beta_{i+1}) = \begin{cases} N_i(x) - \beta_i F(x) & x_{i-2} \le x \le x_i \\ N_i(x) - \beta_{i+1} F(x) & x_i \le x \le x_{i+2} \\ 0 & \text{otherwise} \end{cases} \tag{27}$$

The function F is given over each element by

$$F(x) = \begin{cases} 5(x-x_{i-1})(2(x-x_{i-1})^2/h^2 - 3(x-x_{i-1})/h + 1)/2h & \\ & x_{i-1} \le x \le x_{i+1} \\ 0 & \text{otherwise} \end{cases} \tag{28}$$

These elements used in conjunction with the SS method as defined by (16) allow us to find expressions that give the limits of stability within which the parameters α_i and β_i are to be chosen, and for the constant coefficients case optimal values can be computed that render the exact solution at each nodal point. We must point out, however, that we cannot find one fixed pair of values for α and β (as is the case $\alpha = 1$ for linear elements, for which an upwind scheme is obtained) which will render stability for all values of γ and for computations the optimal values are always used. The situation becomes even more complicated when NR is used, and for these reasons we will not get into further details on these elements. These can be found in reference Heinrich *et al* (1977).

3. TWO DIMENSIONAL PROBLEMS

As stated earlier, we are interested in modelling problems

involving fluid flow, heat transfer, convective-diffusion, etc., which are of immense practical importance. We will restrict our attention here to the solution of three types of problems in two space dimensions and steady state:

a) Flow of viscous incompressible fluids governed by the Navier-Stokes equations which, using inditial notation and the summation convention, can be written together with the continuity equation as

$$\rho u_j \frac{\partial u_i}{\partial x_j} = - \frac{\partial p}{\partial x_i} + 2\mu \frac{\partial}{\partial x_j}\left(\frac{\partial u_i}{\partial x_j} + \frac{\partial u_j}{\partial x_i}\right) + \rho B_i \tag{29}$$

$$\frac{\partial u_i}{\partial x_i} = 0 \tag{30}$$

where ρ stands for the density, μ the viscosity, p the pressure and u_i represents the velocity in the x_i cartesian coordinate. An Eulerian formulation and constant physical properties have been assumed.

b) Convective heat transport where the flow field is assumed independent of the temperature and governed by the convective-diffusion equation

$$\rho c_p u_j \frac{\partial T}{\partial x_j} = \frac{\partial}{\partial x_j} k\left(\frac{\partial T}{\partial x_j}\right) + Q \tag{31}$$

here c_p is the heat capacity at constant pressure and k the termal diffussivity, assumed isothropic.

c) Coupled convective and conductive heat transfer. In this case equations (29)-(31) are coupled through a body force term B_i and possibly through a viscous dissipation term Q of the form

$$B_i = \beta(T-T_o)g_i \tag{32}$$

$$Q = \mu\Phi = \mu \frac{\partial u_i}{\partial x_j}\left(\frac{\partial u_i}{\partial x_j} + \frac{\partial u_j}{\partial x_i}\right) \tag{33}$$

where β is the coefficient of thermal expansion, T_o a reference temperature and g_i the component of gravitational acceleration in the x_i coordinate direction. Φ is called the viscous dissipation function.

In all cases, numerical difficulties arise when the convective acceleration terms are dominant and oscillatory answers are obtained which are quite often meaningless unless some sort of precaution such as a special weighting or the use of extremely refined meshes are taken.

The ideas and procedures introduced in section 2 are readily extended to the two dimensional case by considering the extension to bilinear and biquadratic rectangular four and nine-noded Lagrangian elements in the way familiar to all finite elements users (Gartling, Nickell and Tanner (1977)). In order to compute the weighting parameters we assume that this can be done independently in each coordinate direction. We illustrate the procedure in a four noded bilinear element as shown in figure 1; in this case the weighting functions in isoparametric (ξ,η) coordinates (Zienkiewicz (1977)) are given by

$$\begin{aligned}
\psi^1(\xi,\eta) &= (N^1(\xi) - \alpha_1 F(\xi))(N^1(\eta) - \alpha_4 F(\eta)) \\
\psi^2(\xi,\eta) &= (N^2(\xi) + \alpha_1 F(\xi))(N^1(\eta) - \alpha_2 F(\eta)) \\
\psi^3(\xi,\eta) &= (N^2(\xi) + \alpha_3 F(\xi))(N^2(\eta) + \alpha_2 F(\eta)) \\
\psi^4(\xi,\eta) &= (N^1(\xi) - \alpha_3 F(\xi))(N^2(\eta) + \alpha_4 F(\eta))
\end{aligned} \tag{34}$$

where

$$\left.\begin{aligned}
N^1(\xi) &= \tfrac{1}{2}(1+\xi) \\
N^2(\xi) &= \tfrac{1}{2}(1-\xi)
\end{aligned}\right\} , \quad -1\leq\xi\leq 1 \tag{35}$$

and similar expressions with ξ replaced by η.

The parameters α_i are computed for each side joining nodes i, j using

$$\gamma_i = \frac{V_i h}{2}\,, \quad V_i = \tfrac{1}{2}(\underset{\sim}{V}_i + \underset{\sim}{V}_j)\cdot \underset{\sim}{1}_{ij} \tag{36}$$

where $\underset{\sim}{1}_{ij}$ is unit vector in the direction from i to j and $\underset{\sim}{V}_i$ is a vector whose components are the coefficients of the convective acceleration terms. Further details can be found in Heinrich *et al* (1977) and Heinrich and Zienkiewicz (1977) and Zienkiewicz *et al* (1978). In this last reference it is also pointed out that both for the linear and quadratic case the extension to two dimensions is by no means unique and many possibilities can be explored. In the context of linear convective-diffusion, that is, problems of type b), both bilinear and biquadratic elements together with their corresponding test functions have been successfully applied, as has been reported in Heinrich *et al* (1977), Heinrich and Zienkiewicz (1977) and Zienkiewicz *et al* (1978). These techniques are now extended to problems of type a) and c), which are of nonlinear nature. In the next section we present a selection of examples which illustrate both the need of special procedures and the performance of the present algorithms.

4. EXAMPLES

Some numerical examples have been selected which we believe are representative of a wide class of physical situations recognised for the difficulties they offer to numerical modelling. All computations were performed at the University of Swansea's ICL 1904S computer. In all examples the iterative procedures were stopped when the difference in the Euclidean norm between two successive iterations became less than one percent, i.e.,

$$\left(\sum_{i=1}^{n}(u_i^n - u_i^{n-1})^2\right)^{\frac{1}{2}} \le .01\left(\sum_{i=1}^{n}(u_i^n)^2\right)^{\frac{1}{2}} \qquad (37)$$

where $\{u_i\}_{i=1}^{N}$ represents the solution vector.

Consider first the one dimensional problem defined by equations (1) and (2). Since an analytical solution is available, a careful quantitative evaluation of the results is possible. The problem was solved for $\lambda = 10$ using both linear and quadratic elements using two different grids of 11 and 41 nodes. For linear elements the values $\alpha = 0$, 1 and α opt were used with both the SS and NR algorithms. The results for $h = {}^1/10$ are shown in Table 1 together with the analytical solution.

Although it is clear from the results that for this low value of λ no static oscillations arise, this example shows how the accuracy of the finite element approximation varies with α. The value $\alpha = 1$ introduces a considerable amount of false diffusion while on the other hand the value $\alpha = \alpha$ opt renders a substantial improvement in accuracy, as expected.

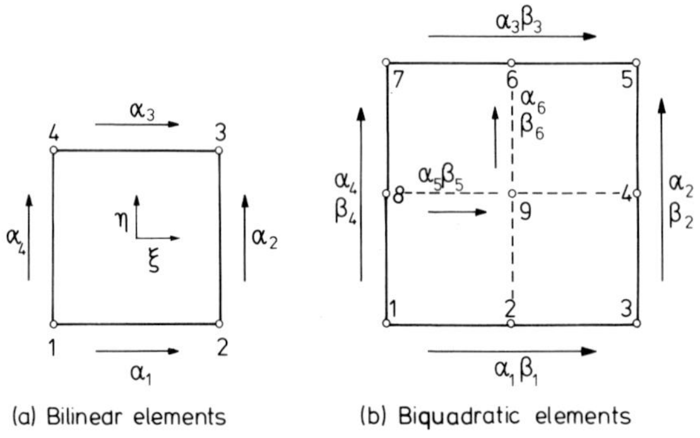

Fig. 1. *Parameters numbering and positive velocity conventions used in two dimensional elements.*

TABLE 1

Results for one dimensional problem using linear elements for $h = {}^1/10$ and $\lambda = 10$.

x	SS α=0	SS α=αopt	SS α=1	NR α=1	NR α=αopt	NR α=1	ANALYTICAL SOLUTION
0.0	1.000	1.000	1.000	1.000	1.000	1.000	1.000
0.2	1.000	0.999	0.996	1.000	0.999	0.996	0.999
0.4	0.997	0.995	0.980	0.997	0.995	0.981	0.995
0.6	0.969	0.963	0.919	0.970	0.964	0.920	0.964
0.8	0.766	0.759	0.689	0.770	0.762	0.693	0.762
0.9	0.460	0.459	0.421	0.464	0.462	0.426	0.462
1.0	0.000	0.000	0.000	0.000	0.000	0.000	0.000

In Table 2 we present the results obtained using quadratic elements for both meshes. In this case only the values $\alpha = \beta = 0$ and $\alpha = \alpha$opt, $\beta = \beta$opt were used.

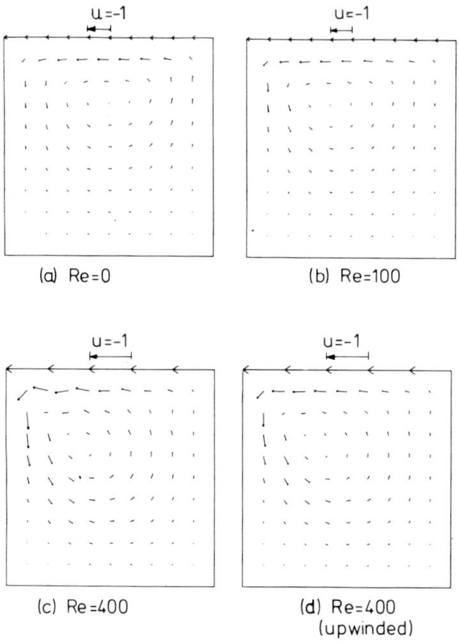

Fig. 2. *Driven cavity flow showing development of oscillations with increasing Reynolds number and smoothing by means of upstream weighting.*

TABLE 2

Results for one dimensional problem using quadratic elements for $h = {}^1/10$ and $h = {}^1/40$, $\lambda = 10$.

	SS(α=β=0)		SS(α=αopt,β=βopt)		NR(α=β=0)		NR(α=αopt,β=βopt)		ANALYTICAL SOLUTION
x	h=.1	h=.025			h=.1	h=.025			
0.0	1.000	1.000	1.000	1.000	1.000	1.000	1.000	1.000	1.000
0.2	0.999	0.999	0.999	0.999	0.999	0.999	0.999	0.999	0.999
0.4	0.995	0.995	0.995	0.995	0.995	0.995	0.995	0.995	0.995
0.6	0.963	0.962	0.963	0.963	0.963	0.964	0.964	0.964	0.964
0.8	0.759	0.758	0.758	0.758	0.763	0.762	0.762	0.762	0.762
0.9	0.456	0.458	0.456	0.458	0.460	0.462	0.461	0.462	0.462
1.0	0.000	0.000	0.000	0.000	0.000	0.000	0.000	0.000	0.000

TABLE 3

Number of iterations needed for convergence

	SS(h=.1)			NR(h=.1)			SS(h=.025)			NR(h=.025)		
λ	α=0	α=αopt	α=1	α=0	α=αopt	α=1	α=0	α=αopt	α=1	α=0	α=αopt	α=1
5	5	5	5	4	4	4	-	-	5	-	-	4
10	6	6	5	5	5	4	-	-	6	-	-	5
10^2	6*	4	5	6*	6	6	-	-	-	-	-	-
10^3	**	4	4	16*	7	7	-	-	4	-	-	8
10^6	-	4	4	**	7	7	-	-	3	-	-	9

In this case we observe no significant differences among the two schemes. This is perhaps due to the fact that the regular ($\alpha=\beta=0$) Galerkin finite element approximations are very accurate to start with. The tendency of the NR method to attain better accuracy is shown better than in the previous case. The problem was also solved using linear elements for a range of values of the parameter λ in order to assess the speed at which the different algorithms converge in Table 3 we give the number of iterations necessary to achieve convergence for λ varying between 5 and 10^6 and using both

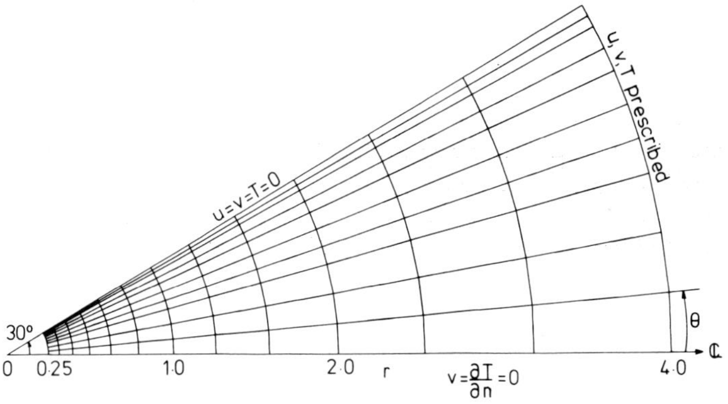

Fig. 3. *Domain and finite element mesh for Jeffery Hamel flow with viscous heat dissipation.*

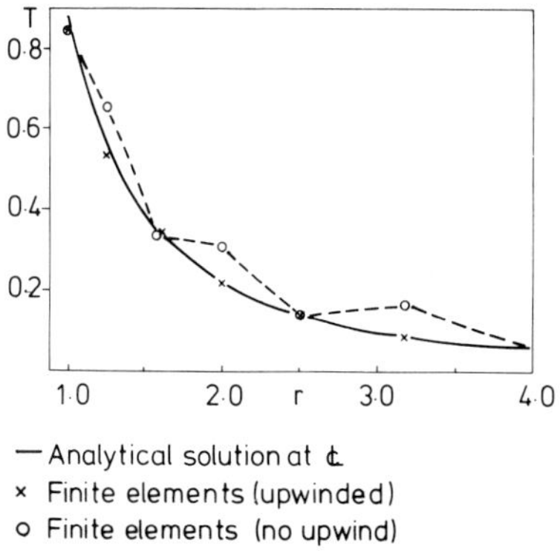

Fig. 4. *Jeffery Hamel flow with viscous heat dissipation temperature profiles along centre line convergent flow at RE=20.*

meshes. The star indicates that an oscillatory solution was obtained, a double star means that no convergence was achieved after 20 iterations, and a dash that we did not run that case. The behaviour of the SS method is quite remarkable taking less iterations to converge for high values of λ; it is also interesting to note that for $\lambda>10$ both methods gave the same solution to within 4 significant digits.

The behaviour of the SS and NR methods observed in the present experiments is in agreement with the observations of Gartling *et al* (1977) for accelerating flow problems. It is clear from the present results that we cannot draw conclusions about which method will be preferable.

The second example that we examine involves 2-dimensional flow of a viscous incompressible fluid in a driven cavity. Our aim here is to show as the Reynolds number increases, oscillations develop in the velocity field as boundary layers grow strong. This is illustrated in Figure 2a, b, c, where solutions for Re = 0, 100 and 400 obtained with a regular 11×11 mesh of bilinear elements are given. In Figure 2d we show an upstream weighted solution using optimal parameters for $R_e = 400$ where it can be clearly observed that oscillations have been eliminated. We will not attempt here to make any further quantitative evaluation of these results since in order to do this there are many other aspects to be considered which are out of the scope of this paper (see, for example, Heinrich *et al*) that should be considered.

The above solutions were obtained using a penalty function formulation (see Heinrich *et al*) and the upwinded solution using the SS method. We must mention that so far we have not been able to reproduce this with the NR method.

Our third example is concerned with coupled convective and conductive heat transfer and it concerns the computation of

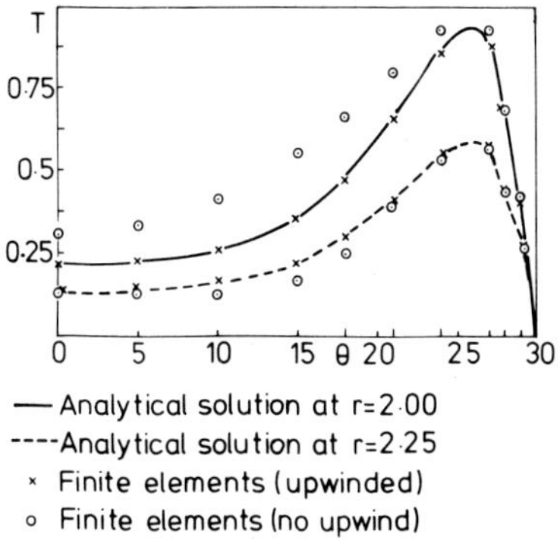

Fig. 5. *Jeffery Hamel flow with viscous heat dissipation temperature profiles along r=2 and r=2.5198 convergent flow at RE=20.*

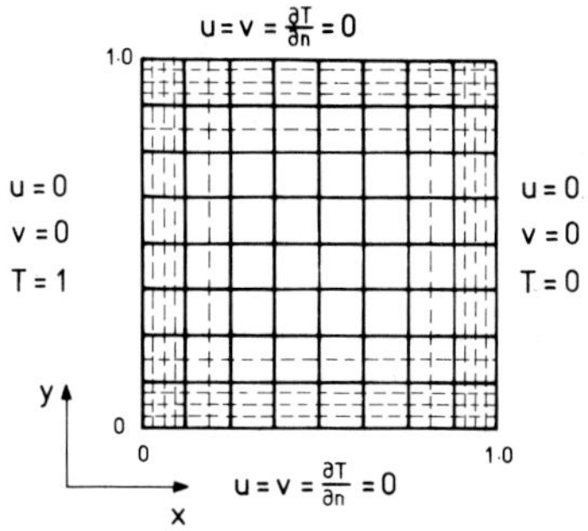

Fig. 6. *Domain, boundary conditions and basic mesh (—) for natural convection in a square. Mesh refinement indicated (----).*

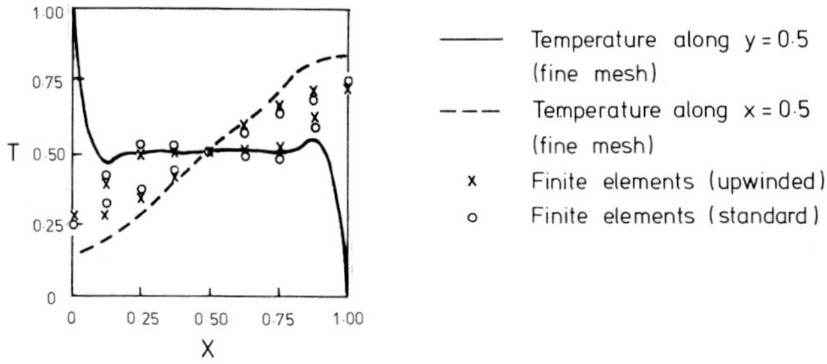

Fig. 7. *Temperature profiles along x=0.5 and y=0.5 for natural convection at Rayleigh number* 10^6 *and Prandtl number 1.*

temperature distributions due to viscous heat dissipation for a Jeffery-Hamel flow in a convergent channel. Details on this problem can be found in Millsaps and Pohlhausen (1953). The problem was solved for convergent flow at Reynolds number 20 and Prandtl number 3 using bilinear elements in the mesh shown in Figure 3, for a half angle of 30°. Since the computation of the flow field did not show any difficulties the special weighting procedure was only applied to the temperature where high oscillations were observed. Strictly speaking, this makes the problem one of linear nature, since the velocities are not dependent on temperature. Nevertheless it is an excellent problem to assess the accuracy of the finite element approximations since an analytical solution is available for both the flow and temperature fields. In Figure 4 we show the solutions obtained for the temperature profile along the centreline and compare with the analytical solution. The same is done in Figure 5 now along two fixed radius $r = 2$ and $r = 2.5198$ for $0 \leq \theta \leq 30°$. The Petrov-Galerkin solution shows an excellent agreement with the analytical solution while the standard Galerkin procedure renders meaningless highly oscillatory answers.

A final example is provided by the problem of natural convection in a square cavity. The domain, finite element meshes and boundary conditions are indicated in Figure 6. The equations governing the flow are the Navier-Stokes equations (29) and (30) with the body force term B_i given by (32) (Boussinesq approximation); and the energy equation (31) where the heat dissipation term Q is neglected. In non-dimensional form, the flow is governed essentially by two parameters, namely the Rayleigh number R_a and Prandtl number P_r (for details see Heinrich *et al* (1978)).

The problem was solved for $R_a = 10^6$ and $P_r = 1$ using

bilinear elements in the coarser regular mesh of size $h = .125$ with and without the use of upstream weighting, the answers showing significant differences. However, when compared with the answers obtained in the much finer irregular mesh shown in Figure 6 using nine noded elements, this third solution is again substantially different from the previous two. This is shown in Figure 7.

It is clear from these results that the problem is extremely sensitive to the model used and that mesh refinement is perhaps unavoidable if accurate answers are desired. Unfortunately, in this case there is just no way of evaluating the quality of the different answers. We choose to accept the solution obtained in the fine mesh with biquadratic elements as the best one. But this solution is about ten times more expensive than any of the other two, and further improvements in the models which would allow us to achieve better accuracy without resorting to such a refinement are needed.

5. CONCLUSIONS

We have shown that finite element schemes of the Petrov-Galerkin type can be very useful in the solution of nonlinear second order differential equations with significant first derivatives when standard Galerkin schemes can produce meaningless answers. In the present work we have used mainly the simplest bilinear elements; further work is in progress on the use of biquadratic elements. Also, this being mainly a qualitative evaluation of the procedure, a more detailed quantitative study is yet to come.

Throughout the examples we have shown that the present schemes can be applied to a variety of complex situations and practical problems in each of which important questions are open to more detailed study. Of particular interest is the behaviour of the successive substitution and Newton-Raphson iterative methods when upstream weighting procedures are used. In the present work we have shown only a small proportion of the possibilities open for investigation and we expect a substantial growth in this field in the future.

ACKNOWLEDGEMENTS

The authors are indebted to SRC for support under grant No. B/SR/9503/4.

REFERENCES

Anderssen, R.S. and Mitchell, A.R. (1977). "The Petrov-Galerkin Method". University of Dundee, Department of Mathematics, Report NA/21.

Barret, K.E. (1973). "The Numerical Solution of Singular Perturbation Boundary-Value Problems", *Q.Jl.Mech.Appl. Math.* 27, 57-68.

Blackburn, W.S. (1976). "Letter to the Editor", *Int.J.Num. Meth.Engng.* 10, 718-719.

Christie, I., Griffiths, D.F., Mitchell, A.R. and Zienkiewicz, O.C., (1976). "Finite Element Methods for Second Order Differential Equations with Significant First Derivatives", *Int.J.Num.Meth.Engng.* 10, 1389-1396.

Gartling, D.K., Nickell, R.E. and Tanner, R.I. (1977). "A Finite Element Convergence Study for Accelerating Flow Problems", *Int.J.Num.Meth.Engng.* 11, 1155-1174.

Griffiths, D.F. and Lorenz, J. (1977). "An Analysis of the Petrov-Galerkin Finite Element Method Applied to a Model Problem", University of Calgary, Department of Mathematics and Statistics Research Paper No. 334, February 1977.

Heinrich, J.C., Huyakorn, P.S., Zienkiewicz, O.C. and Mitchell, A.R. (1977). "An 'Upwind' Finite Element Scheme for Two-Dimensional Convective-Transport Equation", *Int. J.Num.Meth.Engng.* 11, 1831-1844.

Heinrich, J.C. and Zienkiewicz, O.C. (1977). "Quadratic Finite Element Schemes for Two-Dimensional Convective-Transport Problems", *Int.J.Num.Meth.Engng.* 11, 1831-1844.

Heinrich, J.C., Marshall, R.S. and Zienkiewicz, O.C. (1978). "Penalty Function Finite Element Solution of Coupled Convective and Conductive Heat Transfer". Presented at the International Conference on Laminar and Turbulent Flow,

Swansea, July 1978.

Heinrich, J.C., Marshall, R.S. and Zienkiewicz, O.C. (1978). "Solution of Navier-Stokes Equations by a Penalty Function Finite Element Method". University of Swansea, Department of Civil Engineering, Report C/R/308/78.

Miller, J.J.H. (1975). "A Finite Element Method for a Two Point Boundary Value Problem with a Small Parameter Affecting the Highest Derivative". Trinity College, Dublin, School of Mathematics Report TCD-1975-11.

Millsaps, K. and Pohlhausen, K. (1953). "Thermal Distribution in Jeffery-Hamel Flows Between Nonparallel Plane Walls", *J. Aero. Sci.*, 20, 187-196.

Piva, R. and Di Carlo, A. (1975). "Numerical Techniques for Convection/Diffusion Problems". Second Conference on the Mathematics of Finite Elements and Applications, Brunel University.

Tabata, M. (1977). "A Finite Element Approximation Corresponding to the Upwind Finite Difference". *Memoir of Numerical Mathematics* 4, 47-63.

Zienkiewicz, O.C., Gallagher, R.H. and Hood, P. (1975). "Newtonian and Non-Newtonian Viscous Incompressible Flow. Temperature Induced Flows. Finite Element Solutions". Second Conference on the Mathematics of Finite Elements and Applications, Brunel University.

Zienkiewicz, O.C. (1977). *"The Finite Element Method"*. 3rd edition, McGraw-Hill, London.

Zienkiewicz, O.C. and Heinrich, J.C. (1978). The Finite Element Method and Convection Problems in Fluid Mechanics. In: *Finite Elements in Fluids* vol 3. (R.H. Gallagher, J.T. Oden, C. Taylor and O.C. Zienkiewicz, eds). Wiley, London.

HOMOGENISATION OF SECOND ORDER ELLIPTIC EIGENVALUE PROBLEMS

S. Kesavan
IRIA-LABORIA, Domaine de Voluceau Rocquencourt, 78150 Le Chesnay, France

and

School of Mathematics, Tata Institute of Fundamental Research, Bombay, India

ABSTRACT

The aim of this paper is to study the homogenization of the following second-order eigenvalue problem :

$$\begin{cases} -\sum_{i,j=1}^{n} \frac{\partial}{\partial x_i}\left(a_{ij}^{\varepsilon}(x)\frac{\partial u_\varepsilon}{\partial x_j}\right) = \lambda_\varepsilon u_\varepsilon \text{ in } \Omega \\ u_\varepsilon = 0 \text{ on } \Gamma \end{cases}$$

where $\Omega \subseteq \mathbb{R}^n$ is a bounded, open set and Γ its boundary. The main homogenization theorem states that the same operator which serves to homogenize the corresponding static problem works for the eigenvalue problem as well and that the structure of eigenvalues and eigenvectors is in some sense preserved. Formulae for first and second order correctors for eigenvalues are proposed and error estimates are obtained. The results are applied to the case of coefficients with a periodic structure and a simple numerical example is presented. Extensions to other types of boundary conditions are indicated.

1. STATEMENT OF THE PROBLEM

Let $\Omega \subseteq \mathbb{R}^n$ be a bounded open set and let Γ denote its boundary. Let $\varepsilon(>0)$ be a parameter which tends to zero. To each ε is associated n^2 real-valued functions $a_{ij}^{\varepsilon} : \Omega \to \mathbb{R}$. The following hypotheses are made on these functions :

(H1) $\forall \varepsilon > 0,\ \forall 1 \leq i,j \leq n\ ,\ a_{ij}^{\varepsilon} \in L^{\infty}(\Omega)$.

Further there exists $M > 0$, independent of ε, such that

$$\forall\ \varepsilon > 0,\ |a_{ij}^{\varepsilon}(x)| \leq M \text{ a.e.} \tag{1.1}$$

(H2) There exists $\alpha > 0$, independant of ε , such that

$$\forall \xi = (\xi_i)_{i=1}^{n} \in \mathbb{R}^n\ ,\ a_{ij}^{\varepsilon}(x)\xi_i\xi_j \geq \alpha\ \xi_i\xi_i\ . \tag{1.2}$$

(Throughout this paper the convention of summation with respect to repeated indices will be adopted).

(H3) Symmetry : $a_{ij}^{\varepsilon}(x) = a_{ji}^{\varepsilon}(x)$.

To these functions a_{ij}^{ε}, the following second-order partial differential operator can be associated :

$$A^{\varepsilon} = -\frac{\partial}{\partial x_i}\left(a_{ij}^{\varepsilon}(x)\frac{\partial}{\partial x_j}\right)\ . \tag{1.3}$$

Consider the following Dirichlet problems for the operator A^{ε}

(a) *The Static Problem* :

Given f, to find u_{ε} such that

$$\begin{cases} A^{\varepsilon} u_{\varepsilon} = f \text{ in } \Omega \\ u_{\varepsilon} = 0 \text{ on } \Gamma \end{cases} \tag{1.4}$$

(where $(\cdot,\cdot)$ denoted the usual scalar product in $L^2(\Omega)$).

It is well known that the solution u_ε of problem (a) exists uniquely. For problem (**b**) , there exists a sequence of eigenvalues $\{\lambda_\varepsilon^{(\ell)}\}$, which are strictly positive and increasing to $+\infty$ with ℓ. The corresponding sequence of eigenvectors form an orthonormal basis for $L^2(\Omega)$.

In principle these solutions can all be calculated numerically, for instance by the finite element method. However as $\varepsilon \to 0$ if the coefficients become highly oscillatory (cf. the examples below), then it becomes virtually impossible to effect the resolution in practice. The mesh must be extremely fine and this leads to enormous memory requirements. The time of execution is also very large.

Thus arises the need for homogenization. The sequence of operators A^ε is replaced by a "simpler" homogenized operator A so that the solutions u_ε of (a) or $(u_\varepsilon,\lambda_\varepsilon)$ of (b) converge in some sense to the corresponding solutions associated to A.

EXAMPLE 1 (The periodic case).

Let $Y = \prod_{i=1}^{n} [0,y_i]$ be a basic reference call. Then $\mathbb{R}^n$ can be covered by a lattice of calls homothetic to Y by means of the homothecy $x \mapsto x/\varepsilon$. Let $a_{ij} : Y \to \mathbb{R}$ be n^2 functions in $L^\infty(Y)$ taking equal values on opposite faces of Y (such functions are said to be Y-periodic). Then $a_{ij}(x/\varepsilon)$ denotes the extention by periodicity of the function a_{ij} to the whole of $\mathbb{R}^n$. Then $a_{ij}^\varepsilon : \Omega \to \mathbb{R}$ is given by

$$a_{ij}^\varepsilon(x) = a_{ij}(\frac{x}{\varepsilon}). \tag{1.9}$$

EXAMPLE 2 : This is a particular case of the previous example. $Y = (0,1)\times(0,1)$, $\Omega \subseteq \mathbb{R}^2$.

(b) *The Eigenvalue Problem* :

To find $\lambda_\varepsilon \in \mathbb{R}$, and u_ε such that

$$\begin{cases} A^\varepsilon u_\varepsilon = \lambda_\varepsilon u_\varepsilon \text{ in } \Omega \\ u_\varepsilon = 0 \text{ on } \Gamma \end{cases} \tag{1.5}$$

Let $W^{m,p}(\Omega)$ denote the usual Sobolev space of functions in $L^p(\Omega)$ whose distributional derivatives up to order m are also in $L^p(\Omega)$, $1 \le p \le +\infty$. Let the usual Sobolev norm and semi-norm be denoted respectively by $\|\cdot\|_{m,p,\Omega}$ and $|\cdot|_{m,p,\Omega}$. If p=2, the symbol $H^m(\Omega)$ is used to denote $W^{m,2}(\Omega)$. The corresponding norm and semi-norm are denoted by $\|\cdot\|_{m,\Omega}$ and $|\cdot|_{m,\Omega}$ respectively. The closure of $\mathcal{D}(\Omega)$ (the space of C^∞ functions with compact support in Ω) in $W^{m,p}(\Omega)$ is denoted by $W_0^{m,p}(\Omega)$; if p=2 it is denoted by $H_0^m(\Omega)$. Notice that due to Poincaré's inequality, $|\cdot|_{m,p,\Omega}$ is a norm equivalent to that induced by $\|\cdot\|_{m,p,\Omega}$ on the space $W_0^{m,p}(\Omega)$.

Let $a_\varepsilon(\cdot,\cdot)$ be the bilinear form associated to the operator A^ε :

$$\forall\, u,v \in H^1(\Omega) \ , \ a_\varepsilon(u,v) = \int_\Omega a_{ij}^\varepsilon(x) \frac{\partial u}{\partial x_j} \frac{\partial v}{\partial x_i} \, dx \ . \tag{1.6}$$

With these notations, the problems (a) and (b) can be put into their weak forms :

(a) To find $u_\varepsilon \in H_0^1(\Omega)$ such that

$$a_\varepsilon(u_\varepsilon, v) = \langle f,v \rangle_{H^{-1},H_0^1} \ , \ \forall\, v \in H_0^1(\Omega) \tag{1.7}$$

where $f \in H^{-1}(\Omega)$, the dual of $H_0^1(\Omega)$, is given.

(b) To find $(u_\varepsilon, \lambda_\varepsilon) \in H_0^1(\Omega) \times \mathbb{R}$ such that

$$a_\varepsilon(u_\varepsilon, v) = \lambda_\varepsilon(u_\varepsilon, v) \quad \forall\, v \in H_0^1(\Omega) \ , \tag{1.8}$$

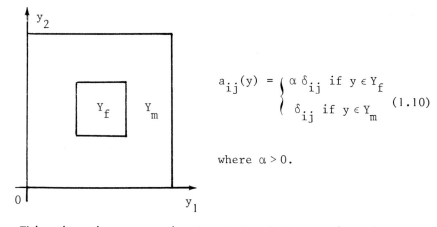

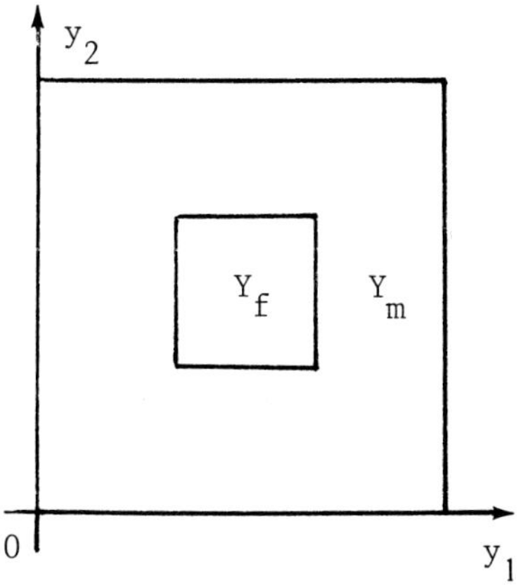

$$a_{ij}(y) = \begin{cases} \alpha\, \delta_{ij} & \text{if } y \in Y_f \\ \delta_{ij} & \text{if } y \in Y_m \end{cases} \qquad (1.10)$$

where $\alpha > 0$.

This situation occurs in the study of the torsion of an elastic bar with reinforcing fibres (Y_f), distributed periodically. By $\varepsilon \to 0$ it is meant that the number of fibres is increased steadily but the ratio of the volume occupied by the fibres to the total volume is kept constant (1/9 as shown by the figure). This operator has been considered by Bourgat and Lanchon (1976) for the static problem and by Kesavan (1978) for the eigenvalue problem to study the effects of homogenization. As suggested by this example, homogenization consists in replacing the heterogeneous material (the reinforced bar) by an "equivalent" homogeneous one.

2. THE STATIC PROBLEM.

The earliest results on the existence and uniqueness of a homogenized operator for the static problem (a) are due to de Giorgi and Spagnolo (1973). Their methods lean heavily on the symmetry hypothesis (H3) and on the maximum principle for second-order problems. Tartar (1977) has proved a more general existence and uniqueness theorem without this assumption. In the periodic case (cf. Example 1) explicit expression is given to the homogenized operator by Bensoussan, Lions and Papanicolaou (1978) and by L. Tartar (cf. Bensoussan et al. (1978)). The former use multiple scales and asymptotic expansions and they need a certain amount of smoothness on the coefficients ; the latter uses energy methods with no regularity other than boundedness, but error estimates are not possible in this case. There are other more physical approaches to this problem due to Sanchez-Palencia and a description of all these can be found in Bensoussan et al.(1978).

The salient features of the homogenization of the static problem will now be recalled. It is assumed throught that the operators A^{ε} (cf. (1.3)) satisfy the hypotheses (H1) and (H2). Let A be an operator of similar type :

$$A = -\frac{\partial}{\partial x_i}\left(q_{ij}(x)\frac{\partial}{\partial x_j}\right) \tag{2.1}$$

The following definition is due to Tartar (1977).

Definition 2.1 : The operator A is said to be the homogenized operator (with respect to the operators A^{ε}) if the following property holds :

Let $v_{\varepsilon} \rightharpoonup v$ in $H^1(\Omega)$ weakly and $A^{\varepsilon}v_{\varepsilon} \to f$ in $H^{-1}(\Omega)$ strongly. Then $\xi_i^{\varepsilon} = a_{ij}^{\varepsilon}(x)\frac{\partial v_{\varepsilon}}{\partial x_j} \rightharpoonup \xi = q_{ij}(x)\frac{\partial v}{\partial x_j}$ in $L^2(\Omega)$ weakly.■

From the above definition it follows immediately that if u_{ε} is the solution of the static problem (a) then u_{ε} converges to u in $H^1(\Omega)$ weakly and hence in $L^2(\Omega)$ strongly, u being the solution of the "homogenized problem" :

$$Au = f \text{ in } \Omega\ ,\ u = 0 \text{ on } \Gamma. \tag{2.2}$$

In the periodic case (Example 1), the operator is calculated as follows : (cf. Bensoussan et al. 1978).

Let Y be the basic cell and let W be the space

$$W = \{v \in H^1(Y) \mid v \text{ is } Y\text{-periodic}\} . \tag{2.3}$$

The bilinear form $a_Y(\cdot,\cdot)$ is defined by

$$a_Y(u,v) = \int_Y a_{ij}(y) \frac{\partial u}{\partial y_j} \frac{\partial v}{\partial y_i} dy , \quad u,v \in W. \tag{2.4}$$

Let $\chi^j, 1 \le j \le n$ denote "the" solution of the following problem :

$$\chi^j \in W , \quad a_Y(\chi^j,v) = a_Y(y_j,v) \quad \forall v \in W, \tag{2.5}$$

where y_j denotes the function

$$y = (y_i,\ldots,y_N) \mapsto y_j . \tag{2.6}$$

The functions χ^j are determined uniquely up to an additive constant. It could be fixed that $\chi^j = 0$ at the vertices of Y or $\int_Y \chi^j = 0$ etc... (As will be seen presently, the choice of this condition does not affect the operator A). The "homogenized coefficients" are now given by

$$q_{ij} = \frac{1}{|Y|} a_Y(\chi^j - y_j, \chi^i - y_i) \tag{2.7}$$

where $|Y|$ is the measure of Y.

Thus the homogenized operator boils down to a constant coefficient operator. In the particular case of Example 2, Bourgat and Dervieux (1978) have shown that $q_{ij} = \gamma\delta_{ij}$ owing to the diverse symmetries of the problem.

The homogenized operator is elliptic with the same ellipticity constant α, though the constant M of hypothesis (H1) may be different. In the symmetric case even this constant remains unchanged and the resulting operator is also symmetric (cf. Tartar (1977)).

The asymptotic expansion in the periodic case leads to the following correctors of first and second order.

$$u_\varepsilon \simeq u - \varepsilon\chi^j\left(\frac{x}{\varepsilon}\right)\frac{\partial u}{\partial x_j} + \varepsilon^2\eta_{ij}\frac{\partial^2 u}{\partial x_i \partial x_j} \tag{2.8}$$

where $\eta_{ij} \in W$ is the solution (upto an additive constant) of the problem

$$a_Y(\eta_{ij},v) = (a_{ij} - q_{ij} - a_{ik}\frac{\partial \chi^j}{\partial y_k} - \frac{\partial}{\partial y_k}(a_{ki}\chi^j),v)\ \forall v \in W. \tag{2.9}$$

Error estimates when these correctors are taken into account are given by Bourgat and Dervieux (1978).

3. THE EIGENVALUE PROBLEM-THEORY.

This section is devoted to the study of the homogenization of the eigenvalue problem (Problem (b)) for the operators A^ε. Unlike Section 2, it is assumed henceforth that all the hypotheses H1, H2 and H3 are valid. Detailed proofs of the results presented here can be found in Kesavan (1978).

Consider the problem :

To find $(u_\varepsilon,\lambda_\varepsilon) \in H^1_o(\Omega) \times \mathbf{R}$ such that, $\forall v \in H^1_o(\Omega)$

$$a_\varepsilon(u_\varepsilon,v) = \lambda_\varepsilon(u_\varepsilon,v) \tag{3.1}$$

A standard argument involving the Green's operator (i.e. inverse of A^ε) reduces the problem to that of finding the characteristic values of a compact, self-adjoint operator mapping $L^2(\Omega)$ into itself. Thus there exists a sequence $\{\lambda^\ell_\varepsilon\}$ of eigenvalues and a sequence of corresponding eigenvectors $\{u^\ell_\varepsilon\}$ in $H^1_o(\Omega)$ such that,

$$0 < \lambda_\varepsilon^1 \le \lambda_\varepsilon^2 \le \ldots \to +\infty \tag{3.2}$$

$$(u_\varepsilon^\ell, u_\varepsilon^m) = \delta_{\ell m} \,. \tag{3.3}$$

The multiplicity of each λ_ε^ℓ is finite (λ_ε^1 is simple, by the maximum principle) and the $\{u_\varepsilon^\ell\}_\ell$ constitute an orthonormal basis for $L^2(\Omega)$. These eigenvalues and eigenvectors can be characterised using the Raleigh Quotient (cf. Courant and Hilbert or Strang and Fix (1972) :

$$\forall v \in H_o^1(\Omega) \;,\quad v \neq 0 \;,\; R_\varepsilon(v) = \frac{a_\varepsilon(v,v)}{(v,v)} \,. \tag{3.4}$$

$$\lambda_\varepsilon^\ell = R_\varepsilon(u_\varepsilon^\ell) = \max_{v \in E_\ell(\varepsilon)} R_\varepsilon(v) = \min_{v \perp E_{\ell-1}^{(\varepsilon)}} R_\varepsilon(v) \tag{3.5}$$

$$= \min_{W \in D_\ell} \max_{v \in W} R_\varepsilon(v)$$

where

$$E_\ell^{(\varepsilon)} = \text{span}\,\{u_2^1, \ldots, u_\varepsilon^\ell\} \tag{3.6}$$

$$D_\ell = \{w \subseteq H_o^1(\Omega) \,|\, W \text{ a subspace of dimension } \ell\}. \tag{3.7}$$

The last relation in (3.5) is the MIN-MAX Principle due to Courant.

If A is the homogenized operator in the sense of definition 2.1, then its eigenvalues and eigenvectors have analogous properties.

The main homogenization theorem reads as follows (cf. Kesavan (1978)) :

THEOREM 3.1 : Let $\{\lambda_\varepsilon^\ell\}$ be the sequence of eigenvalues of A^ε and let $\{u_\varepsilon^\ell\}$ be an orthonormal sequence of eigenvectors satisfying the above mentioned properties. Then given any subsequence ε_n tending to zero a further subsequence can be found (again denoted by ε_n) such that, (i) for every ℓ, $\lambda^\ell_{\varepsilon_n} \to \lambda^\ell$, and $u^\ell_{\varepsilon_n} \to u^\ell$ (in $H^1_0(\Omega)$ weakly and in $L^2(\Omega)$ strongly), with (u^ℓ,λ^ℓ) satisfying the homogenized eigenvalue problem,

(ii) there exists no eigenvalue of the homogenized problem other than the limits obtained above and $\{u^\ell\}$ is a complete orthonormal basis for $L^2(\Omega)$. Consequently, the entire original sequence λ^ℓ_ε converges to λ^ℓ,

(iii) if for any particular ℓ, λ^ℓ is a simple eigenvalue of the homogenized problem, then for ε sufficiently small, λ^ℓ_ε is a simple eigenvalue. In this case given a normalized eigenvector u^ℓ coresponding to λ^ℓ, a normalized eigenvector u^ℓ_ε can be found for λ^ℓ_ε so that $u^\ell_\varepsilon \to u^\ell$ in $H^1_0(\Omega)$ weakly (and in $L^2(\Omega)$ strongly) for the entire original sequence.■

Remark 3.1 : Nothing can be said at this stage regarding convergence of eigenvectors of multiple eigenvalues except for subsequences.■

Remark 3.2 : The above theorem can be generalized to cover types of boundary conditions. Let is be assumed that $a_\varepsilon(\cdot,\cdot)$ is a $H^1(\Omega)$-elliptic bilinear form : for instance

$$a_\varepsilon(u,v) = \int_\Omega (a^\varepsilon_{ij}(x) \frac{\partial u}{\partial x_j} \frac{\partial v}{\partial x_i} + a_o(x)uv)dx \qquad (3.8)$$

where the a^ε_{ij} satisfy the hypotheses (H1)-(H3) and $a_o \in L^\infty(\Omega)$, $a_o(x) \geq \alpha > 0$, a.e. Let the boundary condition be such that the weak formulation is effected with a subspace $V \subseteq H^1(\Omega)$. Then the homogenization theorem still holds.■

ERROR ESTIMATES

The main ideas involved in the estimation of errors for simple eigenvalues will now be outlined. For full details the reader is referred to Kesavan (1978).

For purposes of the rest of this section, the following notations are observed : let λ_o be a simple eigenvalue of the homogenized problem and let $\{\lambda_\varepsilon\}$ be the sequence of corresponding eigenvalues of the operators A^ε so that $\lambda_\varepsilon \to \lambda_o$. As was already observed λ_ε is also simple for ε sufficiently small. Let u_o be a normalized eigenvector for λ_o, chosen once and for all. Let u_ε be a normalized eigenvector for λ_ε so that $u_\varepsilon \to u_o$ in $H_o^1(\Omega)$ weakly and in $L^2(\Omega)$ strongly. Let $\omega_\varepsilon \in H_o^1(\Omega)$ be the solution of the following auxiliary static problem :

$$\omega_\varepsilon \in H_o^1(\Omega),\ a_\varepsilon(\omega_\varepsilon,v) = \lambda_o(u_o,v),\ \forall v \in H_o^1(\Omega). \qquad (3.9)$$

The symmetry of the bilinear forms $a_\varepsilon(\cdot,\cdot)$ (hypothesis H3) leads to a simple and very useful relation :

$$\begin{aligned}\lambda_\varepsilon(u_\varepsilon,\omega_\varepsilon) &= a_\varepsilon(u_\varepsilon,\omega_\varepsilon)\\ &= a_\varepsilon(\omega_\varepsilon,u_\varepsilon) = \lambda_o(u_o,u_\varepsilon) .\end{aligned}$$

Thus it follows that

$$\lambda_\varepsilon(\omega_\varepsilon,u_\varepsilon) = \lambda_o(u_o,u_\varepsilon). \qquad (3.10)$$

The above relations play an important role in the theory of error estimates and correctors.

Remark 3.3 : There is far more flexibility in the relation (3.10). Let λ_ε^ℓ be the ℓth eigenvalue of A^ε and u_ε^ℓ an eigenvector (normalized or not) corresponding to it ; let λ^m be the mth eigenvalue of the homogenized problem and u^m an eigenvector corresponding to it ; let ω_ε be defined as is (3.9) with λ^m as λ_o and u^m as u_o. Then

$$\lambda_\varepsilon^\ell(\omega_\varepsilon, u_\varepsilon^\ell) = \lambda^m(u^m, u_\varepsilon^\ell) \ . \blacksquare \tag{3.11}$$

The relation (3.10) gives an immediate (though rather naive) estimate for eigenvalues. Since $(\omega_\varepsilon, u_\varepsilon) \to 1$, for ε sufficiently small, $(\omega_\varepsilon, u_\varepsilon) \geq \frac{1}{2}$ (say) and consequently

$$\lambda_\varepsilon - \lambda_o = \frac{\lambda_o}{(u_\varepsilon, \omega_\varepsilon)} (u_o - \omega_\varepsilon, u_\varepsilon)$$

leading to the following result :

THEOREM 3.2 : There exists a constant $C > 0$, independent of ε such that, for ε sufficiently small

$$|\lambda_\varepsilon - \lambda_o| \leq C|u_o - \omega_\varepsilon|_{o,\Omega} \ . \blacksquare \tag{3.12}$$

Thus the error for eigenvalues turns out to be the same as that for solutions of static problems. In the periodic case (example 1) if the coefficients are sufficiently smooth, this is of the order $O(\varepsilon)$ (cf. Bensoussan et al. (1978)). Usually a higher rate of convergence for eigenvalues is expected. Indeed Babuska has reported some computational results in the unidimensional case (cf. Osborn (1974)). For the Dirichlet problem he has obtained an order $O(\varepsilon^2)$ but with a Dirichlet-Neuman condition only $O(\varepsilon)$. Now, the procedure adopted to obtain the estimate (3.12) does not depend on the boundary conditions. As long as the homogenization theorem is valid (cf. Remark 3.2), this estimate is always true. So it is possible that this estimate is optimal in a general context.

Eigenvector estimates involve a little more work. Let $f \in L^2(\Omega)$. Let $z_\varepsilon = z_\varepsilon(f)$ be the unique solution of the following problem :

$$\begin{cases} a_\varepsilon(z_\varepsilon,v) - \lambda_\varepsilon(z_\varepsilon,v) = (P_\varepsilon f,v) \quad , \forall v \in H_o^1(\Omega) \\ (z_\varepsilon,u_\varepsilon) = 0 \end{cases} \tag{3.13}$$

where $P_\varepsilon f$ is the projection of f in $L^2(\Omega)$ onto the orthogonal complement of u_ε in $L^2(\Omega)$. (It is assumed that ε is sufficiently small so that λ_ε is simple). Then the following result can be proved.

THEOREM 3.3 : Let $\lambda_o,\lambda_\varepsilon$ be simple. Then there exists a constant $C>0$ independent of ε such that

$$\forall f \in L^2(\Omega), \quad |z_\varepsilon(f)|_{1,\Omega} \leq C|f|_{0,\Omega} \;. \;\blacksquare \tag{3.14}$$

This result can now be applied to the function $\lambda_\varepsilon\omega_\varepsilon-\lambda_o u_o$. Notice that as a result of (3.10), $P_\varepsilon(\lambda_\varepsilon\omega_\varepsilon-\lambda_o u_o) = \lambda_\varepsilon\omega_\varepsilon-\lambda_o u_o$. The corresponding z_ε has the interesting property that $z_\varepsilon + \omega_\varepsilon$ is an eigenvector corresponding to λ_ε. Thus

$$|z_\varepsilon|_{0,\Omega} \leq C_1|z_\varepsilon|_{1,\Omega} \leq C_2|\lambda_\varepsilon\omega_\varepsilon-\lambda_o u_o|_{0,\Omega} \leq C_3|\omega_\varepsilon-u_o|_{0,\Omega} \tag{3.15}$$

The last inequality is a result of a combination of the triangle inequality and the estimate (3.12) for eigenvalues. Thus

$$|z_\varepsilon+\omega_\varepsilon-u_o|_{0,\Omega} \leq C_4|\omega_\varepsilon-u_o|_{0,\Omega} \;.$$

All the constants C_i are independent of ε. Normalizing $z_\varepsilon+\omega_\varepsilon$ does not change the nature of the estimate. Since it is known that u_ε is the eigenvector converging to u_o, it follows that

$$u_\varepsilon = (z_\varepsilon+\omega_\varepsilon)/|z_\varepsilon+\omega_\varepsilon|_{0,\Omega}$$

and the following result is true :

THEOREM 3.4 : <u>There exists a constant</u> $C>0$ <u>independent of</u> ε <u>such that</u>

$$|u_\varepsilon-u_o| \le C|\omega_\varepsilon-u_o| \;.\blacksquare \tag{3.16}$$

The case of multiple eigenvalues is slightly more complex. Theorem 3.3 is not valid in full generality for all $f \in L^2(\Omega)$. It is however valid for particular choices of the type "$\lambda_\varepsilon\omega_\varepsilon-\lambda_o u_o$" which enables the imitation of the preceding argument. The full details are reported in Kesavan (1978).

CORRECTORS

In this section heuristic formulae for correctors will be suggested. These will subsequently be justified by means of error estimates.

Let ω_1,ω_2 be first and second order correctors for ω_ε. In the periodic case these are suggested by the relation (2.8) Let it be assumed that it is possible to expand $\lambda_\varepsilon,u_\varepsilon,\omega_\varepsilon$ asymptotically.

$$\lambda_\varepsilon = \lambda_o + \varepsilon\lambda_1 + \varepsilon^2\lambda_2 +\ldots \tag{3.17}$$

$$u_\varepsilon = u_o + \varepsilon u_1 + \varepsilon^2 u_2 + \ldots \tag{3.18}$$

$$\omega_\varepsilon = u_o + \varepsilon\omega_1 + \varepsilon^2\omega_2 + \ldots \tag{3.19}$$

Let these expression be substituted in (3.10). Comparing coefficients on both sides, and taking into account that u_o is normalized it follows that

$$\lambda_o = \lambda_o \quad (\text{coefficients of } \varepsilon^o) \tag{3.20}$$

$$\lambda_1 = -\lambda_o(u_o,\omega_1) \tag{3.21}$$

$$\begin{aligned}\lambda_2 = &- \lambda_o\{(u_o,\omega_2)+(u_1,\omega_1)\}\\ &- \lambda_1\{(u_o,\omega_1)+(u_o,u_1)\}\end{aligned} \tag{3.22}$$

Substituting (3.18) in the normalizing condition

$$(u_\varepsilon,u_\varepsilon) = 1 \tag{3.23}$$

and comparing coefficients of ε it follows that

$$(u_o,u_1) = 0. \tag{3.24}$$

If u_ε and ω_ε are sufficiently "close", ω_1 could replace u_1 in the second term of (3.22) to get

$$\lambda_2 = -\lambda_o\{(u_o,\omega_2)+(\omega_1,\omega_1)\} - \lambda_1(u_o,\omega_1) \tag{3.25}$$

In the periodic case the expressions for λ_1 and λ_2 read as

$$\lambda_1 = \lambda_o\int_\Omega \chi^j(\tfrac{x}{\varepsilon})u_o(x)\,\frac{\partial u_o}{\partial x_j}\,dx \tag{3.26}$$

$$\left\{\begin{aligned}\lambda_2 = &-\lambda_o\int_\Omega \eta_{ij}(\tfrac{x}{\varepsilon})u_o(x)\,\frac{\partial^2 u_o}{\partial x_i\partial x_j}\,dx - \lambda_o\int_\Omega \Big(\chi^j(\tfrac{x}{\varepsilon})\,\frac{\partial u_o}{\partial x_j}\Big)^2 dx\\ &+ \lambda_1\int_\Omega \chi^j(\tfrac{x}{\varepsilon})u_o(x)\,\frac{\partial u_o}{\partial x_j}\,dx.\end{aligned}\right. \tag{3.27}$$

The following error estimates can be proved (cf. Kesavan (1978)).

THEOREM 3.5 : There exists a constant $C > 0$, independent of ε, such that, for sufficiently small ε,

$$|\lambda_\varepsilon-\lambda_o-\varepsilon\lambda_1| \leq C\{\varepsilon|\omega_\varepsilon-u_o|_{0,\Omega}+|\omega_\varepsilon-u_o|^2_{0,\Omega}+|\omega_\varepsilon-u_o-\varepsilon\omega_1|^2_{0,\Omega}\} \tag{3.28}$$

$$\left\{\begin{aligned} &|\lambda_\varepsilon-\lambda_o-\varepsilon\lambda_1-\varepsilon^2\lambda_2| \leq C\{|\omega_\varepsilon-u_o-\varepsilon\omega_1-\varepsilon^2\omega_2|_{0,\Omega}+\varepsilon|\lambda_\varepsilon-\lambda_o-\varepsilon\lambda_1| \\ &\quad +\varepsilon^2|\omega_\varepsilon-u_o|_{0,\Omega}+\varepsilon^2+|\omega_\varepsilon-u_o|^2_{0,\Omega}\}. \end{aligned}\right. \tag{3.29}$$

In the periodic case, these estimates read as

$$|\lambda_\varepsilon-\lambda_o-\varepsilon\lambda_1| \leq C\{\varepsilon^2+|\omega_\varepsilon-u_o-\varepsilon\omega_1|_{0,\Omega}\} \tag{3.30}$$

$$\left\{\begin{aligned} &|\lambda_\varepsilon-\lambda_o-\varepsilon\lambda_1-\varepsilon^2\lambda_2| \leq C\{\varepsilon^2+\varepsilon|\omega_\varepsilon-u_o-\varepsilon\omega_1|_{0,\Omega} \\ &\qquad\qquad +|\omega_\varepsilon-u_o-\varepsilon\omega_1-\varepsilon^2\omega_2|_{0,\Omega}\} \quad . \end{aligned}\right. \tag{3.31}$$

Remark 3.4 : Notice that the orders of convergence are governed by the "improved" errors $|\omega_\varepsilon-u_o-\varepsilon\omega_1|_{0,\Omega}$ and $|\omega_\varepsilon-u_o-\varepsilon\omega_1-\varepsilon^2\omega_2|_{0,\Omega}$ in (3.30) and (3.31) respectively. This justifies the name "correctors" given to λ_1 and λ_2. ■

4. THE ENGENVALUE PROBLEM - NUMERICAL RESULTS.

For purposes of numerical study, the example considered was indentical with that explained in the Example 2 of Section 1.

The aims of the numerical study were the following :

(i) To compute the first eigenvalue and eigenvector of the real problem and compare them with those of the homogenized problem.

(ii) To study the effect of the correctors on the first eigenvalue.

(iii) To compare $u_\varepsilon, \omega_\varepsilon$ and $u_o+\varepsilon\omega_1+\varepsilon^2\omega_2$.

(iv) To compare multiplicities in case of certain higher eigenvalues.

The sequence $\varepsilon_n \to 0$ was taken to be $\{2^{-n}\}$. The computations were carried out for $\varepsilon=1/2$, $\varepsilon=1/4$. From $\varepsilon=1/8$ onwards the memory requirements became prohibitive thus providing a strong case in favour of homogenization. The problems were all discretized using the finite element method and the resulting finite dimensional eigenvalue problem was solved using an inverse iteration algorithm (cf. Bathe and Wilson (1976)) to get the first eigenvalue and eigenvector. For higher eigenvalues a bisection method due to Peters and Wilkinson (1971) was used. The computations were carried out on an IBM 370/168 computer.

As expected the "homogenized eigenvalue" λ_o approximates the "real" eigenvalue λ_ε better as ε tends to zero. However there is yet another factor which governs the convergence : the coefficient α on the fibres (cf. Example 2, Section 1). If $\alpha \geq 1$, the results are very good and even correctors are not necessary. For instance if $\alpha=10$, the error in the eigenvalue is only about 0.19% if $\varepsilon=1/2$ and 0.04% if $\varepsilon=1/4$. However if $\alpha < 1$, the quality of the results deteriorates steadily as α tends towards zero. The table below shows that for $\varepsilon=1/2$, and $\varepsilon=1/114$, there is an error of 130% in the eigenvalue and this falls to 7% if $\varepsilon=1/4$. It is here that correctors are most useful. For $\varepsilon=1/4$, on using correctors the error falls to 2.5% in the case where $\alpha = 1/114$.

TABLE 1

α	Homogenized eigenvalue (λ_o)	$\varepsilon = 1/2$		$\varepsilon = 1/4$	
		actual eigenvalue (λ_ε)	corrected eigenvalue $\lambda_o+\varepsilon\lambda_1+\varepsilon^2\lambda_2$	actual eigenvalue (λ_ε)	corrected eigenvalue $\lambda_o+\varepsilon\lambda_1+\varepsilon^2\lambda_2$
10	24.17865	24.13341	-	24.18787	-
1/18	16.14206	15.73890	15.62387	16.07114	16.01152
1/30	15.98447	14.97165	15.18859	15.83533	15.78448
1/114	15.80205	6.82241	13.14088	14.76333	15.13542

As far as the first eigenvector is concerned, the homogenized eigenvector u_o approximates u_ε fairly well. Nevertheless, when traversing a fibre (where the differential operator has a coefficient $\alpha < 1$) the solution u_ε shows peaks like those obtained by Bourgat et al. (1978) in the static case. The smaller α is, the more pronounced are the peaks. These peaks can never be approximated by u_o. This is a reflection of the fact that u_ε converges to u_o strongly in $L^2(\Omega)$ but only weakly in $H_o^1(\Omega)$. In order to approximate the gradient as well, correctors are necessary. However the asymptotic expansion as in Bensoussan et al. (1978) yields the same correctors for both u_ε and ω_ε which is hardly realistic. In fact computations show that ω_ε is approximated by $u_o+\varepsilon\omega_1+\varepsilon^2\omega_2$ extremely well and u_ε much less so. A suggestion to approximate the gradient of u_ε comes from the arguments following Theorem 3.3. According to (3.15) ω_ε is "almost" u_ε modulo a small orthogonal complement. Thus in order to approximate the peaks a natural idea is to normalize $u_o+\varepsilon\omega_1+\varepsilon^2\omega_2$ ($\simeq \omega_\varepsilon$).

Regarding higher eigenvalues, the rate of convergence is much slower. Multiplicities show a tendancy to be preserved.

Full details of the numerical experiments are described in Kesavan (1978).

REFERENCES

Bathe, K-J. and Wilson, E.L. (1976). *Numerical Methods in Finite Element Analysis*, Prentice-Hall, Inc.

Bensoussan, A., Lions, J.L., and Papanicolaou, G. (1978). *Asymptotic Methods in Periodic Structures*, North-Holland, Amsterdam.

Bourgat, J.F. and Dervieux A. (1978). Méthode d'Homogénéisation des opérateurs à coefficients périodiques : étude des correcteurs provenant du développement asymptotique, *Rapport LABORIA*, N° 278.

Bourgat, J.F., and Lanchon, H. (1976). Application of the homogenization Method to Composite Materials with Periodic Structure, *Rapport LABORIA* N° 208.

Courant, R. and Hilbert. *Methods of Mathematical Physics*, Interscience Publishers, New-York.

de Giorgi, E. and Spagnolo, S. (1973). Sulla Convergenzia degli integrali dell'energia per operatori ellitici del 2° ordine, *Boll. U. Mat. Ital.*, 8, pp. 391-411.

Kesavan, S. (1978). To appear.

Osborn, J.E. (1974). Spectral Approximation for Compact Operators, *Technical Report* 74-76, University of Maryland.

Peters, G. and Wilkinson, J.H. (1971). Eigenvalues of $Ax=\lambda Bx$ with Band Symmetric A and B, *Computer Journal*, 14, p. 398-404.

Strang, G. and Fix, G.J. (1972). *An analysis of the Finite Element Method*, Prentice-Hall Inc.

Tartar, L. (1977). Cours Peccot, Collège de France (to appear).

COMBINATIONS OF INITIAL AND BOUNDARY VALUE METHODS FOR A CLASS OF SINGULAR PERTURBATION PROBLEMS

J. Lorenz
University of Münster, Institut für Numerische Mathematik, D 44 Münster, W. Germany

ABSTRACT

For many singular perturbation problems a reduced problem is well defined and known a priori. We use this for the numerical solution of a class of nonlinear second order problems and give asymptotic error estimates.

1. INTRODUCTION

Consider the boundary value problem

$$-\varepsilon x'' + \alpha(t,x)x' + \beta(t,x) = 0 \ , \quad t \in [0,1] \qquad (BVP_\varepsilon)$$

$$x(0) = A, \quad x(1) = B$$

for small, positive values of ε and assume that

$$\alpha(t,x) \geq \alpha_0 > 0 \text{ for all } (t,x) \in [0,1] \times \mathbb{R}.$$

Under additional technical conditions to be

described below the problem (BVP_ε) has a unique solution x_ε, and if u is the solution of the reduce problem

$$\alpha(t,u)u'+\beta(t,u)=0,\ t\in[0,1],\ u(0)=A \tag{RP}$$

then the following estimate holds (Howes (1976))

$$|u(t)-x_\varepsilon(t)|=O(\varepsilon)+O(1)e^{(t-1)\alpha_0/\varepsilon},\ t\in[0,1].$$

The $O(\cdot)$-estimates are valid uniformly in t for $\varepsilon\rightarrow 0$. Thus for small $\varepsilon > 0$ the unknown function x_ε will nearly coincide with u on the main part of the interval [0,1] since the exponential term $e^{(t-1)\alpha_0/\varepsilon}$ is small outside a neighbourhood of t=1. In general $u(1)\neq B=x_\varepsilon(1)$, and therefore a boundary layer behaviour of x_ε is expected near the right endpoint t=1. To solve BVP_ε numerically for small values of ε it is suggestive to treat just the reduced problem by an initial value metho on the main part of the interval [0,1] and to trea a boundary value problem in the boundary layer, in case that information about x_ε in the layer is wanted. The boundary value problem to be solved in the boundary layer will be called the boundary layer equation. We shall consider different possibilities for this equation. The whole process then can be described as follows:

1. The interval [0,1] is subdivided in a main part and a boundary layer.

2. Due to the smallness of ε the problem (BVP_ε) is simplified in both regions, leading to the reduced problem and a boundary layer equation.

3. The reduced problem and the boundary layer equation are solved numerically.

Thus in the whole procedure different errors are involved: Errors due to the simplification processes of the equation and errors due to the discretisation of the simplified problems. The purpose of this paper is to study the dependence of these errors upon ε and the chosen subdivision of the domain [0,1]. Especially, we shall see that with the method described the overall error does not grow if ε tends to zero. This is a desirable feature of any method used for problems with a small parameter ε.

The general idea of the described simplification process dates back to Prandtl (1904): If the parameter 1/Re (Re: Reynold's number) in the Navier-Stokes equations is small then the equations simplify to Euler's equations in the main part of the domain, whereas Prandtl's "Grenzschichtgleichungen" have to be solved in the boundary layer. Recently, Flaherty and O'Malley (1977) have used a similar idea very successfully to treat certain stiff systems of ordinary boundary value problems. Also, Hsiao and Jordan (1978) have used this idea to solve linear problems of the form (BVP_ε).

2. THE CONTINUOUS PROBLEM

Boundary value problems of the form (BVP_ε) have been studied by many authors, cf. Coddington and Levinson (1952), Wasow (1956), Dorr, Parter and Shampine (1973), O'Malley (1974), Howes (1976). Therefore we just state the results to be used below and refer to the literature for most of the proofs. For simplicity we shall always take the following global assumptions:

A1: $\alpha,\beta\in C^n([0,1]\times\mathbb{R})$, i.e. all derivatives of orders up to and including n exist and are continuous on $[0,1]\times\mathbb{R}$. n will be specified in the following theorems.

A2: $\alpha(t,x) \geq \alpha_0 > 0 \; \forall \; (t,x)\in[0,1]\times\mathbb{R}$. This assumption excludes turning point problems.

A3: $\beta(t,x)-\beta(t,y) \geq \beta_0(x-y) \; \forall \; t\in[0,1],\ x \geq y$, for some $\beta_0\in\mathbb{R}$. This lower Lipschitz condition on β allows to apply the maximum principle to (BVP_ε). If $\beta_0 \geq 0$ then $\beta(t,\cdot)$ is monotonically increasing.

If a priori estimates of the solutions x_ε are known then our global assumptions can of course be localized.

Theorem 1: (Existence, a priori estimate)
Assume A1 with n=0, A2 and A3. Then there exists $\varepsilon_0=\varepsilon_0(\alpha_0,\beta_0) > 0$ and $C_0=C_0(\alpha_0,\beta_0)$ such that (BVP_ε) has a solution $x_\varepsilon\in C^2[0,1]$ for $0 < \varepsilon \leq \varepsilon_0$. In addition, if $0 < \varepsilon \leq \varepsilon_0$ and if x_ε is any solution of (BVP_ε) then

$$\|x_\varepsilon\|_\infty \leq C_0 \mathrm{Max}\{\|\beta(\cdot,0)\|_\infty, |A|, |B|\}. \tag{1}$$

($\|\cdot\|_\infty$ denotes the maximum norm.)

Proof: (Compare Howes (1976).)
Using a comparison function and the maximum principle the estimate (1) is obtained. Then with Schauder's theorem the existence result is proved.

Theorem 2: (Uniqueness)
Assume A1 with n=1, A2 and A3. Then there exists $\varepsilon_0 > 0$ such that (BVP_ε) has a unique solution $x_\varepsilon \in C^2[0,1]$.

Proof: The arguments used by Coddington and Levinson (1952) can be extended to yield global uniqueness under our global assumptions.

Since the functions x_ε satisfy the a priori estimate (1) we can assume without loss of generality that

A4: α,β and all existing continuous derivatives are globally bounded.

Then the function $\beta(t,u)/\alpha(t,u)$ is globally Lipschitz bounded if $\alpha,\beta \in C^1$. This leads to

Theorem 3: (Reduced problem)
Assume A1 with n=1, A2 and A4. Then (RP) has a unique solution $u \in C^1[0,1]$.

Theorem 4: (Smoothness and estimates)
Assume A1 for some $n \geq 1$, A2 and A4.

Then $x_\varepsilon \in C^{n+2}$, $u \in C^{n+1}$ and

$$|x_\varepsilon(t)-u(t)|=O(\varepsilon)+O(1)E_\varepsilon(t) \tag{2}$$

$$|x'_\varepsilon(t)-u'(t)|=O(\varepsilon)+O(\varepsilon^{-1})E_\varepsilon(t) \tag{3}$$

$$|x_\varepsilon^{(i)}(t)-u^{(i)}(t)|=O(\varepsilon^{2-i})+O(\varepsilon^{-i})E_\varepsilon(t) \tag{4}$$

for i=2,..,n+1, where $E_\varepsilon(t)=\exp((t-1)\alpha_0/\varepsilon)$.

Proof: The estimates for $|x_\varepsilon-u|$ and $|x'_\varepsilon-u'|$ are obtained by Howes (1976). The estimate for $|x''_\varepsilon-u''|$ can be proved similarly, and the estimates of $|x_\varepsilon^{(i)}-u^{(i)}|$ for $i \geq 3$ then simply follow by differentiating the equation $\varepsilon x''_\varepsilon=\alpha(t,x_\varepsilon)x'_\varepsilon+\beta(t,x_\varepsilon)$.

Theorem 4 essentially states that for small ε the function x_ε nearly coincides with u outside a boundary layer. Now consider the problem to determi x_ε within the boundary layer. Relying on O'Malley (1974) we could add to u(t) a boundary layer correct R(t) were R is the solution of

$$-\varepsilon R''(t)+\alpha(1,u(1)+R(t))R'(t)=0,\ t\in(-\infty,1] \tag{5}$$

$$\lim_{t\to-\infty} R(t)=0, \quad R(1)=B-u(1).$$

R solves an autonomous differential equation and can be obtained in explicit form as a function over the boundary layer jump. A method following these lines has been described by Flaherty et al (1977). Though the special structure of the problem (BVP_ε) seems to be best utilized if we proceed as in Flaherty et al (1977) we shall consider simpler methods which treat a boundary layer equation directly by a standard numerical procedure. It seems to be easier to generalize this direct approach to more complex problems.

Let $k > 0$ be chosen so that $k\varepsilon \ll 1$. Consider the feasibilities to approximate $x_\varepsilon(t)$ for $t \in [1-k_\varepsilon, 1]$ by a solution $x(t)$ of one of the following problems:

P1: $-\varepsilon x'' + \alpha(t,x)x' + \beta(t,x) = 0,\ t \in [1-k\varepsilon, 1]$
$x(1-k\varepsilon) = u(1-k\varepsilon),\ x(1) = B.$

P2: $-\varepsilon x'' + \alpha(1,x)x' = 0,\ t \in [1-k\varepsilon, 1]$
$x(1-k\varepsilon) = u(1-k\varepsilon),\ x(1) = B.$

P3: $-\varepsilon y'' + \alpha(t,u(t)+y)y' + \beta(t,u(t)+y) = 0, t \in [1-k\varepsilon, 1]$
$y(1-k\varepsilon) = 0, y(1) = B-u(1),\ x := u+y.$

P4: $-\varepsilon y'' + \alpha(1,u(1)+y)y' = 0,\ t \in [1-k\varepsilon, 1],$
$y(1-k\varepsilon) = 0,\ y(1) = B-u(1),\ x := u+y.$

If P1 is simplified due to the smallness of ε and $k\varepsilon$, then P2 is obtained. (This is clearly seen if the stretched variable $\tau = k + (t-1)/\varepsilon$ is used.) A similar simplification process leads from P3 to P4. Note that P4 is the problem (5) with $\lim_{t \to -\infty} R(t) = 0$ replaced by $y(1-k\varepsilon) = 0$. Thus the simplification process P3 $\to$ P4 can be justified even if $k\varepsilon$ is not small. On the other hand, if e.g. $k = 1/\varepsilon$ then P2 reads

$$-\varepsilon x'' + \alpha(1,x)x' = 0 \text{ in } [0,1],\ x(0)=A,\ x(1)=B,$$

and thus the error $x_\varepsilon - x$ will be $O(1)$, and will not be small for small ε. Thus the simplification proce P1 ➤ P2 cannot be justified for all values of k. In spite of this disadvantage of P2 we shall use P1 or P2 to approximate x_ε within $[1-k\varepsilon,1]$. The reason is that P1 and P2 have a simple advantage over P3 and P4: They do not require approximate values for u inside the layer. (Note that P1 and P are equivalent to P3 and P4 resp. as long as u can be considered to be constant in $[1-k\varepsilon,1]$.) - In order to study the simplification process P1 ➤ P2 we state the following theorem which might be of interest in itself. It is a global stability inequality for a quasilinear equation.

<u>Theorem 5</u> (continuous stability inequality)
Let $\gamma \in C(\mathbb{R})$, $0 < \gamma_0 \le \gamma(x) \le \gamma_1\ \forall\ x \in \mathbb{R}$. Define for $k \ge 1$ the operators

$$Tx = T_k x = -x'' + \gamma(x(t))x' \quad \text{for} \quad x \in C^2[0,k],$$

$$Rx = R_k x = (x(0), x(k)).$$

Assume that $x, y \in C^2[0,k]$ with $x(k)=y(k)$. Then $\|x-y\|_\infty \le C(k\|Tx-Ty\|_\infty + |x(0)-y(0)|)$ with $C=C(\gamma_0,\gamma_1)$ independent of x,y and k.

<u>Proof:</u> Let $z=x-y$, $\eta=z(0)$, $\varphi=Tx-Ty$ and $\Gamma(x)=\int_0^x \gamma(t)dt$, thus $\frac{d}{dt}\Gamma(x(t))=\gamma(x(t))x'(t)$.

Integration of $Tx(s)-Ty(s)=\varphi(s)$ from 0 to t yields:

$$-(z'(t)-z'(0)) + \Gamma(x(t)) - \Gamma(y(t)) = \int_0^t \varphi(s)ds + \Gamma(x(0)) - \Gamma(y(0)).$$

Since $\Gamma(x)-\Gamma(y)=\gamma(\xi)\cdot(x-y)$ it follows that

$$-z'(t)+\gamma(\xi(t))z(t)=\gamma(\xi(0))\eta+\int_0^t \varphi(s)ds-z'(0) =: \tilde{\eta}+\tilde{\varphi}(t)-z'(0). \qquad (6)$$

An integration of this first order equation leads to

$$z(t)=\int_t^k (\tilde{\eta}+\tilde{\varphi}(\tau)-z'(0))E(t,\tau)d\tau \qquad (7)$$

where $E(t,\tau)=\exp(\int_\tau^t \gamma(\xi(s))ds)$. Note that $z(k)=0$ by our assumption. (7) yields $\eta=z(0)=(\tilde{\eta}-z'(0))\cdot \int_0^k E(0,\tau)d\tau+\int_0^k \tilde{\varphi}(\tau)E(0,\tau)d\tau$ and it follows that

$$|z'(0)| \le c(\|\varphi\|_\infty + |\eta|), \quad c=c(\gamma_0,\gamma_1) \text{ for } k \ge 1.$$

Now assume $z'(t_0)=0$. Then (6) yields

$$|z(t_0)| \le C(k\|\varphi\|_\infty + |\eta|)$$

and our assertion follows.

3. THE NUMERICAL TREATMENT

Assume that (BVP_ε) is given. Let A1 with $n \geq 1$, A2 and A4 be satisfied and let $\varepsilon > 0$ be so small that (BVP_ε) has a unique solution $x_\varepsilon \in C^{n+2}[0,1]$. $u \in C^{n+1}[0,1]$ always denotes the solution of the reduced problem (RP). To get a numerical approximat for x_ε we consider methods of the following form:

Step 1: Choose $k \in (0,1/\varepsilon)$ and $n_1, n_2 \in \mathbb{N}$. Define the step-sizes $h_1=(1-k\varepsilon)/n_1$, $h_2=k\varepsilon/n_2$, and the grids $G_1=\{0,h_1,\ldots,1-k\varepsilon\}$, $G_2=\{1-k\varepsilon, 1-k\varepsilon+h_2,\ldots,1\}$.

Step 2: Solve the initial value problem $u'+\beta(t,u)/\alpha(t,u)=0$, $u(0)=A$ on the grid G_1 by an initial value method. This produces a discrete function $u_d \in \mathbb{R}^{G_1}$.

Step 3: Solve the boundary value problem P1 or P2 with $u(1-k\varepsilon)$ replaced by $u_d(1-k\varepsilon)$ on the grid G_2 by a boundary value method.

Together with step 2 we get a grid function on $G=G_1 \cup G_2$ which will be denoted by x_d. Let

$$\|x_d-x_\varepsilon\| := \max \{|x_d(t)-x_\varepsilon(t)| : t \in G\} \tag{8}$$

denote the error in maximum norm on the whole grid. How does $\|x_d-x_\varepsilon\|$ depend on k, n_1, n_2, and ε?

For simplicity we shall assume here that in step 2 the classical Runge-Kutta method and in step 3 the ordinary difference method is used. Also, we assume $\alpha,\beta \in C^4([0,1]\times\mathbb{R})$, and thus $u \in C^5[0,$

Then it holds that

$$\max \{ |u_d(t)-u(t)| : t \in G_1 \} \leq Ch_1^4 \leq Cn_1^{-4}. \tag{9}$$

Here and in the following C denotes a generic constant independent of k, n_1, n_2 and ε, not necessarily the same at successive appearances. The next theorem treats a special case of (BVP_ε) where $\alpha(t,x)=\alpha(t)$, i.e. (BVP_ε) is mildly nonlinear.

<u>Theorem 6</u> (the mildly nonlinear case)
Assume A1 with n=4, A2, A3 with $\beta_0=0$, and A4. Also let $\alpha(t,x)=\alpha(t) \leq \alpha_1$. Assume that in step 3 the problem P1 (with $u_d(1-k\varepsilon)$) is taken. The resulting system of nonlinear equations has a unique solution if $\alpha_1 k/n_2 \leq 2$, and the following asymptotic error estimate holds

$$\|x_d - x_\varepsilon\| \leq C(\varepsilon + e^{-k\alpha_0} + n_1^{-4} + k^2/n_2^2). \tag{10}$$

Before this result is proved we introduce some notation. For $x \in \mathbb{R}^n$ let $|x|$ be the vector with components $|x_i|$. $\delta \in \mathbb{R}^n$ is the vector with $\delta_i=1$ for all i. For $x,y \in \mathbb{R}^n$ let $x \leq y$ iff $x_i \leq y_i$ for all i. Similarly, for matrices A, B let $A \leq B$ iff $a_{ij} \leq b_{ij}$ for all i, j. A is called an M-matrix iff $a_{ij} \leq 0$ for $i \neq j$, A^{-1} exists and $0 \leq A^{-1}$. If A is an M-matrix and D is a nonnegative diagonal matrix then A+D is an M-matrix and $(A+D)^{-1} \leq A^{-1}$. The following lemma is an essential step in the proof of theorem 6.

<u>Lemma 1</u>

Let $0 < \alpha \leq \beta$, $k > 0$, $n \in \mathbb{N}$, $h=k/n$. Furthermore, let $A \in \mathbb{R}^{n+1,\, n+1}$ be a matrix of the form

$$A=h^{-2}\begin{pmatrix} h^2 & & & & \\ -1 & 2 & -1 & & \\ & \ddots & \ddots & \ddots & \\ & & -1 & 2 & -1 \\ & & & & h^2 \end{pmatrix} + \frac{1}{2}h^{-1}\begin{pmatrix} 0 & & & & \\ -\alpha_1 & 0 & \alpha_1 & & \\ & \ddots & \ddots & \ddots & \\ & & -\alpha_{n-1} & 0 & \alpha_{n-1} \\ & & & & 0 \end{pmatrix} \tag{11}$$

with $\alpha \leq \alpha_i \leq \beta$, and let $\eta \in \mathbb{R}^{n+1}$ be given by $\eta_0=\eta_n=0$, $\eta_i=\exp\,((ih-k)\alpha)$ for $i=1,\ldots,n-1$. Then there exists $h_0=h_0\,(\alpha,\beta) > 0$ such that A is an M-matrix and that $A^{-1}\eta \leq (2/\alpha^2)\delta$ if $h \leq h_0$.

<u>Proof:</u> If $h\beta \leq 2$ then A is an M-matrix; use e.g. theorem 1 of Lorenz (1977). Now define the function

$$v_k(\tau) = \frac{1}{\alpha}\left(\frac{1}{\alpha} + k - \tau\right)e^{(\tau-k)\alpha}, \quad \tau\in[0,k],$$

and let $v_{k,h} = (v_k(0),\ v_k(h),\ldots,v_k(k))^t \in \mathbb{R}^{n+1}$. v_k is monotonically increasing and $0 < v_k(\tau) \leq 1/\alpha^2$ for $0 \leq \tau \leq k$. If $\bar{A}$ denotes the matrix in (11) with $\alpha_i=\alpha$ for all i then an elementary calculation shows for $i=1,\ldots,n-1$:

$$(Av_{k,h})_i \geq (\bar{A}v_{k,h})_i$$
$$=\eta_i \{f(h\alpha)+\frac{1}{\alpha}(\frac{1}{\alpha} + k - ih)h^{-2}g(h\alpha)\}$$

with $f(s)=(e^s-e^{-s})/s - \cosh(s)$,

$g(s)=s \sinh(s) - 4\sinh^2(s/2)$.

There exists $s_o > 0$ such that $f(s) \geq 1/2$ and $g(s) \geq 0$ for $0 < s \leq s_o$, and thus $Av_{k,h} \geq \frac{1}{2}\eta$ for $h \leq h_o$. Multiplying with A^{-1} we get our assertion.

Proof of theorem 6:

a. If $t \in G_1$, then

$$|x_d(t)-x_\varepsilon(t)| \leq |u_d(t)-u(t)| + |u(t)-x_\varepsilon(t)|$$

$$\leq C(n_1^{-4}+\varepsilon+e^{-k\alpha_0}) \text{ by (9) and (2).}$$

b. The discrete system which has to be solved in step 3 reads

$$x_o=u_d(1-k\varepsilon),\ x_{n_2}=B$$

$$\varepsilon h_2^{-2}(-x_{i-1}+2x_i-x_{i+1}) + \frac{1}{2} h_2^{-1}\alpha(t_i)(x_{i+1}-x_{i-1})$$

$$+\beta(t_i,x_i)=0 \quad \text{for} \quad i=1,\dots,n_2-1,$$

where $t_i=1-k\varepsilon+ih_2$. Let $h=\frac{1}{\varepsilon}h_2=k/n_2$. Multiply the equations for $i=1,\dots,n_2-1$ by ε and write the resulting system as

$$A_h x + \beta_h(x) = r_h, \quad x \in \mathbb{R}^{n_2+1} \tag{12}$$

with a matrix A_h of the form (11) and a diagonal vector field β_h. If $\alpha_1 h \le 2$ then A_h is an M-matrix and by theorem 13.5.6 of Ortega and Rheinboldt (19 the system (12) has a unique solution $x_h \in \mathbb{R}^{n_2+1}$.

c. Let $x^h \in \mathbb{R}^{n_2+1}$ be the restriction of x_ε to the grid G_2. The vector of consistency errors

$$c^h := |A_h x^h + \beta_h(x^h) - r_h|$$

has components

$$c_0^h = |x_\varepsilon(1-k\varepsilon) - u_d(1-k\varepsilon)| \le C(n_1^{-4} + \varepsilon + e^{-k\alpha_0}),$$

$$c_{n_2}^h = 0,$$

$$c_i^h \le Ch_2^2\,(1 + \varepsilon^{-2}\exp((t_{i+1}-1)\alpha_0/\varepsilon))$$

$$\le Ch^2(\varepsilon^2 + \exp((ih-k)\alpha_0)) \text{ for } i=1,\ldots,n_2-1.$$

(Use the estimate (4).)

d. We can write

$$c^h = |A_h x^h + \beta_h(x^h) - A_h x_h - \beta_h(x_h)|$$

$$= |(A_h + D_h)(x^h - x_h)|$$

with a nonnegative diagonal matrix D_h. Thus

$|x^h - x_h| \le (A_h + D_h)^{-1} c^h \le A_h^{-1} c^h$ and it remains to show that $A_h^{-1} c^h \le C(n_1^{-4} + \varepsilon + e^{-k\alpha_0} + h^2)\delta$. The only difficult part of this estimate stems from the components $\exp((ih-k)\alpha_0)$, and for these lemma 1 yields the assertion.

The implications of the estimate (10) are two-fold. First notice that the bound is not increasing if ε tends to zero. Thus even very small values of ε do not lead to difficulties if the described method is applied. Of course, the method is not appropriate if ε is not small, since then x_ε and u may differ considerably. This fact, that the method gets better if the equation gets stiffer has also been observed in Flaherty et al (1977). Secondly it is important to study the influence of the parameter k. Here the bound (10) indicates that the error is not increasing if k is increased as long as k/n_2 is kept fixed. Notice that k/n_2 is the step length if the ordinary difference method is applied to the boundary layer equation in the stretched variable $\tau = k + (t-1)/\varepsilon$, $\tau \in [0,k]$. In many numerical examples the error terms Ck^2/n_2^2 and $Ce^{-k\alpha_0}$ can clearly be observed and Ck^2/n_2^2 dominates if k is not too small. (See the numerical examples below.)-Let us consider now the equation (BVP_ε) with α actually depending upon x. Also assume that in step 3 the simplified problem P2 (with $u(1-k\varepsilon)$ replaced by $u_d(1-k\varepsilon)$) is solved numerically on G_2 with the ordinary difference method. In the remarks following the description of P1-P4 we have noticed that large values of k must lead to error terms growing with k. Actually, in practice this is not important since a choice $k \approx \log(1/\varepsilon)$ is appropriate because of (2) (compare Hsiao et al (1978)) and $\log(1/\varepsilon)$ will not be large, even for rather small values of ε. But nevertheless a bound like (10) cannot hold if the simplified problem P2 is solved. (To see this choose $k=1/\varepsilon$ and let $n_2 \to \infty$.) Therefore we shall concentrate here on estimates were k is assumed to be fixed, but n_1, n_2 and ε may vary.

<u>Theorem 7</u> (the general quasilinear case)
Assume A1 with n=4, A2, and A4, and let $x_d \in \mathbb{R}^G$ be computed as described above. There exists a constant C independent of $k \geq 1$, n_1, n_2 and ε, and for each fixed k a constant C(k) independent of n_1, n_2 and ε so that

$$\|x_\varepsilon - x_d\| \leq C(k^2\varepsilon + n_1^{-4} + e^{-k\alpha_0}) + C(k)n_2^{-2}.$$

<u>Proof:</u> As in the proof of theorem 5 we have for $t \in G_1$: $|x_d(t)-x_\varepsilon(t)| \leq C(\varepsilon + e^{-k\alpha_0} + n_1^{-4})$. The discrete system which has to be solved in step 3 reads after multiplication of the interior equations with ε (again, $h=\frac{1}{\varepsilon}h_2=k/n_2$):

$$x_0 = u_d(1-k\varepsilon), \quad x_{u_2} = B$$

$$h^{-2}(-x_{i-1}+2x_i-x_{i+1}) + \tfrac{1}{2}h^{-1}\alpha(1,x_i)(x_{i+1}-x_{i-1}) = 0$$

for $i=1,\ldots,n_2-1$,

We abbreviate this as

$$A_h(x)x=r_h, \quad x\in \mathbb{R}^{n2+1}. \tag{13}$$

Now let $y\in C^6[0,k]$ be the solution of

$$-y''(\tau)+\alpha(1,y(\tau))y'(\tau)=0, \quad \tau\in[0,k] \tag{14}$$

$$y(0)=u_d(1-k\varepsilon), \quad y(k)=B$$

i.e. $\tilde{y}(t)=y\ (k+(t-1)/\varepsilon)$, $t\in[1-k\varepsilon,1]$ is the solution of P2 with $u(1-k\varepsilon)$ replaced by $u_d(1-k\varepsilon)$. Then the system (13) is an $O(h^2)$ -discretisation of (14) and it follows from theorem 3.4 of Keller (1975) that

$$\max\ \{|\tilde{y}(t)-x_h(t)|:\ t\in G_2\} \le C(k)h^2.$$

Here x_h denotes that solution of (13) the existence of which is established in Keller (1975). Now let $X_\varepsilon(\tau):=x_\varepsilon(1+\varepsilon(\tau-k))$, $\tau\in[0,k]$. Then the continuous stability inequality of theorem 5 yields

$$\max\ \{|X_\varepsilon(\tau)-y(\tau)|:\ \tau\in[0,k]\} \le C(k^2\varepsilon+n_1^{-4}+e^{-k\alpha_0})$$

to prove the theorem.

4. NUMERICAL EXAMPLES

In the following two simple examples the exact

solutions x_ε and u are known. We have computed numerical approximations for x_ε for different values of ε with the method described. Here P1 and P2 coincide. We always took $n_1=10$ Runge-Kutta steps to solve the reduced problem and made different choices for k and n_2. The following tables list the maximum error $\|x_d-x_\varepsilon\|$ as defined in (10), i.e. the maximum error over the whole grid which contains points outside and inside the boundary layer. Also, we list the difference $x_\varepsilon-u$ at the breaking point $\bar{t}=1-k\varepsilon$.

Example 1:

$$-\varepsilon x''+x'=0 \text{ in } [0,1],\ x(0)=1,\ x(1)=0.$$

$$x_\varepsilon(t)=(1-e^{(t-1)/\varepsilon})\ /\ (1-e^{-1/\varepsilon}),\ u(t)=1.$$

This model problem has been considered by many authors in the study of numerical methods for singular perturbation problems, cf. Bohl (1978), Christie, Griffiths, Mitchell and Zienkiewicz (1976 Griffiths and Lorenz (1978), Hsiao and Jordan (1978

$\varepsilon = 10^{-1}$

k	n_2	$\|x_d-x_\varepsilon\|$	$(x_\varepsilon-u)(\bar{t})$
3	5	.50038(-1)	-.49744(-1)
6	10	.12680(-1)	-.24335(-2)
9	15	.11288(-1)	-.78013(-4)

$\varepsilon = 10^{-10}$

k	n_2	$\|x_d - x_\varepsilon\|$	$(x_\varepsilon - u)(\bar{t})$
3	5	.50079(-1)	-.49787(-1)
6	10	.12711(-1)	-.24788(-2)
9	15	.11319(-1)	-.12341(-3)

Example 2 (see Bohl (1978)):

$$-\varepsilon x'' + x' = e^t \text{ in } [0,1],\ x(0)=x(1)=0.$$

$$x_\varepsilon(t) = (1-\varepsilon)^{-1}\{e^t - (1 - e^{1-1/\varepsilon} + (e-1)e^{(t-1)/\varepsilon})$$

$$/(1-e^{-1/\varepsilon})\},\ u(t) = e^t - 1.$$

$\varepsilon = 10^{-3}$

k \ k/n_2	$\|x_d - x_\varepsilon\|$ 5/4	5/8
5	.98312(-1)	.26729(-1)
10	.94574(-1)	.19672(-1)
20	.94584(-1)	.19652(-1)
40	.94624(-1)	.19690(-1)

k \ k/n_2	$\|x_d - x_\varepsilon\|$ 5/16	5/32	$(x_\varepsilon - u)(\bar{t})$
5	.12138(-1)	.99099(-2)	-.98826(-2)
10	.42042(-2)	.16148(-2)	+.16148(-2)
20	.41767(-2)	.16661(-2)	+.16661(-2)
40	.42090(-2)	.16133(-2)	+.16133(-2)

$\varepsilon = 10^{-6}$

k \ k/n_2	$\|x_d - x_\varepsilon\|$ 5/4	5/8
5	.99527(-1)	.27945(-1)
10	.95779(-1)	.20879(-1)
20	.95768(-1)	.20839(-1)
40	.95768(-1)	.20839(-1)

k \ k/n_2	$\|x_d - x_\varepsilon\|$ 5/16	5/32	$(x_\varepsilon - u)(\bar{t})$
5	.13541(-1)	.11584(-1)	-.11576(-1)
10	.52331(-2)	.13338(-2)	-.76292(-5)
20	.51891(-2)	.12859(-2)	+.17147(-5)
40	.51891(-2)	.12859(-2)	+.17181(-5)

These and other examples clearly demonstrate that the method is insensible to a change of ε; the method works even for very small values of ε. In example 2 the error term Ck^2/n_2^2 of (10) dominates the others for k=10, 20 and 40.

REFERENCES

Bohl, E. (1978). Inverse monotonicity in the study of continuous and discrete singular perturbatio[n] problems. This book.

Christie, I., Griffiths, D. F., Mitchell, A. R. and Zienkiewicz, O. C. (1976). Finite element methods for second order differential equations with significant first derivatives. Int. J. Numer. Meth. Engin. 10, 1389-1396.

Coddington, E. A. and Levinson, N. (1952). A boundary value problem for a nonlinear differen[tial] equation with a small parameter. Proc. Amer. Math. Soc. 3, 73-81.

Dorr, F. W., Parter, S. V. and Shampine, L. F. (1973). Applications of the maximum principle to singular perturbation problems. SIAM Rev. 15, 43-48.

Flaherty, J. E. and O'Malley, Jr., R. E. (1977). The numerical solution of boundary value problems for stiff differential equations. Math. Comp. 31, 66-93.

Griffiths, D. F. and Lorenz, J. (1978). An analysis of the Petrov-Galerkin finite element method. Computer Methods appl. Mech. Engin. 14, 39-64.

Howes, F. A. (1976). Singular perturbations and differential inequalities. Memoirs of the AMS 5, 1-75.

Hsiao, G. C. and Jordan, K. E. (1978). Solutions to the difference equations of singular perturbation problems. University of Delaware, IMS Technical Report 36.

Keller, H. B. (1975). Approximation methods for nonlinear problems with application to two-point boundary value problems. Math. Comp. 29, 464-474.

Lorenz, J. (1977). Zur Inversmonotonie diskreter Probleme. Numer. Math. 27, 227-238.

O'Malley, Jr., R. E. (1974). Introduction to singular perturbations. Academic Press, New York.

Ortega, J. M. and Rheinboldt, W. C. (1970). Iterative solution of nonlinear equations in several variables. Academic Press, New York.

Prandtl , L. (1904). Über Flüssigkeitsbewegung bei sehr kleiner Reibung. Verhandlung d. III. Intern. Math. Kongr. Heidelberg.

Wasow, W. R. (1956). Singular perturbations of boundary value problems for nonlinear differential equations of the second order. Comm. Pure. Appl. Math. 9, 93-113.

SINGULAR PERTURBATIONS, ORDER REDUCTION, AND DECOUPLING OF LARGE SCALE SYSTEMS

R.E. O'Malley, Jr.
Department of Mathematics and Program in Applied Mathematics, University of Arizona, Tucson, Arizona 85721

L.R. Anderson
Department of Aerospace and Mechanical Engineering, University of Arizona, Tucson, Arizona 85721

1. INTRODUCTION

Concepts from singular perturbations have recently suggested very useful new techniques for reduced order modeling, stability analysis, and synthesis of optimal controls (cf., e.g., Kokotović *et al* (1976) and O'Malley (1978)). Their primary advantage is achieved by approximating solutions of systems of the form

$$\begin{cases} \dot{u} = f(u,v,t,\varepsilon) \\ \varepsilon\dot{v} = g(u,v,t,\varepsilon) \end{cases} \tag{1}$$

with solutions of the lower order ("reduced" or "outer") system

$$\begin{cases} \dot{U} = f(U,V,t,0) \\ 0 = g(U,V,t,0) \end{cases} \tag{2}$$

obtained by setting the small positive parameter ε equal to zero. In engineering contexts, analogous procedures have commonly (and necessarily) been used to lower the (often prohibitively high) dimensionality of complex models. These are often explained in terms of neglecting fast as opposed to slow dynamics (cf. Davison (1966), van Ness (1977), and Skira and De Hoff (1977)). Despite remarkable success, these schemes sometimes involve nonsensical use of asymptotic analysis, valid as $\varepsilon \to 0$, for $\varepsilon = 1$ (cf. Calise (1976)). Indeed, such success, so perilously based, requires more careful analysis. In the numerical methods literature such "singularly perturbed" systems of differential equations are called stiff (cf. Willoughby (1974)). Singular perturbations theory has recently contributed to the analysis of such stiff equations (cf., e.g., Miranker (1975), Hemker (1977), Flaherty and O'Malley (1977), and Kreiss (1978)). In physical contexts, however, one is seldom presented with systems in the form (1). The fundamental question usually remains, viz: How does one numerically identify the small parameter(s) involved? This must be answered before one can effectively utilize the singular perturbations machinery, either directly or as an underpinning structure for a numerical approximations procedure.

As the simplest example of such problems, let us consider the linear constant coefficient system

$$\dot{x} = Ax \tag{3}$$

on a finite interval, say $0 \le t \le 1$, when the spectrum $\lambda(A)$ of A has two (or more) time scales (i.e., the eigenvalues of A cluster into two or more sets which are widely separated in magnitude). Specifically, we shall take x to be an n-vector with the spectrum of A being the disjoint sum

$$\lambda(A) = S \cup F \tag{4}$$

where

$$S = \{s_1, s_2, \ldots, s_{n_1}\} \text{ and } |s_i| \le |s_{i+1}| \text{ for each } i,$$

$$F = \{f_1, f_2, \ldots, f_{n_2}\} \text{ with } |f_j| \le |f_{j+1}| \text{ for each } j,$$

$n_1 + n_2 = n$ with $0 < n_1 < n$, and

$$|s_{n_1}| \ll |f_1|. \tag{5}$$

Thus, a small parameter appropriate for our analysis of (3) has been identified as

$$\varepsilon = |s_{n_1}/f_1| \ll 1. \tag{6}$$

More refined partitioning of $\lambda(A)$ (corresponding to several time scales) will often be useful as well, providing several small parameters of decreasing size. Our general approach may even be useful when ε is not so very small, as we've demonstrated numerically.

We shall also assume that

$$\text{Re } |f_j| \gg |s_{n_1}|, \; j = 1, \ldots, n_2. \tag{7}$$

This will allow us to approximate the solutions of (3) by solutions of a lower order (n_1-dimensional) system, except in narrow endpoint boundary layers. We'd have to consider highly oscillatory solutions and use averaging methods if we omitted assumption (7) and allowed some eigenvalues f_j to be purely imaginary (or nearly so) (cf. Hoppensteadt and Miranker (1976) and Chow (1977)).

2. PRELIMINARY MATHEMATICAL ANALYSIS

The solutions of (3) can, of course, be obtained by constructing the matrix exponential e^{At}, or $e^{At}K$ for any nonsingular matrix K. As we'll find, however, its numerical computation is by no means straightforward (cf. Moler and van Loan (1978)). We shall present a particular way of approximating the solutions of (3) by its n_1 slow modes, adding endpoint corrections due to the n_2 fast modes as needed. To proceed, let us suppose x is decomposed into an

n_1 and an n_2 dimensional subvector and A is subdivided in a compatible fashion, i.e.,

$$x = \begin{bmatrix} x_1 \\ x_2 \end{bmatrix} \text{ and } A = \begin{bmatrix} A_{11} & A_{12} \\ A_{21} & A_{22} \end{bmatrix}. \tag{8}$$

We shall transform x to an n-vector y with separated slow and fast subvectors y_1 and y_2. Thus, we set

$$x = \begin{bmatrix} I_{n_1} & -K \\ -L & I_{n_2} + LK \end{bmatrix} y \tag{9}$$

to obtain

$$\dot{y} = By = \begin{bmatrix} B_1 & 0 \\ 0 & B_2 \end{bmatrix} y \tag{10}$$

for

$$B_1 = A_{11} - A_{12}L \text{ and } B_2 = A_{22} + LA_{12} \tag{11}$$

where L is chosen so that

$$\begin{cases} \lambda(B_1) = S = \{s_1, \ldots, s_{n_1}\} \\ \text{and} \\ \lambda(B_2) = F = \{f_1, \ldots, f_{n_2}\}. \end{cases} \tag{12}$$

We note that the transformation (9) is suggested by that commonly used for time-varying systems (cf., e.g., Harris (1973)). Setting

$$y = \begin{pmatrix} y_1 \\ y_2 \end{pmatrix},$$

we have the separate systems

$$\dot{y}_1 = B_1 y_1 \tag{13}$$

and

$$\dot{y}_2 = B_2 y_2. \tag{14}$$

Now the singular perturbations nature of (10), or equivalently (13)-(14), would be made even more obvious if we had multiplied (14) through by the small parameter $\|B_1\|/\|B_2\|$ and compared the result with (1).

We shall obtain the block diagonal system (10) in two steps. If we first take K = 0 in (9), we find that

$$\begin{bmatrix} x_1 \\ y_2 \end{bmatrix}$$

satisfies the decoupled upper block triangular system

$$\begin{pmatrix} \dot{x}_1 \\ \dot{y}_2 \end{pmatrix} = \begin{pmatrix} B_1 & A_{12} \\ 0 & B_2 \end{pmatrix} \begin{pmatrix} x_1 \\ y_2 \end{pmatrix} \tag{15}$$

provided the (non-square) decoupling matrix L satisfies the algebraic Riccati equation

$$LA_{11} - A_{22}L - LA_{12}L + A_{21} = 0 \tag{16}$$

in addition to the eigenvalue placement requirements (11)-(12). (We note that (16) is linear when $A_{12} = 0$. Then, however, A is already in block-triangular form, although not generally decoupled into slow and fast modes.) We next obtain the diagonal system (10) by requiring K to satisfy the linear equation

$$KB_2 - B_1K + A_{12} = 0. \tag{17}$$

This Liapunov equation will have a unique solution since B_1 and B_2 have no eigenvalues in common (cf., e.g., Bellman (1970)). An attractive feature of the nonsingular transformation (9) is that its inverse is simply given by

$$y = \begin{bmatrix} I_{n_1} + KL & K \\ L & I_{n_2} \end{bmatrix} x, \tag{18}$$

so we don't need to numerically perform a matrix inversion to solve (9) for y.

Presuming we can solve for L and K, we can treat the system (3) as the transformed system (10) and then separately integrate (13) for the slowly-varying vector y_1 and (14) for the rapidly-changing vector y_2. If m_1 of the (large) eigenvalues f_j of B_2 have negative real parts, there will be an m_1-dimensional manifold of solutions to (14) which decay rapidly to zero away from t = 0 and an $m_2 = n_2 - m_1$ dimensional manifold of rapidly growing solutions there. Alternatively,

$$e^{B_2(t-1)}$$

has an m_2 dimensional column space of vectors which decay rapidly to zero within [0,1), away from t = 1. Let us use the (not necessarily Jordan) decomposition

$$B_2 = NDN^{-1} \tag{19}$$

where D has the block diagonal form

$$D = \begin{bmatrix} D_1 & 0 \\ 0 & D_2 \end{bmatrix}$$

with D_1 and $-D_2$ being $m_1 \times m_1$ and $m_2 \times m_2$ dimensional stable matrices, and

$$N = [N_1 \quad N_2]$$

where N_1 is $n_2 \times m_1$ dimensional. Then

$$e^{B_2 t} N_1 = N_1 e^{D_1 t}$$

spans all solutions of (14) which decay away from t = 0,

while

$$e^{B_2(t-1)} N_2 = N_2 e^{D_2(t-1)}$$

spans all solutions of (14) which decay away from t = 1. For the matrix

$$C = \begin{bmatrix} I_{n_1} & 0 \\ 0 & N_1 + e^{-B_2} N_2 \end{bmatrix},$$

$$Y = e^{Bt} C = \begin{bmatrix} e^{B_1 t} & 0 \\ 0 & e^{B_2 t} N_1 + e^{B_2(t-1)} N_2 \end{bmatrix} \tag{20}$$

will be a bounded fundamental matrix for the system (10). Our use of the $e^{B_1 t}$ notation is harmless, since any standard numerical code or explicit formulas for the matrix exponential will serve well to produce this slowly-varying solution to the matrix version of (13). Unless B_2 or $-B_2$ is stable, however, $e^{B_2 t}$ is very difficult to compute, since it contains both rapidly growing and rapidly decaying modes (though no slow modes). Less difficulty is involved in separately computing its fast initial and terminal transients $e^{B_2 t} N_1$ and $e^{B_2(t-1)} N_2$. These correspond to integrating (14) on the appropriate m_1 or m_2 dimensional manifold. If we wished to obtain a bounded solution to the initial value problem for (3), to obtain a bounded solution on [0,1] as $\varepsilon \to 0$, we'd need $m_1 = n_2$ or we'd have to restrict the initial vector x(0) to a lower dimensional manifold. We might even consider such problems on $t \geq 0$ if A were stable. Two-point problems, of course, need not have a solution, but we've at least outlined a framework in which we might numerically consider the question. These considerations, of course, parallel those always encountered in using fundamental matrices (cf., e.g., Coppel (1965) and Cole (1968)).

It is convenient to decompose the fundamental matrix $e^{Bt}C$ for (10) as

$$Y = e^{Bt}C = (Y_s \quad Y_{f0} \quad Y_{f1}) \tag{21}$$

with n_1 slow modes

$$Y_s = \begin{pmatrix} Y_{s1} \\ Y_{s2} \end{pmatrix} = \begin{pmatrix} e^{B_1 t} \\ 0 \end{pmatrix}, \tag{22}$$

m_1 fast-decaying initial modes

$$Y_{f0} = \begin{pmatrix} Y_{f01} \\ Y_{f02} \end{pmatrix} = \begin{pmatrix} 0 \\ e^{B_2 t}N_1 \end{pmatrix} \tag{23}$$

and m_2 fast-decaying terminal modes

$$Y_{f1} = \begin{pmatrix} Y_{f11} \\ Y_{f12} \end{pmatrix} = \begin{pmatrix} 0 \\ e^{B_2(t-1)}N_2 \end{pmatrix}. \tag{24}$$

The decomposition (21) implies a corresponding decomposition for a bounded fundamental matrix $e^{At}C$ for our original system (3), i.e.,

$$X = e^{At}\tilde{C} \equiv (X_s \quad X_{f0} \quad X_{f1}) \tag{25}$$

with n_1 slow modes

$$X_s = \begin{pmatrix} X_{s1} \\ X_{s2} \end{pmatrix} = \begin{pmatrix} I_{n_1} & -K \\ -L & I_{n_2} + LK \end{pmatrix} Y_s = \begin{pmatrix} I_{n_1} \\ -L \end{pmatrix} e^{B_1 t} \tag{26}$$

and n_2 fast modes

$$X_{f0} = \begin{pmatrix} X_{f01} \\ X_{f02} \end{pmatrix} = \begin{pmatrix} -K \\ I_{n_2} + LK \end{pmatrix} e^{B_2 t}N_1 \tag{27}$$

and

$$X_{f1} = \begin{pmatrix} X_{f11} \\ X_{f12} \end{pmatrix} = \begin{pmatrix} -K \\ I_{n_2} + LK \end{pmatrix} e^{B_2(t-1)} N_2 \ . \tag{28}$$

(If $m_2 = 0$, e.g., X_{f1} is omitted in (25).) Within (0,1), then, all solutions to (3) are nicely approximated in the column space of the reduced order $n \times n_1$ matrix $X_s(t)$. In the initial "boundary layer," i.e., in a narrow interval near $t = 0$, X_s is nearly constant, so solutions are well approximated by the span of the $n \times (n_1 + m_1)$ matrix $(X_s(0) \quad X_{f0})$, while $(X_s(1) \quad X_{f1})$ is appropriate near $t = 1$. If we're not concerned with the fast endpoint transients, we'd simply use the reduced (or "outer") approximation $X_s(t)$ everywhere. Since $X_s(0)$ has rank $n_1 < n$, while x may be determined by n linearly independent endconditions, we realize that this approximation is usually not correct at $t = 0$ and 1. The approximation of x by a vector multiple of X_s, however, improves as the ratio ε decreases (i.e., the stiffer our system (3), the more negligible are the fast modes X_{f0} and X_{f1} within (0,1)). We note that the separation of growing and decaying modes is common in time-varying stability theory and is formalized in the concept of an exponential dichotomy (cf. Coppel (1978)). It has previously been used by Ferguson (1975) for the numerical solution of singular perturbation problems. Finally, we emphasize the numerical importance of decoupling the slow part and the fast initial and terminal transients. Large integration steps can be used throughout to calculate X_s, while an accurate integration of X_{f0} and X_{f1} (with much smaller stepsizes) can be obtained, if needed, on the appropriate short t-intervals. We intend to primarily emphasize the numerical aspects involved with computing the reduced order model X_s. Readers should note that this primarily involves the decoupling matrix L, although the matrix

K is needed to identify the endconditions appropriate for X_s.

3. THE DECOUPLING MATRICES L AND K

In order to utilize the structure we've developed, we must be able to find an $n_2 \times n_1$ matrix solution L of the algebraic Riccati equation (16) satisfying the eigenvalue placement conditions (11)-(12) as well. We note that the well-studied case when A is a Hamiltonian matrix is very important in applications (then $A_{11} = -A'_{22}$, $A_{12} = A'_{12}$, $A_{21} = A'_{21}$ (with the prime denoting transposition), and the solution of (16) is symmetric). Under assumptions of stabilizability and detectability, (16) has a unique positive semi-definite solution (cf. Kucera (1973)). An analogous approach is required for our problem. Specifically, let us suppose that A has the decomposition

$$A = MJM^{-1} \tag{29}$$

where

$$J = \begin{bmatrix} J_s & 0 \\ 0 & J_f \end{bmatrix}, \tag{30}$$

with the blocks J_s and J_f satisfying

$$\lambda(J_s) = \lambda(B_1) = S, \quad \lambda(J_f) = \lambda(B_2) = F, \tag{31}$$

and the matrix M having the compatible partitioning

$$M = \begin{bmatrix} M_{11} & M_{12} \\ M_{21} & M_{22} \end{bmatrix}. \tag{32}$$

We could take M to be a modal matrix so that J is in Jordan form. Then the columns of M would be right eigenvectors and generalized right eigenvectors of A corresponding to the diagonal entries of J, while the rows of M^{-1} would be corresponding (sometimes generalized) left eigenvectors of A (cf., e.g., Hirsch and Smale (1974)). It's not necessary, however, to be so explicit. All that is needed is the slow-fast block decomposition of J provided by (30)-(31). Indeed, the Jordan form is not particularly well suited to numerical computation (cf. Golub and Wilkinson (1976)).

Since $e^{At}M = Me^{Jt}$ is a fundamental matrix for (3), it follows that the n_1-dimensional space of slowly-varying solutions of (3) coincides with the column span of

$$M\begin{bmatrix} e^{J_s t} & 0 \\ 0 & 0 \end{bmatrix} = \begin{bmatrix} M_{11} \\ M_{21} \end{bmatrix} e^{J_s t}, \tag{33}$$

$e^{J_f t}$ being fast-varying. Comparing with the slowly-varying matrix X_s of (26), it follows that

$$\begin{pmatrix} I_{n_1} \\ -L \end{pmatrix} \text{ and } \begin{pmatrix} M_{11} \\ M_{21} \end{pmatrix}$$

must span the same space. Therefore, if we have a solution L to (16), (11), and (12), M_{11} will be invertible and we must have

$$L = -M_{21}M_{11}^{-1}. \tag{34}$$

Conversely, if $LM_{11} = -M_{21}$ for M_{11} invertible, multiplying out

$$\begin{bmatrix} I_{n_1} & 0 \\ L & I_{n_2} \end{bmatrix} A \begin{bmatrix} I_{n_1} & 0 \\ -L & I_{n_2} \end{bmatrix}$$

and using the decomposition (29) for A produces the lower right entry $LA_{11} - A_{22}L - LA_{12}L + A_{21} = 0$ (after considerable manipulation). Further, if

$$M^{-1} = \begin{bmatrix} Q_{11} & Q_{12} \\ Q_{21} & Q_{22} \end{bmatrix}, \tag{35}$$

we'll have $Q_{22} = (M_{22} - M_{21}M_{11}^{-1}M_{12})^{-1}$ and (34) will hold if and only if

$$L = Q_{22}^{-1}Q_{21}. \tag{36}$$

The formulas (34) and (36) show that the matrix L we are seeking is unique, although M is far from unique and the quadratic equation (16) has many other solutions. Wu and Narasimhamurthi (1976), Narasimhamurthi and Wu (1977), and Anderson (1978) contain alternative and more detailed treatments of these results.

In order to compute L, we first employ a standard eigenanalysis library routine to obtain approximate eigenvalues of the matrix A. This allows us to select n_1, the number of moderate eigenvalues, and thereby determines the number $n_2 = n - n_1$ of eigenvalues with large real parts. We note that multiple and complex conjugate eigenvalues are naturally grouped together in either S or F.

If $n_1 \leq n/2$, we obtain n_1 approximate right (or generalized right) eigenvectors corresponding to the n_1 moderate eigenvalues, thereby forming an approximation

$$\begin{bmatrix} M_{11}^{\circ} \\ M_{21}^{\circ} \end{bmatrix} \text{ to the submatrix } \begin{bmatrix} M_{11} \\ M_{21} \end{bmatrix}$$

of (32). (If the resulting matrix has complex entries, we eliminate them through elementary column operations.) Then we solve the resulting real system

$$\tilde{L}_0 M_{11}^{\circ} = -M_{21}^{\circ}, \tag{37}$$

approximating (34), providing M_{11}° is nonsingular. An efficient method of obtaining an approximate solution L_0 to (37) is the "LU" factorization technique (cf., e.g., Stewart (1973)). If $n_1 > n/2$, it would be more efficient to seek $n_2 < n/2$ approximate left (or generalized left) eigenvectors of A and instead obtain an approximate solution to (36). In either case, we don't need all the eigenvectors of A. For all our computations to date, we have used a modal matrix for M, but we expect to not be so restrictive in later testing.

Thus far, we have used x_1 and x_2 as any n_1 and n_2 dimensional subvectors of x, requiring only that the matrix

$$\begin{bmatrix} M_{11} \\ M_{21} \end{bmatrix}$$

of slow modes (necessarily of rank n_1) have M_{11} nonsingular. This may demand a reordering of the components of x. If we used an ordering for which M_{11} was nearly singular, the approximate solution L_0 would generally be of large norm and, quite likely, difficult to obtain accurately. A small approximate L_0 would, however, result if we kept physical coordinates known to be primarily slowly-varying in x_1, with coordinates dominated by fast boundary-layer variations being kept in x_2. Such a splitting of x would correspond roughly to the splitting

$$y = \begin{pmatrix} y_1 \\ y_2 \end{pmatrix}$$

into slow and fast components. When x_1 and x_2 are weakly coupled like this, the need for the decoupling matrix L in (9) is lessened, so we would expect L to be small. Formally, then, it is natural to introduce a slow-mode coupling ratio

$$\rho_s = \min \left(\|M_{21}\| / \|M_{11}\|\right) \tag{38}$$

where the minimization takes place over all possible reorderings of the rows of the matrix M. If we could obtain $M_{21} = 0$, x_1 and x_2 would be decoupled and we'd have $L = 0$ by (34). The practical significance of such reorderings can be seen in our numerical results and is stressed in the aircraft dynamics examples of Teneketzis and Sandell (1977), though it is certainly not necessary when the original M_{11} matrix is nonsingular.

An efficient method of improving the initial approximation L_0 to the decoupling matrix L can be obtained by using a successive approximations scheme on the algebraic Riccati equation (16) with L_0 as the first iterate. Using (11), we can rewrite (16) as

$$L = B_2^{-1}(LB_1 + A_{21} + LA_{12}L) = B_2^{-1}(LA_{11} + A_{21}). \tag{39}$$

Thus, we naturally define the iterates

$$L_{j+1} = (A_{22} + L_jA_{12})^{-1}(L_jA_{11} + A_{21}),\ j = 0, 1, \ldots, \tag{40}$$

expecting them to converge to the unique L desired. This iteration scheme features rapid convergence provided L_0 is close to L and ε is small enough (cf. Kleinman (1968) who discussed the analogous Hamiltonian problem). Indeed, a proof of convergence follows from a contraction mapping argument. If we introduce

$$D_j = L_{j+1} - L_j, \tag{41}$$

we find that (40) is equivalent to

$$D_j = (A_{22} + L_jA_{12})^{-1}R_j, \quad j = 0, 1, \ldots, \tag{42}$$

for the residual

$$R_j = L_jA_{11} - A_{22}L_j - L_jA_{12}L_j + A_{21}. \tag{43}$$

The procedure terminates when one finds a suitably small R_j. (In particular, if R_0 is judged small enough, we wouldn't begin the iteration scheme (40) or (42).) Further manipulating (42) and (43) implies that

$$D_{j+1} = (A_{22}+L_{j+1}A_{12})^{-1}D_j(A_{11} - A_{12}L_{j+1}). \qquad (44)$$

We note that $A_{22}+L_{j+1}A_{12}$ will be a good approximation to B_2, if L_{j+1} is a good approximation to L. Then, the inverse used in (40)-(44) is legitimate. Likewise, $A_{11} - A_{12}L_{j+1}$ will then be a good approximation to B_1, and since $\|B_1\|/\|B_2\| = O(\varepsilon)$, (44) implies that our scheme ultimately has a rate of convergence of order ε. A closely related approach is presented in Kokotovic (1975) and Chow and Kokotovic (1976). Instead of (37), they used the initial iterate $L_0 = A_{22}^{-1}A_{21}$, appropriate if A_{22} is nonsingular and if the system is sufficiently decoupled so that L is small and thereby easy to compute (cf. (40)). Thus, their convergence criterion is unnecessarily restrictive. Numerical experiments indicate that our scheme is successful even when a rather crude solution to (37) is used for L_0, and when ε is not too small.

After numerically obtaining the decoupling matrix L, we need to solve the linear equation (17) for the matrix K. Rewriting it as

$$K = (B_1K - A_{12})B_2^{-1} \qquad (45)$$

suggests the iteration scheme

$$K_{j+1} = (B_1K_j - A_{12})B_2^{-1}, \quad K_0 = 0, \quad j = 0, 1, 2, \ldots, \qquad (46)$$

converging to its unique solution. A straightforward argument implies a rate of convergence of order ε. Here, unlike for (40), the initial iterate is not crucial. Alternatively, one can use the representations (32) and (35) to show that

$$K = -M_{12}Q_{22} = M_{11}Q_{12} \qquad (47)$$

(cf. Anderson (1978)), though this requires more information about the modal decomposition of A than (46).

4. IMPLEMENTATION

We propose to obtain a reduced order matrix approximation for two-time-scale systems $\dot{x} = Ax$ via the following numerical algorithm:

1) Obtain approximate eigenvalues of A and order them by increasing moduli into sets S and F satisfying (7). Note that different choices are possible for the number n_1 of slow eigenvalues for any given matrix A.
2) Determine an $n \times n_1$ dimensional real matrix

$$\begin{bmatrix} M_{11} \\ M_{21} \end{bmatrix}$$

whose columns span the right eigenspace of A corresponding to the slow eigenvalues. (If $n_1 > n/2$, an alternative based on left eigenvectors is suggested (cf. (36)).
3) (optional) Reorder the variables x so that $\|M_{21}\|/\|M_{11}\|$ is minimized.
4) Find an approximate solution L_0 of $L_0M_{11} = -M_{21}$ (or its alternate).
5) (optional) Improve the accuracy of the approximation L_j by iteration. (a) Stop if $\|R_j\|$ is less than a prescribed tolerance for

$$R_j = L_jA_{11} + A_{21} - L_jA_{12}L_j - A_{22}L_j.$$

(b) Set

$$L_{j+1} = L_j + (A_{22} + L_jA_{12})^{-1}R_j$$

and return to (a) with $j = j + 1$.
6) The approximate reduced order fundamental matrix (appropriate within (0,1)) is

$$X_s^{(j)} = \begin{pmatrix} I_{n_1} \\ -L_j \end{pmatrix} e^{(A_{11}-A_{12}L_j)t}$$

where j is zero or the last index used in 5).

This algorithm has been applied to a number of physical systems with order n ranging up to 32. Two different programs were used for the eigenanalysis of steps 1) and 2). They are the EISPACK subroutines, which are available without cost from the Applied Mathematics Division of the Argonne National Laboratory (cf. Smith *et al* (1976)), and EIGRF, which is part of the proprietary subroutines sold by the International Mathematics and Statistics Library (IMSL) of Houston. The procedure was implemented on the 64 bit CYBER 175 and the 36 bit DEC 10 computers at the University of Arizona Computing Center.

Example 1

This fourth order model of F-8 aircraft longitudinal dynamics (cf. Ektin (1972) and Teneketzis and Sandell (1977)) has slow eigenvalues $s_{1,2} = -0.0075 \pm i0.076$ and fast eigenvalues $f_{1,2} = -0.94 \pm i3.0$, corresponding to the small parameter $\varepsilon = |s_2/f_1| = 0.024$. The physical variables involved are velocity variation x_1 (ft./sec.), flight path angle x_2 (rad.), angle of attack x_3 (rad.), and pitch rate x_4 (rad./sec.). The first two are primarily slow, while the latter two are predominantly fast. The four different orderings of the x coordinates with $n_1 = n_2 = 2$ show that the ratio $\|M_{21}\|/\|M_{11}\|$ and the corresponding $\|L\|$ are loosely related. The numbers are $(1.9 \times 10^{-4}, 1.1 \times 10^{-3})$, $(2.4 \times 10^{-3}, 1.3 \times 10^{3})$, $(5.3 \times 10^{3}, 6.3 \times 10^{3})$, $(4.2 \times 10^{2}, 2.0 \times 10^{4})$ so $\rho_s = 1.9 \times 10^{-4}$ for the ordering suggested above. The success of the iterative procedure is illustrated in Figure 1 where log $\|R_j\|$ is plotted against j. We find that $\|R_{j+1}\|/\|R_j\| \approx \varepsilon$ until the (machine dependent) maximum possible numerical accuracy is achieved. These results also show that the approximate matrix

$$\begin{bmatrix} M_{11} \\ M_{21} \end{bmatrix}$$

obtained using IMSL was better than with EISPACK, since the resulting L_0 was more accurate. Further tests show, however,

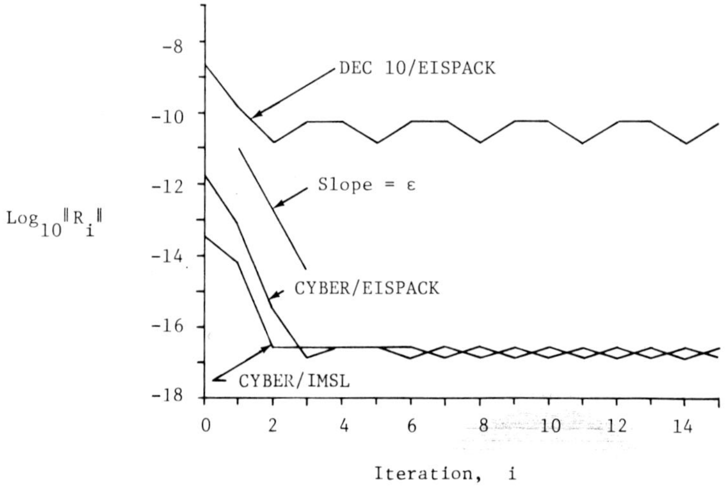

Fig. 1. *The convergence of* L_i *for the F-8 aircraft longitudinal dynamics model.*

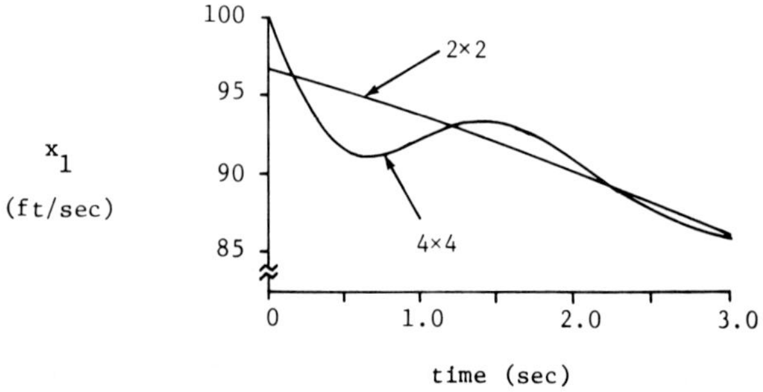

Fig. 2. *Velocity (slow variable) vs. time.*

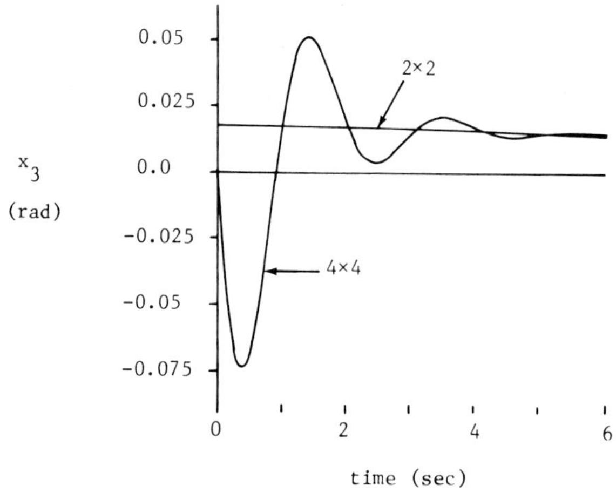

Fig. 3. *Angle-of-attack (fast variable) vs. time.*

that considerable eigenvector inaccuracies may be tolerated with the algorithm still converging to the desired solution when iteration is used. Indeed, the algorithm converges whatever ordering is used for the x components. Most important, however, is the fact that trajectories for initial value problems are well approximated, outside an initial boundary layer region, by the reduced model. Using initial conditions $(x_1(0), x_2(0), x_3(0), x_4(0)) = (100, 0, 0, 0.5)$, the aircraft response for the predominantly slow and fast variables x_1 and x_3 are shown in Figures 2 and 3, both for the full fourth order model and the reduced second order model. After the first few seconds the responses are indistinguishable, so they are not pictured.

Example 2

This 16th order model of an F-100 turbofan engine was used as the theme problem for the recent International Forum on Alternatives for Multivariable Control (cf. Sain (1977)). Two of the x coordinates are shaft speeds, three are

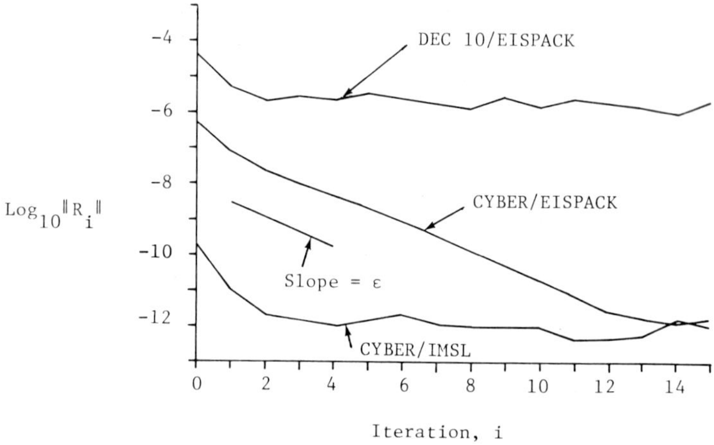

Fig. 4. *The convergence of L_i for decoupling the three slowest modes for the F-100 turbofan engine.*

pressures, and eleven are temperatures. The eigenvalues (ordered in magnitude) are -0.648, -1.906, -2.619, -6.715 ± i1.312, -17.8 ± i4.78, -18.6, -21.3 ± i0.822, -38.7, -47.1, -50.7, -59.2, -175.7, and -577.0. If we take n_1 = 15, 5, and 3, we get ε values 0.304, 0.371, and 0.383, respectively. The choice n_1 = 15 is ruled out because the reduction in dimensionality isn't significant. Even though the remaining ε's aren't very small, our computational algorithm was successful in producing appropriate decoupling matrices L. Results are contained in Figure 4 for n_1 = 3.

ACKNOWLEDGEMENTS

Supported in part by the Office of Naval Research under Contract Number N00014-76-C-0326.

REFERENCES

Anderson, L. (1978). "Decoupling of two-time-scale linear systems," *Proceedings, 1978 Joint Automatic Control Conference.*

Bellman, R. (1970). *Introduction to Matrix Analysis*, Second Edition, McGraw-Hill, New York.

Calise, A.J. (1976). "Singular perturbation methods for variational problems in aircraft flight," *IEEE Trans. on Automatic Control* 21, 345-353.

Chow, J.H. (1978). *Singular Perturbation of Nonlinear Regulators and Systems with Oscillatory Modes*, Ph.D. thesis, University of Illinois, Urbana-Champaign.

Chow, J.H. and Kokotovic, P.V. (1976). "Eigenvalue placement in two-time-scale systems," *Proc. IFAC Symposium on Large-Scale Systems*, 321-326. Udine.

Chow, J.H. and Kokotovic, P.V. (1976). "A decomposition of near-optimum regulators for systems with slow and fast modes," *IEEE Trans. on Automatic Control* 21, 701-705.

Cole, R.H. (1968). *Theory of Ordinary Differential Equations*, Appleton-Century-Crofts, New York.

Coppel, W.A. (1965). *Stability and Asymptotic Behavior of Differential Equations*, D.C. Heath, Boston.

Coppel, W.A. (1978). "Dichotomies in stability theory," *Lecture Notes in Math.* 629, Springer-Verlag, Berlin.

Davison, E.J. (1966). "A method for simplifying linear dynamic systems," *IEEE Trans. Automatic Control* 11, 93-101.

Etkin, B. (1972). *Dynamics of Atmospheric Flight*, Wiley, New York.

Ferguson, W.E., Jr. (1975). *A Singularly Perturbed Linear Two-Point Boundary Value Problem*, Ph.D. Dissertation, California Institute of Technology, Pasadena.

Flaherty, J.E. and O'Malley, R.E., Jr. (1977). "The numerical solution of boundary value problems for stiff differential equations," *Math. of Computation* 31, 66-93.

Golub, G.H. and Wilkinson, J.H. (1976). "Ill-conditioned eigensystems and the computation of the Jordan canonical form," *SIAM Review* 18, 578-619.

Harris, W.A., Jr. (1973). "Singularly perturbed boundary value problems revisited," *Lecture Notes in Math.* 312, 54-64. Springer-Verlag, Berlin.

Hemker, P.W. (1977). *A Numerical Study of Stiff Two-point Boundary Problems*, Ph.D. Dissertation, University of Amsterdam.

Hirsch, M.W. and Smale, S. (1974). *Differential Equations, Dynamical Systems, and Linear Algebra*, Academic Press, New York.

Hoppensteadt, F.C. and Miranker, W.L. (1976). "Differential equations having rapidly changing solutions: Analytic methods for weakly nonlinear systems," *J. Differential Equations* 22, 237-249.

Kleinman, D.L. (1968). "On an iterative technique for Riccati equation computation," *IEEE Trans. Automatic Control* 13, 114-115.

Kokotovic, P.V. (1975). "A Riccati equation for block-diagonalization of ill-conditioned systems," *IEEE Trans. Automatic Control* 20, 812-814.

Kokotovic, P.V., O'Malley, R.E., Jr. and Sannuti, P. (1976). "Singular perturbations and order reduction in control theory - An overview," *Automatica* 12, 123-132.

Kreiss, H.-O. (1978). "Difference methods for stiff ordinary differential equations," *SIAM J. Num. Anal.* 15, 21-58.

Kucera, V. (1973). "A review of the matrix Riccati equation," *Kybernetika* 9, 42-61.

Miranker, W.L. (1975). *The Computational Theory of Stiff Differential Equations*, Instituto per le Applicazioni del Calcolo "Mauro Picone," Rome.

Moler, C.B. and Van Loan, C.F. (1978). "Nineteen ways to compute the exponential of a matrix," *SIAM Review* 20.

Narasimhamurthi, N. and Wu, F.F. (1977). "On the Riccati equation arising from the study of singularly perturbed systems," *Proceedings 1977 Joint Automatic Control Conference*, 1244-1247.

O'Malley, R.E., Jr. ( ). "Singular perturbations and optimal control," *Lecture Notes in Math.*

Sain, M.K. (1977). "The theme problem," *Proceedings International Forum on Alternatives for Multivariable Control*, 1-12.

Skira, C.A. and De Hoff, R.L. (1977). "A practical approach to linear model analysis for multivariable turbine engine control design," *Proceedings International Forum on Alternatives for Multivariable Control*, 29-44.

Smith, B.T., Boyle, J.M., Dongarra, J.J., Garbo, B.S., Ikeke, Y., Klema, V.C. and Moler, C.B. (1976). "Matrix eigensystem routines-EISPACK guide," Second Edition, *Lecture Notes in Computer Science* 6, Springer-Verlag, Berlin.

Stewart, G.W. (1973). *Introduction to Matrix Computations*, Academic Press, New York.

Teneketzis, D. and Sandell, N.R., Jr. (1977). "Linear regulator design for stochastic systems by a multiple time-scales method," *IEEE Trans. Automatic Control* 22, 615-621.

Van Ness, J.E. (1977). "Methods of reducing the order of power system models in dynamic studies," *Proceedings, 1977 IEEE International Symposium on Circuits and Systems*, 858-863.

Willoughby, R. (ed.) (1974). *Stiff Differential Systems*, Plenum Press, New York.

Wu, F.F. and Narasimhamurthi, N. (1976). *A New Algorithm for Modal Approach to Reduced Order Modeling*, Report ERL-M13, Engineering Research Laboratory, University of California, Berkeley.

FINITE DIFFERENCE/FINITE ELEMENT METHODS AT THE PARABOLIC-HYPERBOLIC INTERFACE

A R Mitchell and I Christie
Department of Mathematics, University of Dundee,
Dundee, Scotland

ABSTRACT

Upwinded parabolic and cubic elements are derived on a uniform grid of size h for the finite element Galerkin method applied to the solution of the model convection diffusion problem $\epsilon u'' - Ku' = 0$, $\epsilon, K > 0$ subject to the boundary conditions $u(0) = 1$, $u(1) = 0$. Numerical results are given for this model problem comparing some of the elements derived for a range of $L(= \frac{Kh}{2\epsilon})$, the grid Peclet number.

Application to a nonlinear problem involving Burgers' equation follows together with some numerical results.

1. INTRODUCTION

Much progress has been made recently in developing finite element methods to solve flow problems. Amongst the most difficult problems are convection dominated phenomena like Navier Stokes flows at high Reynolds number and convection-diffusion problems at high Peclet number. Most newly proposed finite element methods have been based on a discretization in space using Galerkin's method with different trial and test functions followed by a time stepping difference algorithm if the problem is transient. Two questions of major importance arise:

(1) How do the new finite element methods compare with existing finite difference schemes?

(2) What are the best time stepping algorithms particularly in the nonlinear case?

This paper will concern itself with attempting to answer question (1), and is a follow up to material contained in Mitchell and Wait (1977), Christie, Griffiths, Mitchell and Zienkiewicz (1976), Heinrich, Huyakorn, Zienkiewicz and Mitchell (1977), and Mitchell and Griffiths (1977, 1978).

2. FINITE ELEMENT GALERKIN METHODS

To set the scene we look first at the two point boundary value problem

$$Lu \equiv \epsilon \frac{d^2u}{dx^2} - K\frac{du}{dx} = 0 , \tag{2.1}$$

with the boundary conditions

$u = 1$ at $x = 0$ and $u = 0$ at $x = 1$,

where ϵ and K are positive constants. The theoretical solution is

$$u = \frac{e^{\frac{K}{\epsilon}x} - e^{\frac{K}{\epsilon}}}{1 - e^{\frac{K}{\epsilon}}} .$$

This problem can be reformulated with <u>homogeneous</u> boundary conditions by putting

$$v = u + x - 1 . \tag{2.1a}$$

The unit interval is now subdivided into n equal elements by the nodes $x = ih$ $(i=0,1,---,n)$ and so $nh = 1$. We shall not consider graded grids (Starius (1978)) because our main object is to compare different forms of the Galerkin finite element method. In the Galerkin methods to follow, the trial and test functions at a node i will be $\phi_i(x)$ and $\psi_i(x)$ respectively.

(i) Linear trial functions. We consider first the standard linear "hat" trial and test functions applied to (2.1). This gives rise to the difference equation

$$(1-L)U_{i+1} - 2U_i + (1+L)U_{i-1} = 0 , \quad (i=1,2,---,n-1) \qquad (2.2)$$

where $L = \frac{hK}{2\epsilon}$ is the grid Peclet number. Oscillations occur in the theoretical solution of (2.2) for $L > 1$. The equivalent of upstream differencing is introduced by modifying the test functions to be

$$\psi_j(x) = \phi_j(x) + \alpha\sigma_2(\tfrac{x}{h} - j) \qquad (j=1,2,---,n-1)$$

where

$$\sigma_2(s) = \begin{cases} 0 & |s| < 1 \\ -3s(1-s) & 0 \le s \le 1 \\ -\sigma_2(-s) , & -1 \le s \le 0 \end{cases}$$

$\phi_j(x)$ $(j=1,2,---,n-1)$ are the linear "hat" functions, and α is an arbitrary parameter. The quadratic perturbing function $\sigma_2(s)$ and the linear "hat" functions are illustrated in Fig 1. This leads to the modified difference formula

$$(1-L+\alpha L)U_{i+1} - 2(1+\alpha L)U_i + (1+L+\alpha L)U_{i-1} = 0 \qquad (2.3)$$

which has been obtained directly from finite difference

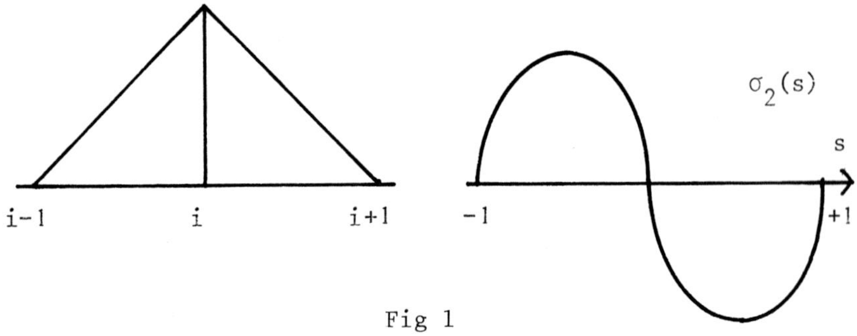

Fig 1

considerations by Hemker (1977) and Miller (1977). Compare also the difference formulae obtained in the more sophisticated vector case of (2.1) by Abramsson, Keller, and Kreiss (1974). Special cases of (2.3) are

$\alpha = 1$ Upstream formula with linear accuracy

$\alpha = 0$ Central formula with quadratic accuracy

$\alpha = \frac{1}{3}L$ Fourth order accuracy

$\alpha = \coth L - \frac{1}{L}$ Complete accuracy (Il'in (1969), Spalding (1972)).

It is easily seen from (2.2) that the effect of introducing asymmetry into the test functions has been to add a dissipation term $\alpha L\delta^2 U_i$ to the resulting difference equation. Alternatively we can add a dissipation term $\frac{1}{2}\alpha hK \frac{d^2u}{dx^2}$ to the original differential equation (2.1) making it

$$(1+\alpha L)\frac{d^2u}{dx^2} - \frac{K}{\epsilon}\frac{du}{dx} = 0 . \tag{2.4}$$

The Galerkin method with linear "hat" trial and test functions applied to (2.4) again leads to (2.3).

For the remainder of the paper we shall use the former method of attacking the original differential equation with different combinations of trial and test functions rather than the alternative method of using standard Galerkin procedures on a modified differential equation.

(ii) Quadratic trial functions. We now look at the Galerkin method with Lagrange quadratic trial and test functions. Applied to (2.1), it gives

$$(1-L)U_{i+1} - 4(2-L)U_{i+\frac{1}{2}} + 14U_i - 4(2+L)U_{i-\frac{1}{2}} + (1+L)U_{i-1} = 0$$

at the integer nodes $i = 1,2,---,n-1$ and

$$(1-\frac{1}{2}L)U_i - 2U_{i-\frac{1}{2}} + (1+\frac{1}{2}L)U_{i-1} = 0$$

at the half integer nodes $i = 1,2,---,n$. Elimination of U at the half integer nodes leads to

$$(1-L+\frac{1}{3}L^2)U_{i+1} - 2(1+\frac{1}{3}L^2)U_i + (1+L+\frac{1}{3}L^2)U_{i-1} = 0 \qquad (2.5)$$

at the integer nodes. The theoretical solution of (2.5) is oscillation free for all L and is fourth order accurate (in h). Unfortunately for large L the solution of (2.5) tends towards the solution of $\delta^2 U_m = 0$ rather than $\nabla U_m = 0$, and so (2.5) requires some modification.

This is carried out by perturbing the trial functions at the integer nodes to give

$$\psi_j(x) = \phi_j(x) - \alpha\sigma_3(\frac{x}{h} - j) - \beta\sigma(\frac{x}{h} - j - 1) \quad (j=1,2,---,n-1) \qquad (2.6)$$

and at the half integer nodes to give

$$\psi_{j-\frac{1}{2}}(x) = \phi_{j-\frac{1}{2}}(x) + 4\gamma\sigma_3(\frac{x}{h} - j) \qquad (j=1,2,---,n) \qquad (2.7)$$

where

$$\sigma_3(s) = \begin{cases} 5s(s+\frac{1}{2})(s+1) & -1 \le s \le 0 \\ 0 & \text{elsewhere} \end{cases} \qquad (2.8)$$

The cubic perturbing function $\sigma_3(s)$ along with the standard Lagrange quadratics are illustrated in Fig.2. The revised

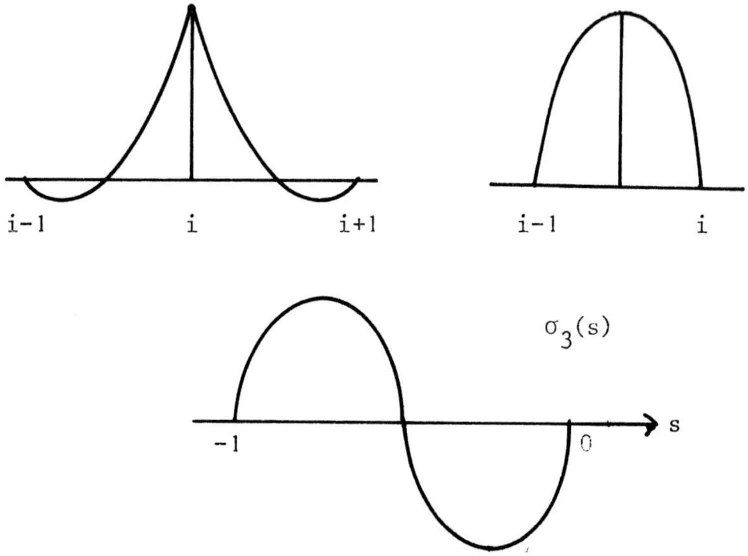

Fig 2

Galerkin method now gives

$$(1-L+\beta L)U_{i+1} - 4(2-L+\tfrac{1}{2}\beta L)U_{i+\frac{1}{2}} + (14+\beta L+\alpha L)U_i - 4(2+L+\tfrac{1}{2}\alpha L)U_{i-\frac{1}{2}} + (1+L+\alpha L)U_{i-1} = 0 \quad (2.9)$$

at the integer nodes and

$$(1-\tfrac{1}{2}L+\tfrac{1}{2}\gamma L)U_i - (2+\gamma L)U_{i-\frac{1}{2}} + (1+\tfrac{1}{2}L+\tfrac{1}{2}\gamma L)U_{i-1} = 0 \quad (2.10)$$

at the half integer nodes. Elimination of U at the half integer nodes leads to

$$[1-(1-\tfrac{1}{2}\gamma)L + (\tfrac{1}{3} - \tfrac{1}{6}\gamma - \tfrac{1}{6}\beta)L^2]U_{i+1} - 2[1 + \tfrac{1}{2}\gamma L + (\tfrac{1}{3} - \tfrac{1}{12}\beta + \tfrac{1}{12}\alpha)L^2]U_i + [1 + (1 + \tfrac{1}{2}\gamma)L + (\tfrac{1}{3} + \tfrac{1}{6}\gamma + \tfrac{1}{6}\alpha)L^2]U_{i-1} = 0 . \quad (2.11)$$

Formula (2.11) has second order accuracy when

$\alpha + \beta = 4\gamma$

and third order accuracy when

$\alpha = \beta = 2\gamma$.

Important values of the triple (α,β,γ) along with the corresponding formulae (2.11) are

$$(8,0,2) \quad U_{i+1} - 2(1+L+L^2)U_i + (1+2L+2L^2)U_{i-1} = 0 \qquad (2.12)$$

$$(\tfrac{4}{3},\tfrac{4}{3},\tfrac{2}{3}) \quad (1 - \tfrac{2}{3}L)U_{i+1} - 2(1+\tfrac{1}{3}L+\tfrac{1}{3}L^2)U_i + (1+\tfrac{4}{3}L+\tfrac{2}{3}L^2)U_{i-1} = 0 \qquad (2.13)$$

$(0,0,0)$ formula (2.5).

Difference equation (2.12) which is the upstream formula with quadratic accuracy has been discovered recently by Axelsson and Gustaffson (1977) using finite difference considerations. Complete accuracy at both integer and half integer nodes can be achieved by a suitable choice of (α,β,γ) but the situation is much more complicated than the complete accuracy case described in (1). The special case of the above procedure when $\alpha = \beta$ is given in Heinrich and Zienkiewicz (1977).

(iii) Cubic trial functions. Consider now the Galerkin method where the trial functions are the standard Lagrange cubics and the test functions are the Lagrange cubics perturbed by a suitable quartic function. This gives test functions of the form

$$\psi_j(x) = \phi_j(x) + \tfrac{1}{2}\alpha_1\sigma_4(\tfrac{x}{h} - j) + \tfrac{1}{2}\alpha_2\sigma_4(\tfrac{x}{h} - j - 1) \qquad (j=1,2,---,n-1)$$

at the integer nodes, and

$$\psi_{j-\frac{2}{3}}(x) = \phi_{j-\frac{2}{3}}(x) + 3\alpha_3\sigma_4(\frac{x}{h} - j)$$

and $\qquad\qquad\qquad\qquad\qquad\qquad\qquad\qquad\qquad\qquad\qquad\qquad (j=1,2,---,n)$

$$\psi_{j-\frac{1}{3}}(x) = \phi_{j-\frac{1}{3}}(x) + 3\alpha_4\sigma_4(\frac{x}{h} - j)$$

at the one third and two thirds integer nodes respectively where

$$\sigma_4(s) = \begin{cases} 21s(s+\frac{1}{3})(s+\frac{2}{3})(s+1) & -1 \le s \le 0 \\ 0 & \text{elsewhere} \end{cases}$$

The quartic polynomial perturbing function $\sigma_4(s)$ along with the standard Lagrange cubics are illustrated in Fig. 3.

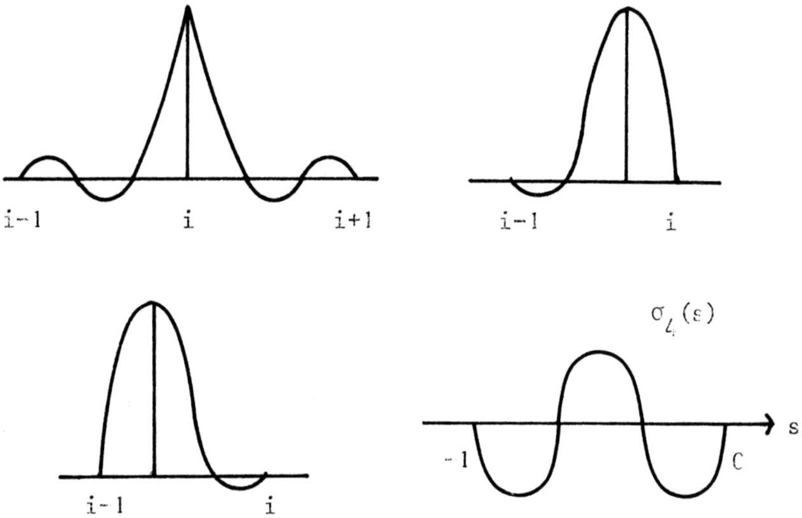

Fig 3

This Galerkin method gives after elimination of U at the intermediate nodes a three point difference formula which we quote for the following special cases of the quadruple

$(\alpha_1,\alpha_2,\alpha_3,\alpha_4)$

$(-14,\frac{1}{2},0,\frac{9}{4})$

$$U_{i+1} - (2+2L+2L^2+\tfrac{4}{3}L^3)U_i + (1+2L+2L^2+\tfrac{4}{3}L^3)U_{i-1} = 0, \qquad (2.17)$$

$(-\frac{27}{5},0,-\frac{9}{10},\frac{9}{5})$

$$(1-\tfrac{1}{2}L)U_{i+1} - (2+L+L^2+\tfrac{1}{3}L^3)U_i + (1+\tfrac{3}{2}L+L^2+\tfrac{1}{3}L^3)U_{i-1} = 0, \qquad (2.18)$$

$(-\frac{3}{2},\frac{3}{2},-\frac{3}{4},\frac{3}{4})$

$$(1-\tfrac{4}{5}L+\tfrac{1}{5}L^2)U_{i+1} - (2+\tfrac{2}{5}L+\tfrac{4}{5}L^2+\tfrac{2}{15}L^3)U_i + (1+\tfrac{6}{5}L+\tfrac{3}{5}L^2+\tfrac{2}{15}L^3)U_{i-1} = 0, \qquad (2.19)$$

$(0,0,0,0)$

$$(1-L+\tfrac{2}{5}L^2-\tfrac{1}{15}L^3)U_{i+1} - 2(1+\tfrac{2}{5}L^2)U_i + (1+L+\tfrac{2}{5}L^2+\tfrac{1}{15}L^3)U_{i-1} = 0. \qquad (2.20)$$

Difference equation (2.17) is the upstream formula with cubic accuracy.

(iv) <u>H^1 Galerkin method</u>. This is a generalised Galerkin method based on the use of the inner product in H^1 rather than in L_2. It was first introduced by Krawchuk in 1932 using global basis functions and developed for finite element methods by de Boor (1966) and by Douglas, Dupont and Wheeler (1974). For the model problem given by (2.1) modified by (2.1a) the H^1 Galerkin method is given by

$$(LV,\psi_j) = (-h^2K,\psi_j) \qquad (j=0,1,---,n) \qquad (2.21)$$

where

$$V(x) = \sum_{i=0}^{n} \gamma_i \phi_i(x)$$

with

$$\psi_j = \phi_j'' . \qquad (j=0,1,---,n)$$

If we take the trial functions to be cubic splines given in Fig (4a), then the test functions are the piecewise linear

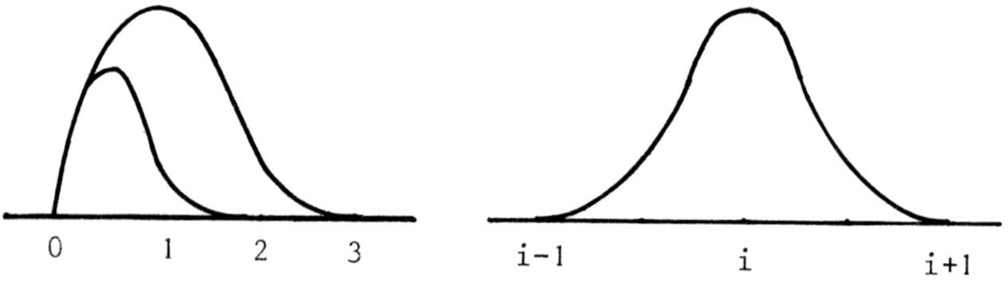

Fig 4(a)

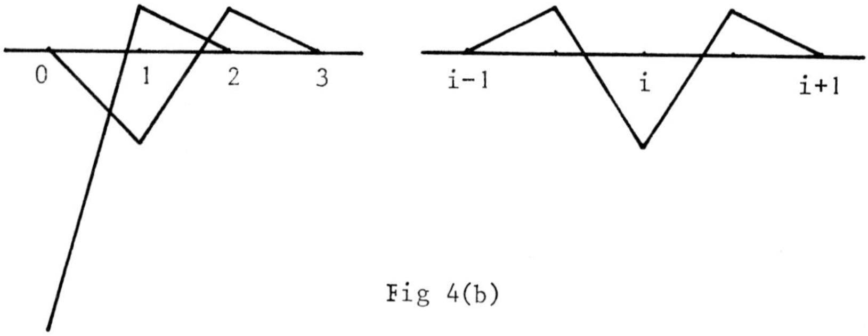

Fig 4(b)

functions shown in Fig (4b). It should be noted that only the trial functions are required to satisfy the homogeneous boundary conditions and at internal nodes

$$U_i = \frac{1}{4}(\gamma_{i+1} + 4\gamma_i + \gamma_{i-1}) \qquad (i=1,2,---,n-1)$$

The numerical implementation is fully explained in Christie (1978) and the numerical results for the model problem are

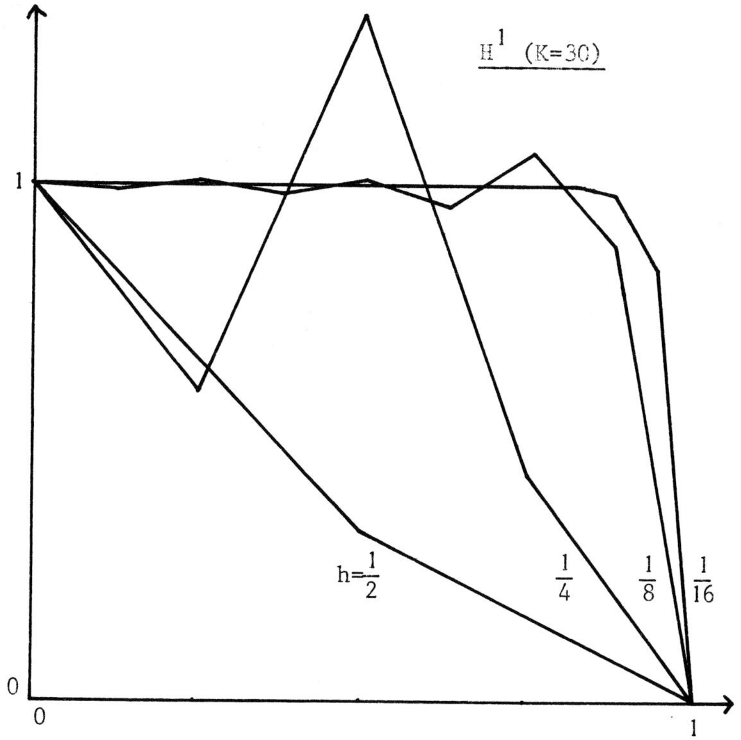

Fig 5

given in Fig 5 for K = 30 with h decreasing from $\frac{1}{2}$ to $\frac{1}{16}$.

(v) <u>H^{-1} Galerkin method</u>. This method was first proposed by Murray in 1943 for global basis functions and follows from the projection of the original differential system in the space H^{-1}. It was originally developed for finite element methods by Rachford and Wheeler (1974). For our model problem, the method is given by

$$(V, L^* \psi_j) = (-h^2 K, \psi_j) \qquad (j=0,1,---,n)$$

where

$$V(x) = \sum_{i=0}^{n} V_i \phi_i(x)$$

with

$$L^* \equiv \epsilon \frac{d^2}{dx^2} + hK \frac{d}{dx}$$

and

$$\phi_i = \psi_i'' . \qquad (i=0,1,---,n)$$

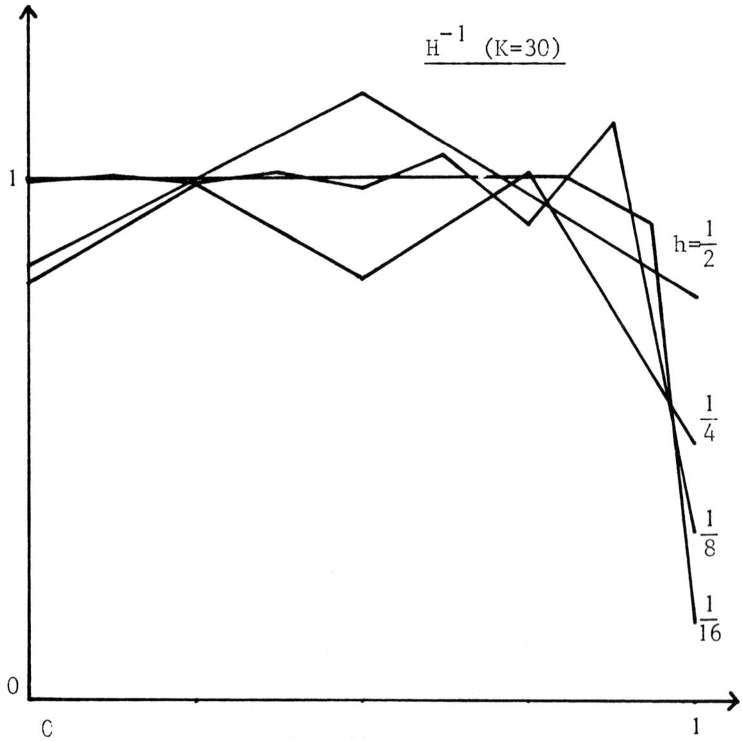

Fig 6

The homogeneous boundary conditions are required to be satisfied by the test functions ψ_j but not by the approximating function $V(x)$. This time we choose the test functions to be the cubic spline functions and so the trial functions are the piecewise linears, leading to

$$U_i = \frac{3}{2}(\delta_{i+1} - 2\delta_i + \delta_{i-1}) \qquad (i=1,2,---,(n-1))$$

Numerical details are given in Lawlor (1976) and Christie (1978), and the results are shown in Fig. 6 again for $K=30$ and decreasing h. Numerical results were also obtained for the H^{-1} Galerkin method using Hermite cubics as test functions, but these were inferior to the results in Fig. 6.

3. NUMERICAL RESULTS

The boundary value problem (2.1) is now solved numerically using finite element Galerkin methods, based on linear, quadratic, and cubic trial functions as described in (i), (ii), and (iii) of the previous section. The methods for which numerical results are given in Table 1 are numbered

1 ((2.3) with $\alpha = 1$)
2 ((2.3) with $\alpha = 0$)
3 (2.12)
4 (2.5)
5 (2.17)
6 (2.20)

All calculations have $\epsilon = 1.0$ and $h = 0.1$ and the theoretical solutions are given for comparison for $K = 100$ and 200 respectively. For the range of values of L covered by the numerical experiments it appears that methods 2 and 6 are rendered useless by the presence of severe oscillations in the solution, whereas method 4, which is good for L

Table 1

L = 5

x	1	2	3	4	5	6	Theoretical
0	-	-	-	-	-	-	-
.1	-	1.0441	-	-	-	-	-
.2	-	0.9779	-	0.9999	-	-	-
.3	-	1.0772	-	0.9998	-	-	-
.4	-	0.9283	-	0.9992	-	-	-
.5	-	1.1517	-	0.9976	-	-	-
.6	0.9999	0.8166	-	0.9916	-	0.9999	-
.7	0.9992	1.3192	-	0.9724	-	1.0009	-
.8	0.9917	0.5654	0.9997	0.9087	-	0.9908	-
.9	0.9091	1.6961	0.9836	0.6977	0.9956	1.0959	0.9999
1.0	0.0000	0.0000	0.0000	0.0000	0.0000	0.0000	0.0000

L = 10

x	1	2	3	4	5	6	Theoretical
0	-	-	-	-	-	-	-
.1	-	1.3451	-	0.9980	-	-	-
.2	-	0.9233	-	0.9942	-	0.9999	-
.3	-	1.4389	-	0.9874	-	1.0002	-
.4	-	0.8087	-	0.9751	-	0.9992	-
.5	-	1.5789	-	0.9526	-	1.0026	-
.6	-	0.6376	-	0.9115	-	0.9916	-
.7	0.9999	1.7881	-	0.8367	-	1.0279	-
.8	0.9977	0.3819	0.9999	0.7005	-	0.9081	-
.9	0.9524	2.1006	0.9954	0.4523	0.9993	1.3031	-
1.0	0.0000	0.0000	0.0000	0.0000	0.0000	0.0000	0.0000

- denotes 1.0000

values in the range $1 \leq L \leq 2$ cannot cope with the higher values of L. This leaves the three upstream based methods 1, 3 and 5 of which 1 (linear accuracy) gives poor results for $L = 5$.

For comparison it should be pointed out that the H^1 and H^{-1} Galerkin methods give exceedingly poor results for L values in the range 1 to 10. (See Figs. 5 and 6). This is mainly due to oscillations in the case of H^1 and to a large boundary error at $x = 1$ in the case of the H^{-1} method.

4. BURGERS' EQUATION

The equation we now examine is the non-linear parabolic equation

$$\frac{\partial u}{\partial t} = \epsilon \frac{\partial^2 u}{\partial x^2} - u \frac{\partial u}{\partial x}, \tag{4.1}$$

where ϵ the coefficient of viscosity is taken to be a parameter. This equation known as Burgers' equation has physical significance both in the theory of shock waves and in turbulence. Its mathematical properties have been studied by Cole (1951) and others. In the theory of shocks, (4.1) provides a model where the velocity u is transported by the fluid motion itself. The higher order term $\epsilon \frac{\partial^2 u}{\partial x^2}$, where ϵ is the viscosity, is the diffusion term, where the velocity is the quantity diffused. Burgers' equation is a good examples of the interaction between non linearity and viscosity.

The weak form of (4.1), where we have assumed homogeneous boundary conditions, is

$$\left(\frac{\partial u}{\partial t}, v\right) + \epsilon \left(\frac{\partial u}{\partial x}, \frac{\partial v}{\partial x}\right) + \left(u \frac{\partial u}{\partial x}, v\right) = 0 \tag{4.2}$$

where $u,v \in H$, an appropriate Hilbert Space. The semi discrete generalised Galerkin method for the solution of (4.2) is:

find

$$U(x,t) = \sum_{i=1}^{n-1} U_i(t)\phi_i(x) \tag{4.3}$$

so that

$$\sum_{i=1}^{n-1} [(\phi_i,\psi_j)\dot{U}_i + \epsilon(\phi_i',\psi_j')U_i] + (\sum_{i=1}^{n-1} U_i\phi_i \sum_{i=1}^{n-1} U_i\phi_i',\psi_j) = 0$$

$$(j=1,2,---,n-1) \tag{4.4}$$

where a dot denotes differentiation with respect to time and a dash with respect to x.

The non linear system of ordinary differential equations (4.4) is now solved for the problem with homogeneous boundary conditions

$$u(0,t) = u(1,t) = 0 \qquad t > 0$$

and the initial condition (4.5)

$$u(x,0) = \sin \Pi x \qquad 0 \leq x \leq 1 .$$

The exact solution can be obtained by transforming (4.1) into a linear diffusion problem. However due to the slow convergence of the Fourier series, the exact solution is not reliable for values of ϵ less than about 0.01. Details of the exact solution are given in Cole (1951).

We have been unable to find many numerical solutions based on finite differences and even fewer based on finite elements for the numerical solution of (4.1) coupled with (4.5). Ames (1978) in his recent text has quoted results due to Miller (1966) who employed both explicit and implicit finite difference methods, the latter depending on a predictor-corrector system. Unfortunately these methods employ a central difference replacement for $\frac{\partial u}{\partial x}$ and so severe oscillations in the numerical solution occur unless a very small grid spacing in x is employed. Frangakis (1978) uses a Galerkin method with linear "hat" trial and test functions which is also equivalent to a central difference for $\frac{\partial u}{\partial x}$.

Our aim is to solve (4.4) numerically for a variety of trial and test functions. In all calculations the grid spacing is $h = 0.1$, and the time step k is sufficiently small to make the integration in time exact to a desired number of decimal places. Also if $x = ih$ and $t = mk$, where i and m are integers, we use the central difference replacements in time in (4.4), viz.

$$\dot{U}_i = \frac{1}{k}(U_i^{m+1} - U_i^m)$$

and $\qquad\qquad (m=0,1,2,---)$

$$U_i = \frac{1}{2}(U_i^{m+1} + U_i^m) ,$$

where $i = 1,2,---,n-1$. The resulting non linear system of equations at each time level $(m+1)k$, $m = 0,1,2,---$ is solved by a standard Newton procedure, where convergence is assumed once an answer consistent to seven decimal places has been obtained.

We now quote some numerical results to four decimal places for equation (4.1) subject to initial and boundary conditions (4.5). The methods used are Galerkin with piecewise linears (method 2) and with piecewise quadratics (method 4). These are shown in Tables 2 and 3 for $\epsilon = 0.1$ and 0.01 respectively. Numerical solutions for $\epsilon = 0.001$ were also attempted by these two methods but the answers were

Table 2

$\epsilon = 0.1$

Method 2

x \ t	0.1	1.0
0	0.0000	0.0000
.1	0.2229 (-5)	0.0665 (+2)
.2	0.4348 (-10)	0.1315 (+3)
.3	0.6238 (-13)	0.1933 (+5)
.4	0.7764 (-13)	0.2487 (+7)
.5	0.8763 (-10)	0.2926 (+7)
.6	0.9040 (-3)	0.3163 (+2)
.7	0.8376 (+7)	0.3073 (-8)
.8	0.6586 (+13)	0.2517 (-20)
.9	0.3667 (+10)	0.1441 (-20)
1.0	0.0000	0.0000

Method 4

x \ t	0.1	1.0	3.0	5.0
0	0.0000	0.0000	0.0000	0.0000
.1	0.2234 (0)	0.0663	0.0117	0.0017
.2	0.4358 (0)	0.1312	0.0225	0.0033
.3	0.6251 (0)	0.1928	0.0314	0.0045
.4	0.7777 (0)	0.2480	0.0375	0.0053
.5	0.8772 (0)	0.2919	<u>0.0402</u>	<u>0.0056</u>
.6	<u>0.9042</u> (0)	<u>0.3161</u>	0.0390	0.0053
.7	0.8369 (0)	0.3081	0.0338	0.0045
.8	0.6573 (0)	0.2537	0.0249	0.0033
.9	0.3658 (0)	0.1461	0.0132	0.0017
1.0	0.0000	0.0000	0.0000	0.0000

Table 3

$\epsilon = 0.01$

Method 2

x \ t	0.1	1.0	1.0 (upwinded $\alpha = \frac{1}{3}R$)
0	0.0000	0.0000	0.0000
.1	0.2353 (-6)	0.0751 (-3)	0.0752
.2	0.4601 (-11)	0.1504 (-2)	0.1502
.3	0.6628 (-15)	0.2273 (+16)	0.2250
.4	0.8303 (-17)	0.2906 (-7)	0.2994
.5	0.9463	0.3965	0.3732
.6	0.9902	0.3945	0.4460
.7	0.9361	0.6083	0.5166
.8	0.7555	0.3996	0.5785
.9	0.4317	0.9284	0.5952
1.0	0.0000	0.0000	0.0000

Method 4

x \ t	0.1	1.0	5.0	10.0	20.0	50.0
0	0.0000	0.0000	0.0000	0.0000	0.0000	0.0000
.1	0.2358 (-1)	0.0754 (0)	0.0188	0.0097	0.0042	0.0003
.2	0.4611 (-1)	0.1506 (0)	0.0375	0.0193	0.0082	0.0005
.3	0.6641 (-2)	0.2257 (0)	0.0563	0.0288	0.0119	0.0007
.4	0.8316 (-4)	0.3003 (0)	0.0751	0.0381	0.0149	0.0008
.5	0.9471	0.3744	0.0938	0.0469	0.0169	0.0009
.6	0.9899	0.4476	0.1124	0.0541	0.0175	0.0008
.7	0.9341	0.5186	0.1299	0.0577	0.0162	0.0007
.8	0.7517	0.5793	0.1414	0.0531	0.0126	0.0005
.9	0.4283	0.5858	0.1192	0.0338	0.0070	0.0003
1.0	0.0000	0.0000	0.0000	0.0000	0.0000	0.0000

unreliable and are not quoted. In fact even for $\epsilon = 0.01$ it can be seen from Table 3 that a large oscillation has appeared at $t = 1.0$ in the numerical solution with piecewise linears. If, however, "upwinding" is introduced in the piecewise linear case according to the law $\alpha = \frac{1}{3}R$ where $R = \frac{U_\ell h}{2\epsilon}$, with U_ℓ the local value of the numerical solution, $h = 0.1$, and $\epsilon = 0.01$, the oscillation is removed, and the results are close to those obtained with piecewise quadratics (see Table 3). This empirical upwinding law is taken from the steady linear conduction convection problem studied by Christie, Griffiths, Mitchell and Zienkiewicz (1976).

The limited numerical results we have displayed in Tables 2 and 3 substantiate in every way the general picture presented by Cole (1951) who stated that "the initial sine wave shows after the first instant a tendency to develop a steep front near $x = 1$, if R is sufficiently large. After a while this steep front broadens and dies out until at the end only a sine wave remains. This sine wave has an amplitude which is smaller than that of the corresponding linear problem because of the increased dissipation over intermediate range of t".

ACKNOWLEDGMENT

The authors are indebted to the Science Research Council for support received from grant number GR/A 39004.

REFERENCES

Abramsson, L.R., Keller, H.B. and Kreiss, H.O. (1974). Difference approximations for singular perturbations of systems of ordinary differential equations. Numer. Math. 22, 367-391.

Ames, W.F. (1978). Numerical Methods for Partial Differential Equations. Second Edition. Nelson. London.

Axelsson, O. and Gustaffson, I. (1977). A modified upwind scheme for convective transport equations and the use of a conjugate gradient method for the solution of non-symmetric systems of equations. Report 77.12R, Computer Sciences, Göteborg, Sweden.

de Boor, C.R. (1966). The method of projections as applied to the numerical solution of two point boundary value problems using cubic splines. Ph.D. Thesis. University of Michigan.

Christie, I., Griffiths, D F., Mitchell, A.R., and Zienkiewicz, O.C. (1976). Finite element methods for second order differential equations with significant first derivatives. Int. J. Num. Meths. in Engng. 10, 1389-1396.

Christie, I. (1978). Finite element Galerkin methods for the conduction-convection problem. Ph.D. Thesis. University of Dundee.

Cole, J D. (1951). On a quasi-linear parabolic equation occurring in aerodynamics. Quart, of App. Maths. 9, 225-236.

Douglas, J., Dupont, T., and Wheeler, M.F. (1974). H^1 Galerkin methods for the Laplace and heat equations. Math. aspects of finite elements. (ed. C. de Boor) Academic Press, New York.

Frangakis, C.N. (1978). Numerical solution of Burgers' equation with finite element methods. (Private communication.)

Heinrich, J.C., Huyakorn, P.S., Zienkiewicz, O.C. and Mitchell, A.R. (1977). An upwind finite element scheme for the two dimensional convective transport equation. Int. J. Num. Meths. in Engng., 11, 131-143.

Heinrich, J.C. and Zienkiewicz, O.C. (1977). Quadratic finite element schemes for two dimensional convective-transport problems. Int. J. Num. Meths. in Engng. 11, 1831-1844.

Hemker, P.W. (1977). A numerical study of stiff two point boundary problems. Mathematisch Centrum, Amsterdam.

Il'in, A.M. (1969). Differencing scheme for a differential equation with a small parameter affecting the highest derivative. Math. Notes of the Head of Sciences of the USSR, 6, 596-602.

Lawlor, F.M.M. (1976). The Galerkin method and its generalisations. M.Sc. Thesis. University of Dundee.

Miller, E L. (1966). Predictor-corrector studies of Burgers' model of turbulent flow. Masters thesis, University of Delaware.

Miller, J.J.H. (1977). Some finite difference schemes for a singular perturbation problem. Proc. Int. Conf. on Constructive Function Theory, Sofia.

Mitchell, A R and Wait, R. (1977). The Finite Element Method in Partial Differential Equations. Wiley, London.

Mitchell, A.R. and Griffiths, D.F. (1977). Generalised Galerkin methods for second order equations with significant first derivative terms. Lecture Notes in Mathematics No.630 (ed. G.A. Watson) 90-104, Springer, Berlin.

Mitchell, A.R. and Griffiths, D.F. (1978). Semi-discrete Galerkin methods for time dependent conduction convection problems. Proc. Brunel Conference (ed. J. Whiteman) Academic Press. (To appear).

Rachford, H.H. and Wheeler, M.F. (1974). An H^{-1} Galerkin procedure for the two point boundary value problem. Math. aspects of finite elements, (ed. C. de Boor) Academic Press, New York.

Spalding, D.B. (1972). A novel finite difference formulation for differential expressions involving both first and second derivatives. Int. J. Num. Meths. in Engng. 4, 552-560.

Starius, G. (1978). Numerical treatment of boundary layers for perturbed hyperbolic equations. Dept of Computer Science Report No.69, University of Uppsala.

HIGHER ORDER SCHEMES OF POSITIVE TYPE FOR SINGULAR PERTURBATION PROBLEMS

M. van Veldhuizen
Wiskundig Seminarium V.U., De Boelelaan 1081, 1081 HV Amsterdam, The Netherlands

ABSTRACT

In this contribution we discuss three related topics. In the first and the second part we discuss methods for the model equation $\varepsilon u'' + a(t)u' = f(t)(a(t) \geq \alpha \geq 0$, with zero boundary conditions). In the last part we try to extend the ideas to elliptic problems of the singular perturbation type in two dimensions. We now give a more detailed survey of the three parts.

In the first place we consider the classical finite element method for the problem $\varepsilon u'' + a(t)u' = f(t)$, where $a(t) \geq \alpha > 0$. The finite element space consists of the piecewise quadratics, and we use Simpson's rule on each subinterval for the approximation of the integrals. We show that this full finite element method is not stable uniformly in ε. However, if we restrict ourselves to the values of the approximation at the nodes (i.e. the points t_i where two subintervals meet) then we do have a stable method uniformly in $\varepsilon \in (0, \varepsilon_0.]$ We also show that the order of accuracy of this method for smooth solutions is $p = 2$ uniformly in $\varepsilon \in (0, \varepsilon_0]$, and not $p = 4$. For related ideas, see Griffiths [3].

Second we use the ideas and techniques obtained by investivating the above scheme in the construction of a *second order* difference scheme of "positive type" (just as the first order upstream difference scheme) for $\varepsilon u'' + a(t)u' = f(t)$. We discuss this scheme. We also give some numerical examples.

In the third place we extend these ideas to schemes for $\varepsilon\Delta u + au_x + bu_y = f$, Δ the Laplacian. We are interested in second order schemes of "positive type". In general the result is negative: such schemes do not exist. However in some cases such schemes do exist. These special cases might be of practical interest in solving numerically a convective-diffusion equation with large convective term. We shall indicate how.

APPROXIMATION BY PIECEWISE QUADRATICS

In this section we consider the boundary value problem

$$\varepsilon u'' + a(t)u' = f(t) \qquad t \in (0,1) \tag{2.1}$$
$$u(0) = u(1) = 0.$$

Here $\varepsilon > 0$ is a small parameter, and a and f are smooth functions on [0,1]. We shall assume $a(t) \geq \alpha > 0$ for all $t \in (0,1)$. I.e. no turning points.

This problem is solved approximately by a finite element method. To this end consider a partitioning Δ of [0,1] consisting of the N+1 nodes

$$0 = t_0 < t_1 < \ldots < t_{N-1} < t_N = 1.$$

The finite element space E_Δ is chosen as the set of functions $u_\Delta \in H_0^1(0,1)$ whose restriction to (t_i, t_{i+1}) is a polynomial of degree at most two, $i = 0,1, \ldots ,N-1$. The condition $u_\Delta \in H_0^1(0,1)$ implies $u_\Delta(0) = u_\Delta(1) = 0$. I.e. $u_\Delta \in E_\Delta$ satisfies the boundary conditions for the problem (2.1). The finite element equations are now given by

$$-\varepsilon(u'_\Delta,\ v'_\Delta) + (au'_\Delta,\ v_\Delta) = (f,v_\Delta) \qquad \forall\ v_\Delta \in E_\Delta, \tag{2.2}$$

with unknown $u_\Delta \in E_\Delta$. Here (f,g) stands for the $L^2(0,1)$ inner product,

$$(f,g) = \int_0^1 f(s)g(s)\ ds.$$

Instead of this continuous inner product we want to use a discrete inner product. This discrete inner product is obtained by approximating an integral over the subinterval $(t_i,\ t_{i+1})$ by Simpson's rule applied to this subinterval, $i = 0,1,\ \dots\ ,N-1$. In this way we obtain the discrete inner product $\langle f,g\rangle_\Delta$,

$$\langle f,g\rangle_\Delta = \sum_{i=0}^{N-1} \frac{h_{i+1}}{6}[f(t_i+0)g(t_i+0) + 4f(t_{i+\frac{1}{2}})g(t_{ir\frac{1}{2}}) + + f(t_{i+1}-0)g(t_{i+1}-0)], \tag{2.3}$$

where $h_{i+1} = t_{i+1} - t_i$, $t_{i+\frac{1}{2}} = \frac{1}{2}(t_{i+1} + t_i)$.

Now we consider the finite element method with numerical quadrature:

$$-\varepsilon\langle u'_\Delta,\ v'_\Delta\rangle_\Delta + \langle au'_\Delta,\ v_\Delta\rangle_\Delta = \langle f,v_\Delta\rangle_\Delta \qquad \forall\ v_\Delta \in E_\Delta. \tag{2.4}$$

This is the discretization we want to investigate. In particular we want to do so without restrictions of the type $\max h_i/\varepsilon$ sufficiently small.

Clearly the discretization (2.4) does not depend on the sign of $a(t)$ on a subinterval. Therefore we call the method (2.4) a *symmetric* method. An asymmetric method is a method like the upstream different scheme and the finite element methods with Radau quadrature, cf Van Veldhuizen (1978). In case of a system of differential equations the use of asymmetric methods is usually a cumbersome process (in the stiff case, of course). For these situations the availability of good symmetric methods is desirable.

Now let us start the investigation of the method (2.4). We want to do so without a restriction of the type $\max_i h_i/\varepsilon$ small. Indeed, we want results for $\varepsilon \in (0,\varepsilon_0]$ and grids Δ unrelated to ε (with one exception: in the boundary-layer an adapted grid might be necessary).

Our arguments are basis-dependent. I.e. we first need a convenient basis in E_Δ. A function $e_\Delta \in E_\Delta$ is a basis element of E_Δ if either e_Δ is a roof function, or e_Δ is given by $(i = 0,1,2, \ldots ,N-1)$

$$\begin{cases} 0, & t \notin (t_i, t_{i+1}) \\ \dfrac{4(t - t_i)(t_{i+1} - t)}{h_{i+1}^2}, & t \in (t_i, t_{i+1}) \end{cases} \tag{2.5}$$

Here $h_{i+1} = t_{i+1} - t_i$. Any $u_\Delta \in E_\Delta$ is written with respect to this basis as a $(2N-1)$-vector

$$u_\Delta \leftrightarrow (u_{\frac{1}{2}}, u_1, \ldots , u_{N-1}, u_{N-\frac{1}{2}})^T \tag{2.6}$$

For integer i we have $u_i = u_\Delta(t_i)$. However we do not have a nodal finite element space. This follows from the relation

$$u_\Delta\left(\frac{t_i + t_{i+1}}{2}\right) = u_{i+\frac{1}{2}} + \frac{u_i + u_{i+1}}{2}, \quad i = 0,1, \ldots , N-1 \tag{2.7}$$

Nevertheless the technique of computing element matrices

cf. Strang and Fix still applies (1973).

We use the notation $a_i = a(t_i)$, $a_{i+\frac{1}{2}} = a(\frac{t_i + t_{i+1}}{2})$, etc. This notation resembles the notation for the coordinates of u_Δ. In this notation and with respect to the basis in E_Δ as described above we find as the element matrix corresponding to the subinterval (t_i, t_{i+1}) the matrix (with $h_{i+1} = t_{i+1} - t_i$)

$$-\frac{\varepsilon}{h_{i+1}}\begin{pmatrix} 1 & 0 & -1 \\ 0 & \frac{16}{3} & 0 \\ -1 & 0 & 1 \end{pmatrix} + \frac{1}{6}\begin{pmatrix} -(a_i + 2a_{i+\frac{1}{2}}) & 4a_i & (a_i + 2a_{i+\frac{1}{2}}) \\ -4a_{i+\frac{1}{2}} & 0 & 4a_{i+\frac{1}{2}} \\ -(2a_{i+\frac{1}{2}} + a_{i+1}) & -4a_{i+1} & (2a_{i+\frac{1}{2}} + a_{i+1}) \end{pmatrix}. \tag{2.8}$$

The contribution of the inhomogeneous term f through $\langle f,v\rangle_\Delta$ on the subinterval (t_i, t_{i+1}) is given by the vector

$$\frac{h_{i+1}}{6}(f_i + 2f_{i+\frac{1}{2}}, 4f_{i+\frac{1}{2}}, 2f_{i+\frac{1}{2}} + f_{i+1})^T. \tag{2.9}$$

Now we apply static condensation, i.e. we eliminate the unknown $u_{i+\frac{1}{2}}$ with fractional index. Then we take together the condensed element matrices in order to obtain the full discretization on [0,1]. The result is a kind of difference scheme for the unknowns $u_0 \overset{=}{D} 0, u_1, u_2, \ldots, u_{n-1}$, and $u_N \overset{=}{D} 0$. This scheme is given by ($i = 1,2, \ldots, N-1$)

$$[\varepsilon \frac{u_{i+1} - u_i}{h_{i+1}} - \varepsilon \frac{u_i - u_{i-1}}{h_i}] +$$

$$+ \frac{1}{6} [(2a_{i-\frac{1}{2}} + a_i)(u_i - u_{i-1}) + (2a_{i+\frac{1}{2}} + a_i)(u_{i+1} - u_i)] +$$

$$+ \frac{a_i}{12\varepsilon} [h_{i+1} a_{i+\frac{1}{2}} (u_{i+1} - u_i) - h_i a_{i-\frac{1}{2}} (u_i - u_{i-1})]$$

$$= \frac{1}{6}[(h_i + h_{i+1}) f_i + 2h_i f_{i-\frac{1}{2}} + 2h_{i+1} f_{i+\frac{1}{2}}] +$$

$$+ \frac{a_i}{12\varepsilon} [h_{i+1}^2 f_{i+\frac{1}{2}} - h_i^2 f_{i-\frac{1}{2}}]. \tag{2.10}$$

This equation is the subject of our investigations. However, we shall make some simplifying assumptions. We do this in order to avoid as much technicalities as possible. We now describe the simplifying assumptions and we comment on them.

Simplifications

In the remainder of this section we assume

(i) $t_i = i/N$, i.e. an equidistant grid with stepsize $h = 1/N$.
(ii) $a(t) = a > 0$, a constant.

The first assumption is not very realistic. Usually the solution has a boundary-layer near t=0. One may expect that a symmetric method as considered here will not do very well unless a sufficiently small stepsize is used in the boundary layer region. In fact, this is so for the method (2.10). Hence our only justification for this simplification is of a didactical character: without this condition the essential idea is obscured by many technicalities. Also, once the idea has been explained, it is easily seen that the same idea works for *any* grid Δ, and for all $\varepsilon \in (0,\varepsilon_0]$.

The same argument applies in justifying the assumption $a(t) = a > 0$. In fact, the only essential condition is

$a(t) \geq \alpha > 0$ for all $t \in [0,1]$. I.e. no turning points.

Adopting the above simplifications the scheme (2.10) may be written as

$$\left(\frac{\varepsilon}{h} - \frac{a}{2} + \frac{a^2h}{12\varepsilon}\right) u_{i-1} - 2\left(\frac{\varepsilon}{h} + \frac{a^2h}{12\varepsilon}\right) u_i + \left(\frac{\varepsilon}{h} + \frac{a}{2} + \frac{a^2h}{12\varepsilon}\right) u_{i+1} =$$

$$= \frac{h}{3}(f_i + f_{i+\frac{1}{2}} + f_{i-\frac{1}{2}}) + \frac{h^2a}{12\varepsilon} (f_{i+\frac{1}{2}} - f_{i-\frac{1}{2}}). \tag{2.11}$$

Observe that the equation (2.11) consists of linear combinations of central difference operators and suitable right-hand sides. This again is a reason to call it a symmetric method. Put

$$\mathcal{u}_N = (u_1, u_2, \ldots, u_{N-1})^T. \tag{2.12}$$

Since (2.11) holds for $i = 1,2, \ldots, N-1$ we have $N-1$ linear equations in the $N-1$ unknowns $u_1, u_2, \ldots, u_{N-1}$. The matrix of this set of linear equations will be called A_N.

Lemma 2.1. For all $\varepsilon > 0$ and all $h = 1/N$ the matrix A_N is an irreducible and weakly diagonally dominant matrix.

Proof. Clearly, A_N is a tridiagonal matrix. Hence, A_N is irreducible if its codiagonals contain non-zero elements only. The codiagonals contain $\frac{\varepsilon}{h} \pm \frac{a}{2} + \frac{a^2h}{12\varepsilon}$. It is easily seen that these two expressions do not vanish for all $\varepsilon > 0$, $h > 0$, and $a \in \mathbb{R}$. Hence A_N is irreducible.

Since $\frac{\varepsilon}{h} \pm \frac{a}{2} + \frac{a^2h}{12}$ do not vanish for all $\varepsilon > 0$, $h > 0$ and $a \in \mathbb{R}$ these expressions are of constant sign. Then obviously,

$$\frac{\varepsilon}{h} \pm \frac{a}{2} + \frac{a^2h}{12\varepsilon} > 0.$$

Now consider a row of A_N, but not the first row or the last one. Then the sum of the two off-diagonal elements equals the modulus of the diagonal element. Only in the first row and the last one the modulus of the diagonal element dominates the one off-diagonal element. This proves the weakly diagonal dominance of A_N. □

Corollary 2.2. For all $\varepsilon > 0$, all $h = 1/N$, $N \in \mathbb{N}$, and all $a \in \mathbb{R}$ the inverse A_N^{-1} exists and $A_N^{-1} < 0$.

Observe that $A_N^{-1} \leq 0$ (i.e. each of the coefficients is not positive) follows by standard arguments. The somewhat stronger conclusion $A_N^{-1} < 0$ is a result of the tridiagonal form of A_N.

Henceforth we shall assume $a > 0$. This is not a restriction since all results depend on the above Corollary 2.2., and this corollary holds true for all $a \in \mathbb{R}$. However, in the construction of barriers we have to distinguish between $a > 0$ and $a < 0$ (for technical reasons only). So we adopt $a > 0$, without any restriction.

Now we construct a *barrier* (comparison function).

Put

$$z(t) = \frac{1}{2a}(t-1)(t+3-\frac{2\varepsilon}{a}). \tag{2.13}$$

Then

$$\varepsilon z''(t) + az'(t) = 1 + t. \tag{2.14}$$

Also $z(t) \leq 0$ for $t \in [0,1]$, $z(1) = 0$. Hence z is a good barrier for the differential equation. Since z is a quadratic polynomial we expect that the vector $z_N = (z_1, \ldots, z_{N-1})^T$, $z_i = z(t_i)$, is a good barrier for the matrix A_N. Indeed, this is so, because with $z_i = z(t_i) = z(ih)$ we have for $i = 1, \ldots, N-1$

$$(\frac{\varepsilon}{h} - \frac{a}{2} + \frac{a^2h}{12\varepsilon})\ z_{i-1} - 2(\frac{\varepsilon}{h} + \frac{a^2h}{12\varepsilon})\ z_i + (\frac{\varepsilon}{h} + \frac{a}{2} + \frac{a^2h}{12\varepsilon})\ z_{i+1} =$$

$$= h(1 + t_i) + \frac{h^3a}{12\varepsilon} \tag{2.15}$$

By standard arguments we now obtain from Corollary 2.2. and (2.15) the following fundamental result.

Theorem 2.3. Assume

(i) $\varepsilon \in (0, \frac{3}{2}a)$ and $a > 0$.

(ii) $A_N u_N = hv_N + \frac{h^3a}{12\varepsilon} w_N$.

(iii) For $i = 1,2, \ldots ,N-1$ we have $|v_i| \leq \alpha(1 + ih)$, $|w_i| \leq \alpha$.

Then $|u_i| \leq \alpha|z(ih)|$ for $i = 1,2, \ldots ,N-1$.

The new feature of the above theorem is the *comparison on two levels* of the right-hand side of the given equations with the right-hand side resulting from the barrier z_N. These two levels are determined by ε, but not uniquely. A given right-hand side is to be interpreted as $hv_N + \frac{h^3a}{12\varepsilon} w_N$, and in doing so there is some freedom of choice. Of course, one should try to find an optimum distribution over v_N and w_N.

It is obvious that the simplifications (i.e. $a(t) = a$ constant and constant stepsize $h = 1/N$) are not essential in obtaining the result of Theorem 2.3. In fact, an analogue of Theorem 2.3 may be established under more general conditions.

Now we must face one of the consequences of the simplifications. We want to estimate the *global error* of the method (2.11). However, because of the constant stepsize we can do this with a realistic result only in case of a smooth solution, i.e. we must assume the absence of a boundary-layer. Of course, it is possible to include a solution with a boundary-layer, but then we should not use an equidistant grid. The point is this: The stability result Theorem 2.3

remains true in case of a varying stepsize, without any condition on the stepsize variation. Hence we have no problems with the stability. The other source of possible problems is the local error. However, by suitable gradation of the grid we can control this local error. So we are able to estimate the global error, not only in case of an equidistant grid, but in all situations, including a boundary-layer. However, these general situations require much more computations, e.g. in obtaining the local error. Therefore we restrict ourselves to equidistant grids and smooth solutions u. We assert that the more realistic situations can be dealt with similarly, at the expense of some tedious computations.

It follows from this sketch that we are interested in the local error of the scheme (2.11) for a smooth solution u. We say that u is a smooth solution if u and its derivatives up to order seven are bounded on [0,1] uniformly in $\varepsilon \in (0,\varepsilon_0]$. This excludes boundary-layers. Now with u a smooth solution we replace u_i in (2.11) by $u(t_i)$ and we compute the residual $h\delta_i(h)$, $\delta_i(h)$ the local error. We find

$$\begin{aligned}\delta_i(h) = {} & \varepsilon[u'' + \frac{h^2}{12}u^{(4)} + O(h^4)] + [au' + \frac{h^2}{6}au''' + O(h^4)] + \\ & + \frac{a^2h^2}{12\varepsilon}[u'' + \frac{h^2}{12}u^{(4)} + O(h^4)] + \\ & - [f + \frac{h^2}{12}f'' + O(h^4)] - \frac{ah^2}{12}[f' + \frac{h^2}{24}f''' + O(h^4)].\end{aligned} \tag{2.16}$$

All evaluations of u', u'', f, f' etc. should take place at $t = t_i$. We use the $O(h^p)$-symbol as follows: $g(h) = O(h^p)$ if $|g(h)| \le C.h^p$ for all $h \in (0,h_o]$ with C and h_o independent of ε. I.e. we use very essentially the smoothness of u, and the absence of a boundary-layer. We simplify the expression for $\delta_i(h)$ by using the three equalities:

$$\varepsilon u'' + au' = f$$
$$\varepsilon u''' + au'' = f'$$
$$\varepsilon u^{(4)} + au''' = f''.$$

The result is (with evaluations at $t = t_i$):

$$\delta_i(h) = \varepsilon O(h^4) + O(h^4) + \frac{ah^2}{12\varepsilon}\left[-\frac{h^2}{24} f''' + \frac{h^2}{12} au^{(4)} + O(h^4)\right]. \tag{2.17}$$

The expression between [ ] is $O(h^2)$. Now apply Theorem 2.3. We want results uniformly in $\varepsilon \in (0,\varepsilon_0]$. So we choose for w_N the vector corresponding to the expression between [ ] in (2.17). Then we find the following error estimate.

Theorem 2.4. The method (2.11) is a method of order $p = 2$ *uniformly* in $\varepsilon \in (0,\varepsilon_0]$.

In fact we only show that $p \geq 2$, p = order of accuracy. However for $u(t) = t^3(t-1)/24$ we can compute the global error explicitly. This confirms $p = 2$.

Remark 2.5. In the classical situation, i.e. ε kept fixed, we have $p = 4$. This is the celebrated superconvergence result of Douglas and Dupont (1974). Consequently, even in case of a problem with constant coefficients there is a difference between the order of accuracy for $h \to 0$, ε kept fixed and the order of accuracy for $h \to 0$ uniformly in $\varepsilon \in (0,\varepsilon_0]$.

Remark 2.6. Some experimental evidence for $p = 2$ (and not $p = 4$) uniformly in $\varepsilon \in (0,\varepsilon_0]$ is given in Hemker (1977). In particular, the results for the LOB2-method in Table 3.7.2 in Hemker (1977) are quite convincing.

At this moment we know that the finite element method (2.4) has a unique solution u_Δ for all $\varepsilon \in (0,\varepsilon_0]$ and all grids Δ (only with a proof in a simplified situation). We also have $u_\Delta(t_i) - u(t_i) = O(h^2)$ for smooth solutions u. Now we consider the error between the nodes. For $t \in (t_i,t_{i+1})$ we then have to take into account the coordinate $u_{i+\frac{1}{2}}$, given by

$$u_{i+\frac{1}{2}} = \frac{h^2}{8\varepsilon}\left(a\,\frac{u_{i+1} - u_i}{h} - f_{i+\frac{1}{2}}\right)$$

Hence, $|u_{i+\frac{1}{2}}|$ is bounded uniformly in $\varepsilon \in (0,\varepsilon_0]$ if and only if

$$a\,\frac{i_{i+1} - u_i}{h} - f_{i+\frac{1}{2}} = O(\varepsilon), \qquad \varepsilon \to 0$$

In general we may not expect this behaviour. Hence in general $|u_{i+\frac{1}{2}}| \to \infty$ for $\varepsilon \to 0$.

Conclusion 2.7. The full finite element method (2.4) is not stable uniformly in $\varepsilon \in (0,\varepsilon_0]$, although the condensed method (2.11) (or rather (2.10)) is.

Let us mention one final result. We modify the inner product $\langle .,. \rangle_\Delta$, cf. (2.3), by using two point Gaussian quadrature instead of Simpson's rule. This yields a slightly modified method (2.10). The ideas of this section apply to this modified method as well. For this modified method, with Gaussian points, we find $u(t_i) - u_i = O(h^4)$, uniformly in $\varepsilon \in (0,\varepsilon_0]$ and for smooth u. Hence, with Gaussian points we do not loose the superconvergence, cf. Theorem 2.4.

The method (2.10) is also considered in Griffith's [3]. See also Mitchell and Christie [5].

A SECOND ORDER METHOD OF POSITIVE TYPE

As an introduction to this section let us consider the boundary value problem

$$\varepsilon u'' + a(t)u' = f(t), \qquad t \in (0,1) \tag{3.1}$$
$$u(0) = u(1) = 0$$

where $a(t) \geq \alpha > 0$, $t \in [0,1]$. The functions a and f are smooth. Then the solution u exists and is unique. It also has a nice monotonicity property: u(t) decreases if we increase the function f, and this is so for all $t \in (0,1)$.

We want to construct a difference scheme of second order with this monotonicity property. First, let us consider the upstream difference scheme for (3.1) on a grid with constant stepsize $h = \frac{1}{N}$. Then we obtain the scheme:

$$\varepsilon \frac{u_{i+1} - 2u_i + u_{i-1}}{h^2} + a_i \frac{u_{i+1} - u_i}{h} = f_i, \quad i = 1,2, \ldots ,N-1 \tag{3.2}$$

In matrix form we may write it as $B_N u_N = f_N$, where B_N is a tridiagonal matrix with $B_N^{-1} < 0$. The vector f_N is given by

$$f_N = (f_1, \ldots ,f_{N-1})^T.$$

Hence $u_N = B_N^{-1} f_N$ depends on f_N in the desired way: if we increase f_N (i.e. we increase at least one of the coordinates of f_N without decreasing one) then u_N decreases. We shall say that the scheme (3.2) is of positive type. This is so for all $\varepsilon > 0$ and all $h = 1/N$, provided $a_i > 0$ for all i. Observe that the scheme (3.2) is of first order of accuracy only.

The condensed method (2.11) is a second order method. However, this method is *not* of positive type for all $\varepsilon > 0$ and all $h = 1/N$. The proof is rather technical: it is based on the occurrence of $f_{i+\frac{1}{2}} - f_{i-\frac{1}{2}}$ in the right-hand side of (2.11). In the proof one needs to consider the interaction between A_N^{-1} and this right-hand side. This requires tedious computations. We just mention the result: for h/ε large the scheme (2.11) is not of positive type.

The purpose of this section is the construction of a

second order scheme of positive type for the problem (3.1). These properties (second order, positive type) should hold uniformly in $\varepsilon \in (0,\varepsilon_o]$.

Observe that the scheme (2.11) may be obtained from the central difference scheme by adding some suitable difference operators (a second order one in the left-hand side, and a first order one in the right-hand side). This is the technique we shall adopt in relation to the scheme (3.2). The choice of what is added to this scheme is made by considering the local error. For easier explanation we restrict ourselves to constant stepsizes $h = 1/N$.

The local error of the scheme (3.2) equals the residual after substitution of $u_i = u(t_i)$, where u is a smooth solution as in section 2. Then we find for the scheme (3.2):

$$\delta_i(h) = a_i \frac{h}{2} u''(t_i) + O(h^2) \tag{3.3}$$

Now we add to the scheme a suitable discretization for $\frac{a_i h}{2} \{\frac{au' - f}{\varepsilon}\}$. We try to do this in such a way that the $a_i \frac{h}{2} u''(t_i)$ - term from $\delta_i(h)$ in (3.3) and this discretization together reduce the global error. No reduction of the local error is expected. In fact, we add an $O(1/\varepsilon)$ - level to the scheme. So we expect an increase in the local error. But by a two-level technique as in Theorem 2.3 this might result in a change of order of accuracy from $p = 1$ to $p = 2$ for the global error. We propose the scheme (i.e. scheme (3.2) plus additional discretization)

$$\varepsilon \frac{u_{i+1} - 2u_i + u_{i-1}}{h^2} + a_i \frac{u_{i+1} - u_i}{h} + a_i \frac{a_{i+\frac{1}{2}}}{2\varepsilon} (u_{i+1} - u_i) =$$

$$= f_i + a_i \frac{h}{2\varepsilon} f_{i+\frac{1}{2}} \tag{3.4}$$

One may replace $f_{i+\frac{1}{2}}$ and $a_{i+\frac{1}{2}}$ by $(f_i + f_{i+1})/2$ and $(a_i + a_{i+1})/2$ respectively. The theory for this slightly modified scheme is similar and gives similar results. If the model (3.1) should be interpreted as the Frechet derivative of a non-linear problem (Newton's method), then the use of fractional indices might be problematic. Then the slightly modified scheme seems appropriate. Henceforth we restrict ourselves to the scheme (3.4). We have the following two results.

Theorem 3.1. Assume $a(t) \geq \alpha > 0$ for all $t \in [0,1]$. Then the scheme (3.4) is of positive type for all $\varepsilon \in (0,\varepsilon_o]$ and all $h = 1/N$.

Theorem 3.2. In addition to $a(t) \geq \alpha > 0$, let u be a smooth solution of problem (3.1). Then the global error $u_i - u(t_i)$ is of order $O(h^2)$, $h \to 0$, uniformly in $\varepsilon \in (0,\varepsilon_o]$.

These results correspond to Theorem 2.3 and Theorem 2.4 respectively. We do not give the proofs of the above theorems because of the great similarity with the proofs of the corresponding results from section 2. The scheme (3.4) is closely related to the scheme

$$\frac{\varepsilon}{1 + \frac{h}{2\varepsilon} a_i} \frac{u_{i+1} - 2u_i + u_{i-1}}{h^2} + a_i \frac{u_{i+1} - u_i}{h} = f_i \qquad (3.5)$$

This scheme may be found in Axelsson and Gustafsson (1977). The scheme is of positive type for all ε and all h provided that $a(t) > 0$ for all t. The order of accuracy of the scheme is $p = 2$. However, uniformly in $\varepsilon \in (0,\varepsilon_o]$ the order of accuracy for a smooth solution u equals $p = 1$ and not $p = 2$. This is easily seen by considering the problem $\varepsilon u'' + u' = f$ with smooth solution $u(t) = t^2$; in this case an explicit expression for the global error may be derived from which one obtaine $p = 1$ and not $p = 2$ uniformly in $\varepsilon \in (0,\varepsilon_o]$.

The scheme (3.4) has another interesting property. Consider the differential equation $\varepsilon u'' + u' = 0$ (i.e. $a(t) = 1$ and $f(t) = 0$ for all $t \in [0,1]$). The scheme (3.4) applied to this differential equation yields a recursion with constant coefficients. It is easily seen that any solution of this recursion is given by

$$u_i = \alpha + \beta . q(\frac{h}{\varepsilon})^i \tag{3.6}$$

where $\alpha, \beta \in \mathbb{R}$ are constants, and where q is a function given by

$$q(z) = \frac{1}{1 + z + z^2/2} \tag{3.7}$$

It is also easily seen that the central difference scheme, the method (2.11) and the method (3.2) when applied to $\varepsilon u'' + u' = 0$ all yield a recursion any solution of which is of the form (3.6). The function q will depend on the scheme. We have the following table:

scheme	q(z)
central difference scheme	$\frac{1 - z/2}{1 + z/2}$
scheme (2.11)	$\frac{1 - z/2 + z^2/12}{1 + z/2 + z^2/12}$
scheme (3.2)	$\frac{1}{1 + z}$
scheme (3.4), (3.5)	$\frac{1}{1 + z + z^2/2}$

From these functions q(z) as listed above it is obvious that the two symmetric schemes, i.e. the central difference scheme (2.11), are very sensitive to a proper gradation of the grid in the boundary-layer. The first-order scheme (3.2), and even more so the second order scheme (3.4) are not very sensitive to a proper gradation in the boundary-layer. In fact, these last two schemes might be used on a course grid, without resolving the boundary-layer. This is a desirable property if we want to generalise our ideas to problems in two or more dimensions.

The $q(z)$ listed above are entries in the Padé table for $\exp(-z)$. See Mitchell and Christie (1979), and Van Veldhuizen (1978) for schemes that correspond to other Padé approximations.

In the table below we give some results for the scheme (3.5), and the upstream difference scheme. We use an equidistant grid. In the table we list the maximum error at the nodes, i.e. with $h = 1/N$.

$$E_\Delta = \max_{1 \le i \le N-1} |u(ih) - u_i|$$

The problem is

$$\varepsilon u'' + \left[\frac{2\varepsilon}{1+t} + \frac{2}{(1+t)^2}\right] u' = f(t)$$

with exact solution

$$u(t) = \cos\left(\frac{\pi s}{2}\right) + \frac{\exp(-1/\varepsilon) - \exp(-s/\varepsilon)}{1 - \exp(-1/\varepsilon)}, \quad s = \frac{2t}{1+t}$$

For $\varepsilon = 10^{-5}$ the results are given in the table.

The results for the new method (3.4) are better than the results for the other two methods. The method (3.5) behaves like the upstream difference scheme (3.2) for h/ε large. If h/ε decreases, then the performance of the method (3.5) improves relative to the upstream difference scheme.

However, even for moderate h/ε the results of the method (3.5) are not as good as the results for the new method (3.4).

N	method (3.4)	method (3.5)	method (3.2)
20	7.6 E - 4	1.1 E - 2	1.1 E - 2
40	2.3 E - 4	7.3 E - 3	6.2 E - 3
80	6.3 E - 5	4.3 E - 3	3.6 E - 3
160	1.6 E - 5	2.3 E - 3	2.0 E - 3
320	4.1 E - 6	1.2 E - 3	1.1 E - 3
640	2.0 E - 5	6.2 E - 4	2.6 E - 3
1280	8.1 E - 5	3.8 E - 4	6.1 E - 3

A TWO-DIMENSIONAL PROBLEM

Let $L = \frac{\partial^2}{\partial x^2} + \frac{\partial^2}{\partial y^2}$ denote the Laplacian. In this section we are interested in difference schemes of positive type and of order $p \geq 2$ for problems of the type

$$\begin{aligned} \varepsilon Lu + a\frac{\partial u}{\partial x} + b\frac{\partial u}{\partial y} &= f \qquad (x,y) \in \Omega \\ u &= 0 \qquad (x,y) \in \delta\Omega \end{aligned} \tag{4.1}$$

Here Ω is a domain in $\mathbb{R}^2$. A problem of the type (4.1) is a generalization to dimension two of the one-dimensional problem (3.1).

We want to construct difference schemes on a grid in Ω such that for all $\varepsilon \in (0,\varepsilon_0]$ the method is of order of accuracy $p \geq 2$ and such that the method is of positive type. We shall attempt the construction of such methods by generalising the idea of Section 3.

For sake of simplicity let Ω be a square with sides parallel to the coordinate axes. The points $(ph, qh) \in \mathbb{R}^2$ are the gridpoints, p,q integer. The meshwidth h should be such that Nh is the length of a side of the square Ω. For P a gridpoint, we have the four neighbours N,E,S and W, see Fig. 1. The value of a gridfunction u at P is denoted by u_P; the value of a function from Ω to $\mathbb{R}$ at P is denoted by u(P).

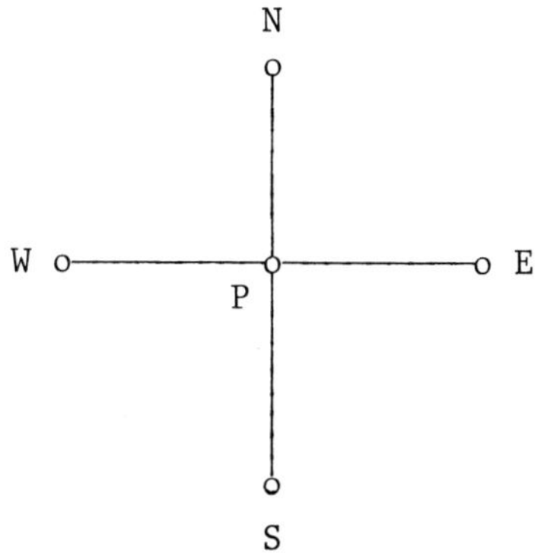

Fig. 1. *The point P and four neighbours.*

As in Section 3 we start with the upstream difference scheme. Assume a > 0, b > 0 and a and b are constants, cf. (4.1). Then the upstream difference scheme is given by

$$\frac{\varepsilon}{h^2}[u_N + u_E + u_S + u_W - 4u_P] + \frac{a}{h}[u_E - u_P] + \frac{b}{h}[u_N - u_P] = f(P). \tag{4.2}$$

For smooth solutions u the local error of this scheme is given by

$$\delta_P(h) = \varepsilon O(h^2) + \frac{ah}{2}u_{xx}(P) + \frac{bh}{2}u_{yy}(P) + O(h^2). \tag{4.3}$$

As in Section 3, we want to add difference quotients to (4.2) such that the global error goes from O(h) for the scheme (4.2) to $O(h^2)$ for the new scheme. The idea of Section 3 is the introduction of an $O(\frac{1}{\varepsilon})$-level in formula (4.2). In view of the form of the local error (4.3) it is obvious to approximate on the $O(\frac{1}{\varepsilon})$-level the problem

$$\frac{a+b}{2\varepsilon}\, h\, [\varepsilon Lu + au_x + bu_y - f] = 0. \tag{4.4}$$

The O(h)-terms in the local error (4.3) should be considered part of the discretization for (4.4). Consequently, we should add difference quotients to the left-hand side of (4.2) which approximate in a discrete sense the differential operator

$$\frac{bh}{2}\, u_{xx} + \frac{ah}{2}\, u_{yy} + \frac{a+b}{2\varepsilon}\, h\, [au_x + bu_y],$$

and we should add to the right-hand side of (4.2) a suitable approximation to $\frac{a+b}{2\varepsilon}$ hf. Since we want a scheme of positive type, we shall require that the left-hand side of the new scheme is of the form

$$\alpha_P u_P + \sum_{Q \text{ neighbour of } P} \beta_{P,Q} u_Q \tag{4.5}$$

with $\alpha_P < 0$, $\beta_{P,Q} \geq 0$ for all $\varepsilon \in (0,\varepsilon_0]$, all h and all P, and Q. Hence, the matrix corresponding to the set of difference equations is weakly diagonally dominant. It's inverse has non-positive coefficients. If we can achieve this (and we can) then the only resulting question concerns the order of accuracy. We have the following negative results.

Theorem 4.1. Let the left-hand side of the scheme be of the form (4.5) with $\alpha_P < 0$, $\beta_{P,Q} \geq 0$ for all $\varepsilon \in (0,\varepsilon_0]$, all h and all P,Q. Then the scheme is of order of accuracy $p \leq 1$, unless $ab = 0$ or $a = \pm b$.

This negative result is a consequence of the non-existence of a second order scheme of the type (4.5) with $\alpha_P < 0$ and $\beta_{P,Q} \geq 0$ for the hyperbolic problem $au_x + bu_y$ unless $ab = 0$ or $a = \pm b$ (we have a square grid).

If a = 0 or b = 0 or a = b, then order of accuracy p = 2 is possible. In fact, we shall describe a scheme of order p = 2 and of positive type (both properties uniformly in ε) for the elliptic problem

$$\varepsilon[au_{xx} + bu_{yy} + cu_x + du_y] + u_x = f, \tag{4.6}$$

where $a(x,y) \geq \alpha > 0$, $b(x,y) \geq \beta > 0$ for $(x,y) \in \Omega$. Moreover, a,b,c,d and f should be smooth functions. Then a second order scheme of positive type for this problem (on a grid as in Fig. 1) is given by

$$\varepsilon[a(P)\,\frac{u_E - 2u_P + u_W}{h^2} + b(p)\,\frac{u_N - 2u_P + u_S}{h^2} + c(P)\,\frac{u_E - u_W}{2h} +$$

$$+\ d(P)\,\frac{u_N - u_S}{2h}] + \frac{u_E - u_P}{h} +$$

$$+\ \frac{h}{2\varepsilon a(P)}\,[\varepsilon b(P)\,\frac{u_N - 2u_P + u_S}{h^2} + \varepsilon c(P)\,\frac{u_E - u_W}{2h} + \varepsilon d(P)\,\frac{u_N - u_S}{2h} +$$

$$+\ \frac{u_E - u_P}{h}]$$

$$= f(P) + \frac{h}{2\varepsilon a(P)}\,\frac{f(P) + f(E)}{2}\ . \tag{4.7}$$

One easily recognizes the upstream difference scheme and the discretizations added on the $O(\frac{1}{\varepsilon})$-level. It is obvious that this scheme is of positive type; we shall not prove that this scheme is of order p = 2 for smooth solutions. However, this proof strongly resembles the proof of Theorem 2.4.

The above scheme applies to very special problems only. The special character of the problem may be removed to some extent by observing that these special problems may be obtained from rather general problems by a suitable transformation. Let us consider the elliptic problem

$$\varepsilon[u_{xx} + u_{yy}] + a(x,y)u_x + b(x,y)u_y = f(x,y) \tag{4.8}$$

with $a^2(x,y) + b^2(x,y) \neq 0$ for all $(x,y) \in \Omega$. We want to bring this problem in a suitable form by the introduction of the new variables ξ and η, $\xi=\xi(x,y)$, $\eta=\eta(x,y)$. In the new variables the problem (4.8) becomes

$$\varepsilon[(\xi_x^2 + \xi_y^2)u_{\xi\xi} + 2(\xi_x\eta_x + \xi_y\eta_y)u_{\xi\eta} + (\eta_x^2 + \eta_y^2)u_{\eta\eta}] +$$

$$+ \varepsilon[(\xi_{xx} + \xi_{yy})u_\xi + (\eta_{xx} + \eta_{yy})u_\eta] +$$

$$+ [(a\xi_x + b\xi_y)u_\xi + (a\eta_x + b\eta_y)u_\eta] = f.$$

Now we determine ξ and η by the (hyperbolic) differential equations

$$a\eta_x + b\eta_y = 0$$

$$-b\xi_x + a\xi_y = 0.$$

Hence, on Ω we have,

$$\xi_x\eta_x + \xi_y\eta_y = 0.$$

Consequently, the transformed problem is of the form (4.6) (with ξ,η - coordinates).

This simple example shows what may be done. The new scheme (4.7) might be very useful for elliptic problems which are easily brought in a form like (4.6). In most cases this means that no transformation is required (the problem is already of the form (4.6)) or the computation of the new coordinates can be done analytically. If the new coordinates are to be determined numerically, then the use of a scheme like (4.7) might be impractical. However, much remains to be done in order to determine the merits of the new scheme relative to other schemes.

REFERENCES

Axelsson, O. and Gustafsson, I. (1977). A Modified Upwind Scheme for Convective Transport Equations etc., Technical Report 77.12 R, Chalmers University of Technology and the University of Göteborg, Göteborg, Sweden.

Douglas, J.,Jr. and Dupont, T. (1974). Galerkin Approximations for the Two Point Boundary Value Problem using Continuous Piecewise Polynomial Spaces. *Numer.Math.* 22, 99-109.

Griffiths, D.F. Toward Time Stepping Algorithms for Convective Diffusion. These proceedings.

Hemker, P.W. (1977). A Numerical Study of Stiff Two-Point Boundary Value Problems. Mathematical Centre Tracts 80, Mathematisch Centrum, Amsterdam.

Mitchell, A.R. and Christie, I. Contribution to this conference. These proceedings.

Strang, G. and Fix, G.J. (1973). An Analysis of the Finite Element Method. Prentice Hall, Englewood Cliffs N.J.

Veldhuizen, M. van. Higher Order Methods for a Singularly Perturbed Problem. *Numer.Math.* 30, 267-279 (1978).

Part II

CONTRIBUTED PAPERS

ITERATIVE SOLUTION OF SINGULAR PERTURBATION 2ND ORDER BOUNDARY VALUE PROBLEMS

O. Axelsson and I. Gustafsson
Chalmers University of Technology and the University of Göteborg, Department of Computer Sciences, Fack, S-402 20 Göteborg, Sweden

ABSTRACT

Discretizations of singular perturbation problems lead to strongly unsymmetric matrices. Iterative methods seem to have been used mostly for positive definite symmetric matrix problems.

An iterative method will be described, which works for unsymmetric matrix problems that are (weakly) diagonally dominant, that is, in this context, for M-matrices. As is well known, there exists several discretizations which lead to M-matrices for singular perturbation, 2nd order problems on the form

$$-\varepsilon\Delta u + \sum_{1}^{n}\beta_i(x)\frac{\partial u}{\partial x_i} + \sigma u = f,\ x \in \Omega \subset R^n,\ \sigma > 0. \qquad (0.1)$$

We present a preconditioned minimal residual conjugate gradient algorithm which is based on a modified incomplete factorization of the given matrix.

1. INTRODUCTION

We will present an iterative method which is applicable also for unsymmetric matrices for instance for the discrete analogue of the problem (0.1).

In order to ensure that the method will work we assume that the matrix is (weakly) diagonally dominant, that is, in this context, an M-matrix. In section 2 some discretizations leading to such matrices are presented. The iterative method is based on approximate factorization of the matrix in combination with a minimal residual conjugate gradient procedure. The latter method will be presented in section 3 and in section 4 the incomplete factorization algorithms are described. In section 5 we present numerical results and give some concluding remarks.

2. STABLE DISCRETIZATION SCHEMES

We will present some stable discretization methods for (0.1) leading to (weakly) diagonally dominant coefficient matrices. For simplicity but with no loss of generality we restrict ourselves to n=1, that is, to the one-dimensional problem

$$- \varepsilon u_{xx} + \beta(x)u_x + \sigma u = f, \ \sigma > 0.$$

The following schemes will give an appropriate approximation in the above sense.

(a) *The usual upwind (downwind) scheme*

$$\beta(x)u_x = \begin{cases} \beta(x)\dfrac{u(x+h) - u(x)}{h} & \text{if } \beta(x) < 0 \\ \beta(x)\dfrac{u(x) - u(x-h)}{h} & \text{if } \beta(x) > 0. \end{cases}$$

This scheme, however, gives only O(h) accuracy. The following two scemes give $O(h^2)$ accuracy for fixed ε.

(b) *The modified upwind scheme*

$$- \varepsilon u_{xx} + \beta(x)u_x = \begin{cases} - \dfrac{\varepsilon}{1-\delta} u_{xx} + \beta(x)\dfrac{u(x+h) - u(x)}{h} & \text{if } \beta(x) < 0 \\ - \dfrac{\varepsilon}{1+\delta} u_{xx} + \beta(x)\dfrac{u(x) - u(x-h)}{h} & \text{if } \beta(x) > 0 \end{cases}$$

where $\delta = \dfrac{h\beta(x)}{2\varepsilon}$.

This scheme was considered by Samarskii, see Gushchin and Shchennikov (1974), see also Axelsson and Gustaffson (1977a) and Kelogg and Tsan (1978).

(c) *The Il'in scheme*

$$- \varepsilon u_{xx} + \beta(x)u_x = -\varepsilon z \coth(z)u_{xx} + \beta(x)\frac{u(x+h) - u(x-h)}{2h},$$

where $z = \beta(x)h/2$.

This scheme was proposed by Il'in (1969) and by Barrett (1974), see also Kellogg *et al* (1978).

In all these three schemes the central differences are used for u_{xx}.

3. THE MINIMAL RESIDUAL CONJUGATE GRADIENT (MRCG) METHOD

Let $Au = f$ be the system of equations (of order N) corresponding to the discrete analogue of (0.1) obtained by one of the schemes of the preceding section.

We first observe that since the matrix A is unsymmetric $(r^TA^{-1}r)^{1/2}$ is not a norm and the classical conjugate gradient method, which for positive definite matrices would minimize $r^TA^{-1}r$, is inappropriate. (Here r stands for the residual in each iteration step.) However, a modified conjugate gradient method, see Axelsson and Munksgaard (1977b), which minimizes r^Tr is useful.

The preconditioned variant of this method (for the solution of $C^{-1}Au = C^{-1}f$) can be summarized in the following algorithm.

Let u^0 be arbitrary and calculate

$r^0 = C^{-1}(Au^0 - f)$, $\rho_0 = (r^0,r^0)$, $d^0 = -r^0$,

$d1^0 = C^{-1}Ad^0$ and $\delta_0 = (d1^0,d1^0)$.

Then for $\ell = 1,2,\ldots$ calculate

$\lambda_\ell = -(r^{\ell-1},d1^{\ell-1})/\delta_{\ell-1}$,

$u^\ell = u^{\ell-1} + \lambda_\ell d^{\ell-1}$, $r^\ell = r^{\ell-1} + \lambda_\ell d1^{\ell-1}$,

$\rho_\ell = (r^\ell,r^\ell)$, $r1^\ell = C^{-1}Ar^\ell$,

$\beta_\ell = (r1^\ell,d1^\ell)/\delta_{\ell-1}$, $d^\ell = -r^\ell + \beta_\ell d^\ell$,

$d1^\ell = -r1^\ell + \beta_\ell d1^\ell$ and $\delta_\ell = (d1^\ell,d1^\ell)$.

Here $(x,y) = x^Ty$ is the usual inner product in R^N. As stopping criterion we have used $\rho_\ell \leq \varepsilon_0^2\rho_0$.

It is shown in Axelsson (1978), see also Axelsson (1973b), that if the eigenvalues of $C^{-1}A$ are situated in an ellipse in the complex plane (see Fig. 1) with the semiaxes α and β satisfying $0 \leq \beta < \alpha < 1$ then this method converges to a relative error ε_0 in at most

$$p_0 = \mathrm{Ent}\left(\frac{\ln 1/\varepsilon_0}{\ln(1+\sqrt{1-\alpha^2-\beta^2}) - \ln(\alpha+\beta)} + 1\right)$$

number of iterations.

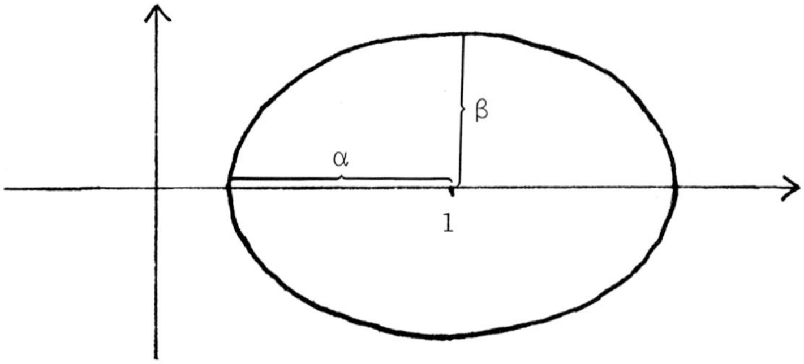

Fig. 1. *The ellipse containing the eigenvalues of* $C^{-1}A$.

It is easily seen that for our problem with C=I we have $\alpha = 1 - c_1h^2$ and $\beta = c_2h$, where c_1 and c_2 are independent of h (but depend on ε). Elementary calculations then give

$$p_0 = O(h^{-1}\ln 1/\varepsilon_0)$$

independent of ε.

In the following section this result will be improved by the use of a proper preconditioning matrix C.

4. INCOMPLETE FACTORIZATIONS

An incomplete factorization is obtained by neglecting elements during the factorization in triangular factors of the given matrix. We then get A = LU + R, where R is the defect matrix.

The following two strategies can be used.

(a) *Neglect small elements*

For instance a new element $a_{lm}^{(k)}$ of row l and column m is accepted as fill-in in $A^{(k)}$ in the k´th elimination step <u>iff</u>

$$|a_{lm}^{(k)}| \geq c\{\max_{j\geq k}|a_{lj}^{(k-1)}|,\ \max_{i}|a_{im}^{(k-1)}|\},$$

where $c > 0$ is a parameter.

This strategy was used by e.g. Tuff and Jennings (1973).

(b) *Neglect elements in certain positions*

This idea is useful for problems with well defined structure like PDE problems and was considered by Meijerink and van der Vorst (1977). Earlier references are Varga (1960), Stone (1968), Dupont, Kendall and Rachford Jr. (1968), Axelsson (1973a) and Saylor and Bracha-Barak (1973).

In the modified incomplete factorization methods the elements are not neglected but moved (and added) to the diagonal of U. In addition a small positive number (of order $O(h^2)$ for Dirichlet PDE problems) is added to the diagonal.

This modification preserves (almost) the rowsum of the matrix, that is,

$$\text{rowsum}(A) \approx \text{rowsum}(LU).$$

For symmetric second order PDE problems the modified incomplete factorizations give $\mathcal{K}((LL^T)^{-1}A) = O(h^{-1})$ and the required number of CG iterations (for the solution of $(LL^T)^{-1}Au = (LL^T)^{-1}f$) is $O(h^{-.5})$.

The unmodified incomplete factorizations, however, give $\mathcal{K}((LL^T)^{-1}A = O(h^{-2})$ (that is, actually the same order as for $\mathcal{K}(A)$) and $O(h^{-1})$ number of CG iterations.

For a detailed study of these methods see Gustafsson (1977) and Gustafsson (1978).

Here we will only give an example. Let A be the usual 5-point matrix,

A = 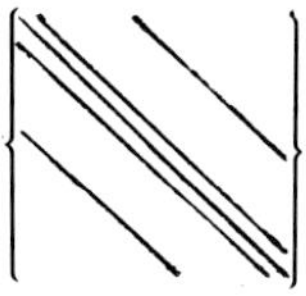 . Then the MIC(0) method is defined by

L = 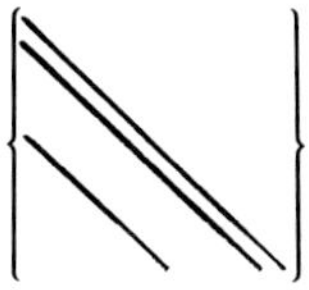 and U = 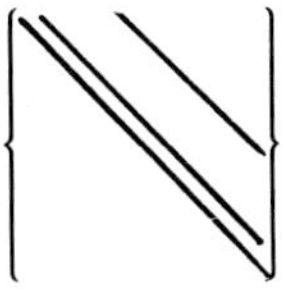 and the MIC(2)

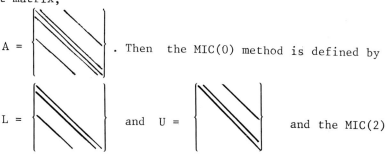

method is defined by

L = 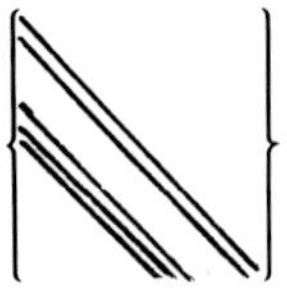 and U = 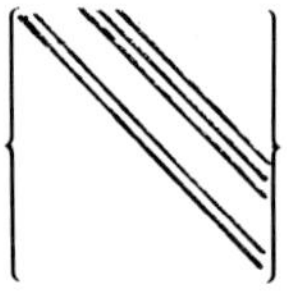 .

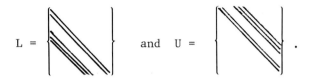

We always have MIC(d) with $d = O(1)$. Thus the storage requirements for these methods is of optimal order $O(N)$, where N is the number of unknowns.

By comparing A, $C = LU$ and R the formulas for computing the elements of L and U are easily derived, see Gustafsson (1977).

It is proved by Meijerink *et al* (1977) that the (unmodified) incomplete factorization algorithms are stable, that is, the elements of diag(U) are positive, when A is an M-matrix. For the modified methods stability is proved for diagonally dominant matrices (in this context for M-matrices), see Gustafsson (1978)

For our unsymmetric problem we have that with a modified incomplete factorization the eigenvalues of $(LU)^{-1}A$ are situated in an ellipse, see Fig. 1, with $\alpha = 1 - c_1 h$ and $\beta = c_2 h$, where c_1 and c_2 are independent of h. It then follows that the number of MRCG iterations is $O(h^{-.5})$ for fixed ε. For details see Gustafsson (1978).

5. NUMERICAL RESULTS AND CONCLUSIONS

The following two-dimensional problem, where Ω is the unit square was solved by the MICMRCG(2) method

$$\begin{cases} -\Delta u + \beta(u_x - u_y) = 1 & \text{in } \Omega \\ u = 0 & \text{on } \partial\Omega \end{cases} .$$

The required number of iterations to reduce the relative error by a factor $\varepsilon_0 = 10^{-3}$ is given in TABLE I for different values of h and β.

Similar results have been obtained for various problems even for problems with turning points, see Axelsson *et al* (1977a).

TABLE I

The number of iterations with the MICMRCG(2) method, $\varepsilon = 10^{-3}$, modified upwind scheme.

β =	0	5	10	100	1000	10^5
$h^{-1}=8$	3	3	3	3	3	3
$h^{-1}=16$	4	5	4	4	4	4
$h^{-1}=32$	6	7	6	5	5	5

Here the modified upwind scheme was used to obtain the discrete system of equations. However, the same or almost the same rate of convergence was observed for the two other schemes in section 2.

A slight increase in the number of iterations is observed as β increases. However for very large β the number of iterations will decrease since then the first derivative term will dominate yielding a quotient between the largest and smallest eigenvalues of A of order $O(h^{-1})$.

The factorization work and the work per iteration for this method are 17N and 21N respectively.

In conclusion we notice that the MIC(MR)CG methods require $O(N^{5/4} \ln 1/\varepsilon_0)$ operations for two-dimensional symmetric or unsymmetric PDE problems.

For three-dimensional problems the corresponding result is $O(N^{7/6} \ln 1/\varepsilon_0)$ operations.

The storage requirement is of optimal order O(N).

REFERENCES

Axelsson, O. (1973a).A generalized SSOR method. *BIT*. 12, 443-467.

Axelsson, O. (1973b). Notes on the numerical solution of the biharmonic equation. *J. Inst. Math. Applics.* 11, 213-226.

Axelsson, O. and Gustafsson, I. (1977a). A modified upwind scheme for convective transport equations and the use of a conjugate gradient method for the solution of non-symmetric systems of equations. *Report 77.12 R, Department of computer sciences, Chalmers University of Technology,*

Göteborg, Sweden.

Axelsson, O. and Munksgaard, N.(1977b). A class of preconditioned conjugate gradient methods for the solution of a mixed finite element discretization of the biharmonic operator. *Report No NI-77-14, Inst. for Num. Anal., Technical Univ. of Denmark, Lyngby, Denmark.*

Axelsson, O. (1978). On preconditioned conjugate gradient methods. (*lecture notes*).

Barrett, K.E. (1974). The numerical solution of singular-perturbation boundary-value problems. *J. Mech. Appl. Math.* 27, 57-68.

Dupont, T., Kendall, R. and Rachford, H.H.,Jr. (1968). An approximate factorization procedure for solving self-adjoint elliptic difference equations. *SIAM J. Numer. Anal.* 5, 559-573.

Gushchin, V.A. and Shchennikov, V.V. (1974). A monotonic difference scheme of second order accuracy. *U.S.S.R. Comp. Math. and Math. Phys.* 14, 252-256

Gustafsson, I. (1977). A class of first order factorization methods. *Report 77.04 R, Department of computer sciences, Chalmers University of Technology, Göteborg, Sweden.*

Gustafsson, I.(1978). Stability and rate of convergence for modified incomplete factorization methods. (*in progress*).

Il´in, A.M. (1969). Differencing scheme for a differential operator with a small parameter affecting the highest derivative. *Mat. Zametki* 6, 237-248 - *Math. Notes* 6, 596-602.

Kellogg, R.B. and Tsan, A. (1978). Analysis of some difference approximations for a singular perturbation problem with turning points. (*to appear*).

Meijerink, J.A. and van der Vorst, H.A. (1977). An iterative solution method for linear systems of which the coefficient matrix is a symmetric M-matrix. *Math. Comp.* 31, 148-162.

Saylor, P. and Bracha-Barak, A. (1973). A symmetric factorization procedure for the solution of elliptic boundary value problems. *SIAM J. Numer. Anal.* 10, 190-206.

Stone, H.L. (1968). Iterative solution of implicit approximations of multidimensional partial difference equations. *SIAM J. Numer. Anal.* 5, 530-558.

Tuff, A.D. and Jenning, A. (1973). An iterative method for large systems of linear structural equations, *Int. J. Numer. Meth. in Engng.* 7, 175-183.

Varga, R.S. (1960). Factorization and normalized iterative methods. In: *Boundary problems in differential equations.* (R.E. Langer ed.), pp 121-142. Proceedings, The University of Wisconsin Press, Madison.

ON THE THICKNESS OF THE BOUNDARY LAYER IN ELLIPTIC ELLIPTIC SINGULAR PERTURBATION PROBLEMS

J. Baranger
Mathématiques, Université de Lyon 1,
69621 Villeurbanne, France

ABSTRACT

We propose a method to define the thickness of the boundary layer. It is based on L^2 estimates outside the boundary layer which are valid for a domain Ω with Lipschitzian boundary, for second order operators with L^∞ coefficients and for some inequalities too.

1. INTRODUCTION

We consider the problem $-\varepsilon \Delta u_\varepsilon + u_\varepsilon = u$ in Ω, $u_\varepsilon /_\Gamma = 0$ the first reference on this topic seems to be Rothe (1933). The most considered (and more complicated) problem in the litterature is $-\varepsilon \Delta u_\varepsilon + (u_\varepsilon)_x = u$. Wasow (1944) proved the pointwise convergence outside a part Γ^- of Γ near which there is a boundary layer. Denote by $d(x, \Gamma)$ the distance from $x \notin \Omega$ to $\Gamma = \partial\Omega$. The asymptotic expansion $u_\varepsilon = u + v_0 + z$ where v_0 is more or less of the form $\exp(-\mu\, d(x, \Gamma) / \varepsilon)$ and $z = O(\varepsilon)$ uniformly in Ω, or part of it, has been studied by Levinson (1950), Visik and Lyusternik (1957), Eckhaus and Dejager (1966) and Grisvard (1978). The analagous expansion for our problem is $u_\varepsilon = u + \exp(-\mu\, d(x, \Gamma) / \sqrt{\varepsilon}) + O(\varepsilon)$; it is not proved in the papers quoted above excepted in Grisvard (1978) where $n = 2$ and Γ is a square. We may consider that the thickness of the boundary layer is obtained when the two terms which express the difference $w = u_\varepsilon - u$

are of the same order in ε, that is to say :

$$\exp(-\mu\, d(x, \Gamma) / \sqrt{\varepsilon}) = C\varepsilon$$

so : $d(x, \Gamma) = \frac{\sqrt{\varepsilon}}{\mu} \text{Log } (C\varepsilon)^{-1} = O(\varepsilon^{1/2} \text{ Log } \varepsilon^{-1})$. We believe that this is a nice way to define the thickness of the boundary layer but it imposes strong regularity asumptions on Γ (going from C^1 to C^6 !). We propose a method based on L^2 estimates outside the boundary layer -see Baranger (1976)- which is valid for a Lipschitzian boundary for second order operators with L^∞ coefficients and for some inequalities too.

2. MODEL PROBLEM

We look first to the model problem :

$$(1) \begin{cases} -\varepsilon\, \Delta u_\varepsilon + u_\varepsilon = u & \text{in } \Omega \\ u_\varepsilon / _\Gamma = 0 \end{cases}$$

It is well known that $u_\varepsilon \to u$ in $L^2(\Omega)$ when $\varepsilon \to 0$ and that $u_\varepsilon \not\to u$ in $H^1(\Omega)$ or in $L^\infty(\Omega)$ if $u /_\Gamma \neq 0$. This is the boundary layer phenomenon. The greatest "speed of convergence" in $L^2(\Omega)$ is :

$$(1') \quad ||u - u_\varepsilon||_{L^2(\Omega)} \leq C\varepsilon^{1/4}$$

This result, due to Lions (1973), is optimal as is readily seen from the one dimensional exemple :
$-\varepsilon\, u''_\varepsilon + u_\varepsilon = \text{ch } x,\ u_\varepsilon(\pm 1) = 0$.

Nevertheless, if we remove a strip along the boundary, we obtain -for $u \in H^2_{loc}(\Omega) \cap H^{-1}(\Omega)$- in the remainding open set $\Omega' \subset\subset \Omega$ (outside the boundary layer) the estimate :

$$(2) \quad ||u - u_\varepsilon||_{L^2(\Omega')} \leq C\varepsilon$$

This result is also optimal ; see Baranger (1976).

We now study the connection between the thickness of the

strip to be removed and ε in order that (2) be valid.

For that purpose, consider the variational form of (1) :

$$(3) \quad \varepsilon \int \nabla u_\varepsilon \nabla v + \int u_\varepsilon v = \int u v \qquad \forall v \in H_o^1(\Omega)$$

From now on Ω is a bounded domain with Lipschitzian boundary and $u \in H^2(\Omega)$. Put $w = u_\varepsilon - u$ and $v = \phi^2 w$ where ϕ is a sufficiently regular function to be precised latter. We get :

$$\varepsilon \int \nabla w \int \nabla \phi^2 w = - \varepsilon \int \nabla u \nabla \phi^2 w = \varepsilon \int \Delta u \quad \phi^2 w$$

This gives us the basic estimate :

$$(4) \quad \varepsilon \int \phi^2 \nabla w^2 + \int \phi^2 w^2 \leq \varepsilon \alpha \int \phi^2 \nabla w^2 + \frac{\varepsilon}{\alpha} \int w^2 \nabla \phi^2$$

$$+ \frac{\varepsilon^2 \beta}{2} \int \phi^2 \Delta u^2 + \frac{1}{2 \beta} \int \phi^2 w^2$$

where α and β are positive numbers. Now , we choose :

$$\phi(x) = \text{Min } \{1, w_\varepsilon \left[\exp \frac{\mu \, d(x, \Gamma)}{\Theta_\varepsilon} - 1 \right]\}$$

We shall see that the choices for the parameter Θ_ε, w_ε and μ are very limited ; this will allow us to define the thickness of the boundary layer. (A function like ϕ with $w_\varepsilon = \varepsilon^{1/4}$ and $\Theta_\varepsilon = \varepsilon^{1/2}$ was introduced by Tartar (1973)).

$$\phi(x) = 1 \text{ iff } d(x, \Gamma) \geq \frac{\Theta_\varepsilon}{\mu} \text{Log } (w_\varepsilon^{-1} + 1)$$

It is easy to check that :

$$|\nabla \phi(x)|^2 \leq 2 \, \Theta_\varepsilon^{-2} \mu^2 \left[\phi(x)^2 + w_\varepsilon^2 \right] \text{ p.p.}$$

Using (4) with $\alpha = 1$, we get :

$$(1 - \frac{1}{2\beta}) \int \phi^2 w^2 \leq 2\,\varepsilon\,\Theta_\varepsilon^{-2}\,\mu^2 \int \phi^2 w^2 + 2\,\varepsilon\,\Theta_\varepsilon^{-2}\,\mu^2\,w_\varepsilon^2 \int w^2$$

$$+ \varepsilon\,\frac{2\beta}{2} \int \phi^2\,\Delta u^2$$

So, we had to choose $\Theta_\varepsilon = \varepsilon^{1/2}$. Then μ and β are 0 (1) and we can choose $\mu^2 = 1/8$ and $\beta = 2$ for example. Then by (1') we can take $w_\varepsilon = \varepsilon^{3/4}$.

So $\phi(x) = 1$ in Ω', the open set obtained from Ω after we have removed the strip $d(x, \Gamma) \geq \varepsilon^{1/2}/8 \text{ Log } (\varepsilon^{-3/4} + 1)$.

We suggest to call thickness of the boundary layer every function $e(\varepsilon)$ such that $\phi(x) = 1$ iff $d(x, \Gamma) \geq Ce(\varepsilon)$ for some constant C and all $\varepsilon \in [0, \varepsilon_o]$. For the model problem we obtain $e(\varepsilon) = \varepsilon^{1/2} \text{ Log } \varepsilon^{-1}$.

Remark 1 : If $e_1(\varepsilon) = O(e(\varepsilon))$, e_1 is also a thickness of the boundary layer. If in place of (1'), we use the coarser estimate $\int w^2 \leq Cte$ we choose $w_\varepsilon = \varepsilon$ and this does not change $e(\varepsilon)$.

Remark 2 : As we pointed out in the introduction our definition is consistent with the "L^∞ theorie" of singular perturbation but uses much weaker hypothesis.

Remark 3 : The same study can be made in the L^p norm -see Baranger (1978)- the thickness of the boundary layer is the same.

3. GENERAL SECOND ORDER PROBLEM

We consider now the more general problem where :

$$A = - \sum_{i,j} \partial_i (a_{i,j}(x)\,\partial_j) + \sum_j b_j(x)\,\partial_j + b_o(x)$$

take the place of Δ. We suppose that the bilinear form a(u, v) associated to A is continuous coercive on $H_o^1(\Omega)$, and that u is given in D(A), the domain of the unbounded operator in

$L^2(\Omega)$. Then, we have :

$$\varepsilon\, a(w, \phi^2 w) + \int \phi^2 w^2 = -\varepsilon\, a(u, \phi^2 w) = \varepsilon(Au, \phi^2 w)$$

The right member is less than $\varepsilon^2 \frac{\beta}{2} \int \phi^2 Au^2 + \frac{1}{2\beta} \int \phi^2 w^2$.
We calculate :

$$a(w, \phi^2 w) = \int \phi^2 \Big[\sum_{i,j} a_{ij}(x)\, \partial_i w\, \partial_j w + \sum_j b_j(x) \partial_j w . w + b_o(x)\, w^2\Big] + 2 \int \phi \Big[\sum_{i,j} a_{ij}(x)\, w\, \partial_j \phi\, \partial_i w\Big]$$

We suppose in addition that the first term is positive ; the opposite of the second can be majorized by :
$C \sup_{i,j} ||a_{ij}||_{L^\infty(\Omega)} \int \phi\, |\nabla w|\, w\, |\nabla \phi|$. The method of § 2 remains valid and the thickness of the boundary layer is the same as for Δ.

4. A VARIATIONAL INEQUALITY

The following variational inequality as a unique solution (see for example Lions (1973), Chapitre 2). Convergence in L^p norm has been studied by Celli (1976).

$$\varepsilon \int \nabla u_\varepsilon\, \nabla(v - u_\varepsilon) + \int u_\varepsilon (v - u_\varepsilon) \geq \int u(v - u_\varepsilon)\ \forall\, v \in K$$

$$u_\varepsilon \in K \text{ with } K = \{v \in H^1(\Omega)\ ;\ v /_\Gamma = 0\}$$

Using the same notations as before, we have :
$v = u_\varepsilon - \phi^2 w \in K$, so for $u \in H^2(\Omega)$, we get :

$$\varepsilon \int \nabla w\, \nabla(\phi^2 w) + \int \phi^2 w^2 \leq -\varepsilon \int \nabla u\, \nabla \phi^2 w = \varepsilon \int \Delta u\, \phi^2 w$$

Then, we can perform the same calculations as in § 2. this gives us $||u_\varepsilon - u||_{L^2(\Omega')} \leq \varepsilon\, ||u||_{H^2(\Omega)}$ in the domain $\Omega' = \{x \in \Omega\ ;\ \phi(x) = 1\}$ and the thickness of the boundary layer is the same as for the equation.

Acknowledgements : the author aknowledges L.S. Frank for an interesting discussion.

5. REFERENCES

1. Baranger J. (1976) "Estimations d'erreur à l'intérieur pour un problème de couche limite" in : *Singular Perturbations and Boundary Layer Theory* (Lyon 1976) Springer Verlag 594 (1977).

2. Baranger J. (1978) "Evaluation de l'épaisseur de couche limite", *Rapport de recherche,* Université de Lyon.

3. Celli A. (1976) "Un problema di perturbazione singolare", *Bollettino UMI* (to appear).

4. Eckhaus W. and Dejager E.M. (1966) "Asymptotic solutions of singular perturbations problems for linear differential equations of elliptic type", *Arch. Rat. Mech. Anal.* 23, 26-86.

5. Grisvard P. (1978) "Perturbations singulières". Exposé au séminaire d'Analyse Numérique de Lyon - Saint-Etienne Ecole Centrale de Lyon.

6. Levinson N. (1950) "The first boundary value problem for $\varepsilon\Delta u + A(x, y)u + B(x, y)u + C(x, y)u = D(x, y)$ for small ε", *Annals of Mathematics,* 51, n° 2, 428-445.

7. Lions J.L. (1973) "*Perturbations singulières dans les problèmes aux limites et en contrôle optimal*", Springer Verlag.

8. Rothe E. (1933) "Ueber asymptotische Entwicklungen bei Randwertaufgaben elliptischer partieller Differentialgleichungen", *Mathematische Annalen,* 108, 578-594.

9. Tartar L. (1973) Quoted in Lions J.L. (1973) p 129.

10. Wasow W. (1944) "Asymptotic solution of boundary value problems for the differential equation $\Delta U + \lambda \partial U / \partial x = \lambda f(x, y)$", *Duke Math. J.,* 11, 405-415.

A MINIMAX PRINCIPLE FOR THE NAVIER-STOKES EQUATIONS

K.E.Barrett, G.Demunshi and D.N.Shields

Mathematics Department, Lanchester Polytechnic,
Coventry, U.K.

ABSTRACT

Optimal control methods have been used successfully to solve transonic flow problems. The technique can be made very efficient as the solution is found by solving a series of Poisson problems and an interlaced sequence of constrained optimisation problems. The two basic ingredients are an Euler-Lagrange equation and an optimal control problem. In this paper it will be shown that for non self-adjoint problems such as the Navier-Stokes equations there is an underlying functional from which an Euler-Lagrange equation and a corresponding optimal control problem can be derived simultaneously.

INTRODUCTION

Millikan(1929) and Finlayson(1972) have shown that there is no straightforward variational principle for the Navier-Stokes equations involving solely the velocity field $\underline{v}$ and the pressure p unless the cross product $\underline{v}$ x curl $\underline{v}$ is identically zero. On the other hand, a number of authors (Slattery(1964), Finlayson(1972a), Deshpande(1974), Usher and Craik(1974)) have shown that, if additional pseudo-velocity and pressure fields are introduced alongside the adjoint differential equation system then the extended system does have a variational principle. The process has been termed an embedding method by Sewell(1977). Usher and Craik have used this technique to produce significant results by analysing a nonlinear wave stability problem. Collins(1976) has derived dual variation principles for a large class of non self-adjoint problems by the same method. In this paper a variational principle for the Navier-Stokes equations is obtained and related to the optimal control method currently being used by the French school(1976a,b) to analyse transonic flows.

MODEL PROBLEM

The Navier-Stokes momentum equations express a balance between diffusion and convection and form part of a non self-adjoint system. It is convient first to discuss some simple one dimensional model problems, the easiest being

$$T'' - U T' = 0,\ U = \text{constant},\ T(0) = 0,\ T(1) = 1, \tag{2.1}$$

which describes convective heat flow between two walls. This problem has no variational principle in which the Lagrangian is a polynomial in T and its derivatives. Consider however the problem of finding stationary values of the functional of two variables T_1 and T_2

$$J(T_1,T_2) = \int_0^1 (T_1'^2 - T_2'^2 - UT_2T_1' + UT_1T_2')dx, \tag{2.2}$$

among the class of functions T_1 and T_2 which are sufficiently continuous and satisfy the Dirichlet conditions

$$T_1(0) = T_2(0) = 0,\ T_1(1) = T_2(1) = 1. \tag{2.3}$$

If the first variation in J is zero then

$$T_1'' = UT_2', \tag{2.4}$$

$$T_2'' = UT_1'. \tag{2.5}$$

The difference function

$$\bar{T} = T_1 - T_2, \tag{2.6}$$

satisfies the homogeneous boundary value problem

$$\bar{T}'' + U\bar{T}' = 0, \; \bar{T}(0) = \bar{T}(1) = 0, \tag{2.7}$$

so $\bar{T} = 0$, i.e. $T_1 = T_2$. If T_1 and T_2 are equal both equations (2.4) and (2.5) reduce to (2.1) and finding the stationary point for the functional (2.2) is equivalent to solving the original problem. It may be noted that if T_2 is fixed, J has a minimum w.r.t. T_1 and if T_1 is fixed, J has a maximum w.r.t. T_2, i.e. J is a saddle functional. The solution is obtained from

$$\max_{T_2} \; \min_{T_1} \; J(T_1,T_2)$$

where, in the extremum processes, the trial functions satify the appropriate Dirichlet conditions. The following algorithm could be used to obtain a solution

1. Choose T_2 to satisfy $T_2(0) = 0$, $T_2(1) = 1$.
2. Minimise J w.r.t T_1 where $T_1(0) = 0$, $T_1(1) = 1$.
   i.e. solve $T_1'' = UT_2'$, $T_1(0) = 0$, $T_1(1) = 1$.

At the end of stage 2 the functions T_1 and T_2 are known and the corresponding value of J may be shown to be

$$\begin{aligned} J(T_1(T_2),T_2) &= - \int_0^1 ((T_1'-T')^2 + (T_2'-T')^2) \, dx \\ &= - \int_0^1 (T_1' - T_2')^2 dx \end{aligned} \tag{2.8}$$

where T is the solution being sought. J provides a direct measure of the error being committed.

3. Maximise J given by (2.8) with respect to T_2 and thus obtain the unknown parameters involved in T_2. Table (2.1) shows the results obtained by dividing the range into 10 subintervals and using piecewise linear approximations for T_2.

As a second model problem consider the nonlinear problem

$$T'' - RTT' = 0, \; T(0) = 0, \; T(1) = 1 \tag{2.9}$$

This has the exact solution

$$T = \pi \beta \tan(\pi \beta x/2)/R \tag{2.10}$$

where $1 = \pi\beta \tan(\pi\beta/2)/R$. If this problem is to be embedded in an extended system involving two variables T_1 and T_2 there are at least two ways of achieving this result. The skew symmetric functional

$$J(T_1,T_2) = \int_0^1 (T_1'^2 - T_2'^2 - RT_1T_2T_1' + RT_2T_1T_2')\,dx \qquad (2.11)$$

leads to the two Euler-Lagrange equations

$$T_1'' - R/2(T_1 + T_2)T_2' = 0,$$
$$T_2'' - R/2(T_2 + T_1)T_1' = 0,$$

which have the unique solution $T_1 = T_2 = T$. On the other hand the functional

$$J(T_1,T_2) = \int_0^1 (T_1'^2 - T_2'^2 + RT_1T_2T_2' - R/2T_2^2T_1' - R/2T_2^2T_2')\,dx, \qquad (2.$$

leads to two Euler-Lagrange equations

$$T_1'' - RT_2T_2' = 0, \qquad (2.13)$$
$$T_2'' - RT_2T_1' = 0, \qquad (2.14)$$

which again have the unique solution $T_1 = T_2 = T$. However at

TABLE 2.1

Linear model problem

u	10	20	30	50
$T_2(0.5)$	0.0247	0.0670	Oscillations in solution	
$T_1(0.5)$	-0.0149	-0.0773		
exact	0.0067	0.000045		
J	-0.4003	-2.8868	-8.5042	-29.6612

the second stage of the solution algorithm (2.12) is

$$J(T_1(T_2),T_2) = -\int_0^1 (T_1' - T_2')^2\,dx \qquad (2.15)$$

just as for the linear example. This result can be proved as follows

$$J(T_1(T_2),T_2) = \int_0^1 \left(-(T_1'-T_2')^2 + 2T_1'^2 - 2T_1'T_2' + RT_1T_2T_2' - \frac{R}{2}T_2^2T_1' - \frac{R}{2}T_2^2T_2'\right)dx,$$
$$= \int_0^1 \left(-(T_1'-T_2')^2 - 2T_1''T_1 + 2T_1''T_2 + RT_1T_2T_2' - \frac{R}{2}T_2^2T_1' - \frac{R}{2}T_2^2T_2'\right)dx,$$

by integration by parts and since $T_1 = T_2$ at 0 and 1.

$$J(T_1(T_2),T_2) = \int_0^1 (-(T_1'-T_2')^2 - R/2(T_2^2T_1)' + R/2(T_2^3)')\,dx,$$

TABLE 2.2

Nonlinear model problem

R	10	50	100	200
$T_2(0.5)$	0.2022	0.0570	-0.2005	Oscillates
exact	0.2027	0.0569	0.0294	
J	-0.0857	-5.3082	-21.9295	-89.9755

by (2.13). A final integration gives (2.15). The functional has a minimum w.r.t. T_1 for any given T_2 but does not appear to have a maximum w.r.t. T_2 for all T_1. The solution corresponds to

$$\max_{T_2} \min_{T_1} J(T_1, T_2)$$

where the trial functions satisfy the appropriate Dirichlet conditions during the extremum process. Table 2.2 shows the results obtained in this case using the same algorithm as in the linear problem.

The French school would solve (2.10) by constructing an equation

$$T_1'' = R\, f(T_2) \text{ where } f(T_1) = T_1 T_1'$$

and solving it with $T_1(0) = 0$, $T_1(1) = 1$ to determine $T_1(T_2)$. They would then construct a suitable optimal control principle which forced T_1 and T_2 to be close to each other. In this paper an underlying principle has been found which leads to a suitable optimal control principle and automatically associates it with a differential equation constraint.

NAVIER-STOKES EQUATIONS

In the previous section a simple nonlinear convective diffusion problem has been shown to be associated with two variational principles, the second leading to a least squares optimal control problem. It will now be shown that the Navier -Stokes equations have a similar underlying principle behind them. It must be stressed that it is purely a mathematical construction with no apparent physical meaning. The Navier -Stokes equations for a steady viscous incompressible flow are

$$\nu \frac{\partial^2 v_i}{\partial x_j^2} = \frac{1}{\rho}\frac{\partial q}{\partial x_i} + v_j \frac{\partial v_i}{\partial x_j}, \quad \frac{\partial v_i}{\partial x_i} = 0 \text{ in } \Omega, \tag{3.1}$$

where v_i are the velocity components, q is the pressure, ρ is the density and ν the kinematic viscosity. Suppose the problems being considered are subject to the boundary conditions

$$v_i = V_i \text{ on } \partial\Omega \tag{3.2}$$

where V_i is a prescribed velocity field. Consider the functional of two pseudo-velocity and two pseudo-pressure fields

$$J(u_i, w_i, p, r) = \int_\Omega \left\{ \frac{\nu}{2}\left(\frac{\partial u_i}{\partial x_j}\right)^2 - \frac{\nu}{2}\left(\frac{\partial w_i}{\partial x_j}\right)^2 + u_i\frac{\partial p}{\partial x_i} - w_i\frac{\partial r}{\partial x_i} + w_i u_j \frac{\partial w_i}{\partial x_j} - w_i w_j \frac{\partial u_i}{\partial x_j} \right\} d\tau \tag{3.3}$$

The Euler-Lagrange equations for this functional, where arbitrary variations are possible in all variables, except that

$$u_i = V_i, \; w_i = V_i \text{ on } \partial\Omega, \tag{3.4}$$

and $p = r$ on $\partial\Omega$,

are

$$\delta u_i: \quad \frac{\partial^2 u_i}{\partial x_j^2} = \frac{\partial}{\partial x_i}\left(p + \frac{1}{2} w_j^2\right) + \frac{\partial}{\partial x_j}(w_i w_j), \tag{3.5}$$

$$\delta p: \quad \frac{\partial u_i}{\partial x_i} = 0, \tag{3.6}$$

$$\delta w_i: \quad \frac{\partial^2 w_i}{\partial x_j^2} = \frac{\partial r}{\partial x_i} + w_j\frac{\partial u_j}{\partial x_i} + \frac{\partial}{\partial x_j}(w_i u_j) + w_j\frac{\partial u_i}{\partial x_j} - u_j\frac{\partial w_i}{\partial x_j}, \tag{3.7}$$

$$\delta r: \quad \frac{\partial w_i}{\partial x_i} = 0. \tag{3.8}$$

These equations have the same solution as the Navier-Stokes equations when

$$u_i = w_i = V_i \tag{3.9}$$

and

$$p = r = \frac{q}{\rho} - \frac{1}{2}v_i^2 \tag{3.10}$$

it has not yet proved possible to establish that (3.5)-(3.8) only have this solution. Suppose now that w_i is chosen to satisfy the boundary conditions (3.4) and to be divergence free and then u_i, p and r are allowed to vary freely subject to p = r on $\partial\Omega$ and $u_i = V_i$ on $\partial\Omega$. J is stationary when (3.5) and (3.6) are satisfied. Once u_i and p have been determined J can be evaluated and it can be shown from (3.5) and Green's theorem that

$$J(u_i(w_i),w_i,p(w_i),r(w_i)) = -\frac{\nu}{2\Omega}\int\left(\frac{\partial u_i}{\partial x_j} - \frac{\partial w_i}{\partial x_j}\right)^2 d\tau \tag{3.11}$$

i.e. the functional evaluated at the point $u_i(w_i)$, w_i etc. once more gives a direct estimate of the error for any given initial solenoidal field w_i.

The following algorithm is suggested

1. Choose a solenoidal velocity field w_i satisfying $w_i=V_i$ on the boundary.
2. Make J stationary subject to $u_i=V_i$ on the boundary. The functional can then be computed from (3.3) or (3.11).
3. Minimise J given by (3.11) with respect to w_i to the unknown parameters involved in w_i.

REFERENCES

Collins,W.D.(1976). Dual extremal principles for dissipative systems. *Maths.Res.Center Rep.* 1624 U. of Wisconsin.

Deshpande,S.M.(1973). On the connection between the elliptic equations of the Navier-Stokes type and the theory of harmonic functionals. *Quart.App.Math.* 31,43-52

Finlayson,B.A.(1972a). On the existence of variational principals for the Navier-Stokes equations. *Phys. Fluids.* 15,963-7.

Finlayson,B.A.(1972b).*The method of weighted residuals and variational principles*. Academic,New York

Glowinski,R., Periaux,J. and Pirronneau,O.(1976a). Use of optimal control theory for the numerical simulation of transon flow by the method of finite elements. In: *Lecture notes in Physics*, 59,205-211

Glowinski,R., Periaux,J. and Pirronneau,O.(1976b). Transonic flow simulation by the finite element method via optimal contr *Finite element methods in flow problems*. ICCAD, 247-261.

Millikan,C.B.(1929). On the steady motion of viscous incompres ible fluids; with particular reference to a variational princ- ipal. *Phil. Mag.*,7, 641-662.

Sewell,M.(1977). Degenerate duality, catastrophes and saddle functionals. *Ossolineum Publishing House of Polish Academy*

Slattery,J.C.(1964). A widely applicable type of variational integral. *I.Chem.Eng.Sci.*,19,801-806.

Usher,J.R. and Craik,A.D.(1974). Nonlinear wave interactions in shear flows. Part 1. A variational formulation, *J.F.M*,66, 209-223.

SINGULAR PERTURBATIONS AND BOUNDARY LAYER CORRECTORS FOR A SYSTEM OF TWO QUASI-LINEAR PARABOLIC EVOLUTION EQUATIONS WITH TWO PARAMETERS ε AND η

A. Bourgeat - R. Tapiéro

Centre de Mathématiques, 303, I.N.S.A.

20, Avenue A. Einstein, 69621 Villeurbanne Cedex, France

ABSTRACT

The studied problem of reaction-diffusion in a plasma leads to a system of two quasi-linear evolution equations including two perturbation parameters ε and η. A boundary layer, near time $t = 0$ and near the space boundary, appears. In order to calculate the boundary layer correctors, inner and outer asymtotic expansions, adapted to the non-linearities, are constructed. Two cases are studied. Firstly ε tends to zero with fixed η , then η tends to zero with $\varepsilon = 0$. Secondly ε and η tend to zero independently. These expansions give good error estimates but are efficient only at first order, the terms of greater order giving no better estimates.

1. INTRODUCTION

The studied problem represents evolution of the distribution profile of atomic and molecular ions in a plasma composed of a unique gas (Sadeghi, 1974). We have the non-dimensional system of two quasi-linear parabolic equations :

$$P_1 u + Cu + a_0 u(u+v) + a_1 u(u+v)^2 = 0 \tag{1-1}$$

$$\eta P_2 v - \eta Cu + v(u+v) = 0 \tag{1-2}$$

in $Q = \Omega \times (0,T)$

$$u|_{\Sigma} = v|_{\Sigma} = 0 \quad , \Sigma = \partial\Omega \times (0, T) \tag{1-3}$$

$$u(0) = u_{in} \in H^1_o(\Omega) \cap L^{\infty}(\Omega) \quad , \quad u_{in} \geq 0 \tag{1-4}$$

$$v(0) = \varepsilon v_{in} \in H^1_o(\Omega) \cap L^{\infty}(\Omega) \quad , \quad v_{in} \geq 0 \tag{1-5}$$

$$||u_{in}||_{\infty} = ||v_{in}||_{\infty} = 1 \tag{1-6}$$

where :

$\Omega = \{(x, y, z) \in R^3 / x^2 + y^2 < 1, |z| < 1\}$

P_i is the heat operator $P_i = \frac{\partial}{\partial t} - d_i\Delta$, $i = 1,2$.

d_1, d_2, C, a_0, a_1 are positive constants $\simeq 1$.

$0 < \varepsilon, \eta << 1$.

We have the following result (Bourgeat and Tapiéro, 1977):

For fixed positive ε and η, there exists a unique solution of (1-1) - (1-5) : $(u_{\varepsilon\eta}, v_{\varepsilon\eta}) \in \{W(0,T) \cap L^{\infty}(Q)\}^2$ where $W(0,T) = L^2(0,T ; H^2(\Omega) \cap H^1_o(\Omega)) \cap H^1(0,T ; L^2(\Omega))$ and such that :

$$0 \leq u_{\varepsilon\eta} \leq 1 \quad \text{a.e. in } Q \tag{1-7}$$

$$0 \leq v_{\varepsilon\eta} \leq C(\varepsilon + \eta) \quad \text{a.e. in } Q \tag{1-8}$$

The physical sense of ε and η does not allow to let $\eta \to 0$ without $\varepsilon \to 0$. So the perturbation problems we will study here are the convergence and asymptotic expansions of $u_{\varepsilon\eta}$ and $v_{\varepsilon\eta}$ when $\varepsilon \to 0$ with fixed η which is a regular perturbation; then $\eta \to 0$ with $\varepsilon = 0$. Otherwise we study the convergence and asymptotic expansions when ε and $\eta \to 0$ independently.

2. CONVERGENCE AND ASYMPTOTIC EXPANSIONS WHEN $\varepsilon \to 0$, η FIXED

(a) Convergence

By (1-7), (1-8) $u_{\varepsilon\eta}$ and $v_{\varepsilon\eta}$ are bounded in $L^{\infty}(Q)$ and by a compacity argument and Lebesgue's theorem we prove that : when $\varepsilon \to 0$, $(u_{\varepsilon\eta}, v_{\varepsilon\eta})$, solution of (1-1) - (1-5), tends to $(u_{o\eta}, v_{o\eta})$ in $W(0, T)$, the unique positive solutions of

(1-1) - (1-4) where (1-5) is replaced by :

$$v(0) = 0 \tag{2-1}$$

Moreover :

$$0 \leq u_{o\eta} \leq 1 \tag{2-2}$$

$$0 \leq v_{o\eta} \leq C\eta \tag{2-3}$$

and we have the error estimates :

$$||u_{\varepsilon\eta} - u_{o\eta}||_{W(0,T) \cap L^{\infty}(Q)} \leq K\varepsilon \tag{2-4}$$

$$||v_{\varepsilon\eta} - v_{o\eta}||_{W(0,T) \cap L^{\infty}(Q)} \leq K\varepsilon \tag{2-5}$$

To prove (2-4), (2-5) we make the change of functions : $U = (u_{\varepsilon\eta} - u_{o\eta})/\varepsilon$, $V = (v_{\varepsilon\eta} - v_{o\eta})/\varepsilon$; for fixed $\varepsilon > 0$, $(U,V) \in \{W(0,T) \cap L^{\infty}(Q)\}^2$ and are solutions of a system :

$$P_1 U + \alpha_1 U + \alpha_2 V = 0$$

$$\eta P_2 V - \beta_1 U + \beta_2 V = 0$$

$$U(0) = 0, \ V(0) = v_{in}$$

where α_1, α_2, β_1, β_2 are positive functions depending on $u_{\varepsilon\eta}$, $v_{\varepsilon\eta}$ but are bounded in $L^{\infty}(Q)$ independently of ε. So, we have the following result (Ladyženskaja, Solonnikov, Ural'ceva 1968): $||U||_{L^{\infty}(Q)} + ||V||_{L^{\infty}(Q)} \leq K$ which gives (2-4),(2-5).

(b) Asymptotic expansion

We can expand the solutions at every order n :

$$u_{\varepsilon\eta} = u_{o\eta} + \varepsilon u_{1\eta} + \dots + \varepsilon^n u_{n\eta} + R^n_{\varepsilon\eta}$$

$$v_{\varepsilon\eta} = v_{o\eta} + \varepsilon v_{1\eta} + \dots + \varepsilon^n v_{n\eta} + S^n_{\varepsilon\eta}$$

where $u_{n\eta}$, $v_{n\eta}$ are the unique solutions in $W(0,T) \cap L^{\infty}(Q)$ (Ladyženskaja, et al., 1968) of the linear system :

$$P_1 u + a_n(x\ ;\ t)u + b_n(x\ ;\ t)\ v = f_n(x\ ;\ t)$$

$$\eta P_2 v + c_n(x\ ;\ t)u + d_n(x\ ;\ t)\ v = g_n(x\ ;\ t)$$

where a_n, b_n, c_n, d_n, f_n, g_n are bounded functions in $L^{\infty}(Q)$.

The remaining terms $R^n_{\varepsilon\eta}$, $S^n_{\varepsilon\eta}$, as in paragraph (a), verify a linear system with bounded coefficients and we obtain by the same technique :

$$\|R^n_{\varepsilon\eta}\|_{L^\infty(Q)\cap W(0,T)} \leq K\varepsilon^{n+1} \tag{2-6}$$

$$\|S^n_{\varepsilon\eta}\|_{L^\infty(Q)\cap W(0,T)} \leq K\varepsilon^{n+1} \tag{2-7}$$

3. CONVERGENCE AND ASYMPTOTIC EXPANSIONS WHEN $\eta\to 0, \varepsilon = 0$

(a) *Convergence*

(2-2), (2-3) show that $u_{o\eta}$ is bounded in $L^\infty(Q)$ and $v_{o\eta}\to 0$ in $L^\infty(Q)$ as $\eta \to 0$.

(i) Limit of $u_{o\eta}$ *and error estimate*. When $\eta \to 0$, $u_{o\eta}\to u_{oo}$ in W(0,T) where u_{oo} is the unique positive solution in W(0,T) of the reduced problem :

$$P_1u + Cu + a_o u^2 + a_1 u^3 = 0 \tag{3-1}$$

$$u(0) = u_{in} \tag{3-2}$$

Moreover we have the properties (Bourgeat and Tapiéro, 1977) :

$$\|u_{o\eta} - u_{oo}\|_{W(0,T)} \leq K\eta \tag{3-3}$$

$u_{\varepsilon\eta} \geq u_{oo} \geq \beta_k > 0$ a.e. in $\Omega_k \times (0,T)$ for every compact set $\Omega_k \subset \Omega$. (3-4)

(ii) Error estimate for $v_{o\eta}$. (1-2) and (2-2)-(2-3) show that P_2v is bounded in $L^2(Q)$ so $v_o \rightharpoonup 0$ weakly in W(0,T) and by compacity $v_{o\eta}\to 0$ in $V = L^2(0,T\,;\,H^1_o(\Omega))$.

Making the scalar product in $L^2(Q)$ of (1-2) by $v_{o\eta}$ we obtain the estimate :

$$\|v_{o\eta}\|_V \leq K\eta^{1/2} \tag{3-5}$$

(b) *Asymptotic expansions*

There is no boundary layer at order 0 ($v_{oo} \equiv 0 \in C^\infty(\overline{Q})$ $v_{oo}|_\Sigma = 0$, $v_{oo}(0) = 0$). But, if we make formally an expansion at order 1 : $u_{o\eta} \simeq u_{oo} + \eta u_{o1}$, $v_{o\eta} \simeq \eta v_{o1}$, (1-2) gives :

$$P_2v_{oo} + v_{o1}\,u_{oo} = Cu_{oo}$$

and, as $v_{oo} \equiv 0$, $u_{oo} > 0$ a.e. inside Ω, $v_{o1} \equiv C$

and a boundary layer appears at t=0 and at the boundary. To find the boundary layer functions we make an inner and outer expansion of each function $u_{o\eta}$ and $v_{o\eta}$:

$$u_{o\eta} = u(\eta;\ x\ ;\ t) + u'(\eta;\frac{x}{\sqrt{\eta}}\ ;\ \frac{t}{\eta})$$
$$v_{o\eta} = v(\eta;\ x\ ;\ t) + v'(\eta;\frac{x}{\sqrt{\eta}}\ ;\ \frac{t}{\eta})$$

with u and v solutions of (1-1), (1-2) (outer expansion) and u', v' represent the boundary layer and are solutions of the ramainding equations :

$$P_1'u' + C\eta u' + \eta a_o\ \{(2u + v + v')\ u' + u'^2 + uv'\}$$
$$+\eta a_1\ \{(3u^2 + 2u(v+v') + (v+v')^2)u' + (3u + v + v')u'^2$$
$$+u'^3 + (u^2 + 2uv)v' + uv'^2\} = 0 \tag{3-6}$$
$$P_2'\ v' + (2v + u + u')v' + v'^2 + (v-C\eta)u' = 0 \tag{3-7}$$

where P_i', i = 1,2 are the heat operators in the variables $x' = \frac{x}{\sqrt{\eta}}$, $t' = \frac{t}{\eta}$.

The boundary and initial conditions are taken in the easiest way such that :

$$(u+u')\ (0) = u_{in} \qquad (v+v')\ (0) = 0$$
$$(u+u')|_\Sigma = (v+v')|_\Sigma = 0$$

We expand each of these four functions in the powers of η and find :

(i) Order 0

. u_{oo} is the solution of (3-1), (3-2).

. $v_{oo} \equiv u'_{oo} \equiv v'_{oo} \equiv 0$

(ii) Order 1

. u_{o1} is the unique solution in $W(0,T) \cap L^\infty(Q)$ of the linear heat equation :

$$P_1 u + (C+2a_o u_{oo} + 3a_1 u_{oo}^2)u = -Cu_{oo}(a_o + 2a_1 u_{oo}) \tag{3-8}$$
$$u(0) = 0 \tag{3-9}$$

$.u'_{o1} \equiv 0 \qquad v_{o1} \equiv C$

$.v'_{o1}$ is the boundary layer function at order 1 solution of :

$$\eta P_2 v + u_{oo}\, v = 0 \tag{3-10}$$

$$v(0) = -C \tag{3-11}$$

$$v|_{\Sigma} = -C \tag{3-12}$$

which has a unique solution such that $v'_{o1} + C \in W(0,T) \cap L^{\infty}(Q)$ and $-C \leq v'_{o1} \leq 0$ a.e. in Q.

(iii) Error estimates at order 1. We obtain :

$$||u_{o\eta} - u_{oo} - \eta u_{o1}||_V + ||v_{o\eta} - \eta(C + v'_{o1})||_V \leq K\eta \tag{3-13}$$

(iv) We can continue the expansions at orders greater than one but we cannot have better estimates than (3-13). The reason is that, in the equations verified by the remaining terms, we will have estimates of the preceding order in the second member.

4. CONVERGENCE AND ASYMPTOTIC EXPANSIONS WHEN ε AND $\eta \to 0$ INDEPENDENTLY

(a) Convergence

When ε and $\eta \to 0$, $u_{\varepsilon\eta} \to u_{oo}$ in $W(0,T)$ where u_{oo} is the solution of (3-1), (3-2) ; $v_{\varepsilon\eta} \to 0$ in $L^{\infty}(Q) \cap V$ and we have the error estimates :

$$||u_{\varepsilon\eta} - u_{oo}||_{W(0,T)} + ||v_{\varepsilon\eta}||_{L^{\infty}(Q)} \leq K(\varepsilon+\eta) \tag{4-1}$$

$$||v_{\varepsilon\eta}||_V \leq K(\varepsilon+\eta)^{1/2} \tag{4-2}$$

(b) Asymptotic expansions

We use the same technique of inner and outer expansions as in section 3-(b), but we expand u, u', v, v' in the powers of ε and η : $u = u_{oo} + \varepsilon u_{1o} + \eta u_{o1} + \ldots$, etc... We find :

(i) Order 0 .

. u_{oo} is the solution in W(0,T) of (3-1), (3-2).

. $v_{oo} \equiv u'_{oo} \equiv v'_{oo} \equiv 0$

(ii) Order 1 in ε, 0 in η.

. $u_{1o} \equiv v_{1o} \equiv u'_{1o} \equiv 0$

. v'_{1o} is the boundary layer function in W(0,T) solution of :

$$\eta P_2 v + u_{oo} v = 0 \tag{4-3}$$

$$v(0) = v_{in} \tag{4-4}$$

$$v|_{\Sigma} = 0 \tag{4-5}$$

Moreover :

$$0 \le v'_{1o} \le 1 \text{ a.e. in } Q$$

(iii) Order 0 in ε, 1 in η .

. u_{o1} is the solution in W(0,T) of (3-8), (3-9).

. $v_{o1} \equiv C$

. $u'_{o1} \equiv 0$

. v'_{o1} is the solution of (3-10)-(3-12).

(iv) Error estimates .

$$||u_{\varepsilon\eta} - u_{oo} - \eta u_{o1}||_V \le K(\varepsilon+\eta) \tag{4-6}$$

$$||v_{\varepsilon\eta} - \varepsilon v'_{1o} - \eta(C+v'_{o1})||_V \le K \frac{(\varepsilon+\eta)^2}{\eta} \tag{4-7}$$

$$||v_{\varepsilon\eta} - \varepsilon v'_{1o} - \eta(C+v'_{o1})||_{L^2(0,T\,;\,L^2_{loc}(\Omega))} \le K \frac{(\varepsilon+\eta)^2}{\eta^{1/2}} \tag{4-8}$$

(4-8) is obtained by using property (3-4).

(v) Further orders. As in section 3-(b)-(iv) and for the same reasons we cannot get better estimates than (4-6) - (4-8) at further orders.

REFERENCES

Bourgeat A. and Tapiéro R. (1977). *Thesis*. Université Claude Bernard - Lyon I.

Ladyženskaja O.A., Solonnikov V.A. and Ural'ceva N.N. (1968). *Linear and quasi-linear equations of parabolic type*. Translations of mathematical monographs. Vol. 23. A.M.S. Providence, Rhode Island.

Sadeghi Kharrazi N. (1974). *Thesis*. Université Scientifique et médicale de Grenoble.

SINGULAR PERTURBATION THEORY IN CHEMICAL KINETICS

G.M. Côme
Institut National Polytechnique de Lorraine
E.R.A. n° 136 du C.N.R.S.
1, rue Grandville, 54042 NANCY (France)

ABSTRACT

Mathematical models in chemical kinetics often consist of stiff ordinary differential equations with initial values. A boundary layer type method is proposed for solving these problems. The conditions of use of such a method are precised. Splitting the variables into stiff and non-stiff ones and determining small perturbing parameters are a priori achieved on reactivity considerations. Adimensional criteria allow the determination of boundary layer lengths corresponding to various approximations (Reactant stationarity, pseudo and quasi-stationary state, long chain approximations). Both the inner and outer regions are solved by standard numerical procedures.

1. INTRODUCTION

The mathematical model of an homogeneous isothermal batch reactor consists of a system of ordinary differential equations

$$dc_j/dt = R_j(c_1, c_2, \ldots, c_q) \qquad (1.1)$$

with initial values :

$$c_j\,(o) = c_j^{\circ} \quad ; \quad j = 1, 2, \ldots, q \tag{1.2}$$

where :

$$R_j = \sum_{i=1}^{s} \nu_{ij}\, r_i \tag{1.3}$$

$$r_i = k_i \prod_{\ell=1}^{q} c_\ell^{\,n_{i\ell}} \tag{1.4}$$

$$n_{i\ell} = (|\nu_{i\ell}| - \nu_{i\ell})/2 \tag{1.5}$$

The stoichiometric coefficients ν_{ij} are integer

$$|\nu_{ij}| \leqslant 2 \tag{1.6}$$

The rate constants k_i and the concentrations c_j are real positive. The reaction time t increases from 0 to a finite value.

The differential system (1.1) has been shown to be stiff by Côme (1977) and therefore difficult to deal with numerically. Since most methods for stiff systems tend to degrade as the stiffness increases, it would be interesting to have recourse to a boundary layer type method, which improves as the stiffness increases.

Miranker (1973) has proposed such a method, which can be summarized as follows : a) The boundary layer nature of the solution is assumed b) A purely numerical procedure is used to split the variables into stiff and non-stiff c) The method does not need to identify a priori the small parameter ε d) The solution is given as a first order outer approximation, i.e. the boundary layer is supposed to be over e) There is no estimate of the precision, because ε is not known.

It is perfectly clear, from a user's point of view, that <u>a priori</u> answering questions a) to e) could avoid both bad approximations and a lot of computer time. This is the purpose of this work, for a particular system, while Miranker's method is for general purposes.

2. BOUNDARY LAYER TYPE PROBLEMS

The subclass of stiff problems to which the boundary layer methods are designed to be applied is such that the eigenvalues of the Jacobian associated to the stiff variables have negative real parts. Using Hadamard-Gerschgorin's theorem, one obtains the following sufficient conditions for the system (1.1) :

$$n_{ij} \left(\nu_{ij} + \sum_{\ell \neq j} |\nu_{i\ell}|\right) \leqslant 0, \quad i = 1, 2, \ldots, s \qquad (2.1)$$

These equations cannot be met for all values of j, but only for those variables which are stiff.

3. PARTITIONING OF VARIABLES

This partitioning can be achieved on the basis of physicochemical considerations. Four groups of variables are defined :

Non-determining radicals	$x = (c_1, c_2, \ldots, c_m)$,	$x(o) = 0$
Determining radicals	$y = (c_{m+1}, c_{m+2}, \ldots, c_n)$,	$y(o) = 0$
Reactants	$u = (c_{n+1}, c_{n+2}, \ldots, c_p)$,	$u(o) \neq 0$
Products	$v = (c_{p+1}, c_{p+2}, \ldots, c_q)$,	$v(o) = 0$

x and y are stiff and u and v non stiff variables. Therefore, for the time being, we shall assume that $u \simeq u(o)$ and $v \simeq v(o)$. Furthermore, the inequalities (2.1) will be assumed to be granted for j = 1 to n. (1.1) becomes :

$$\frac{dc_i}{dt} = 2\, a_i + \sum_{j=1}^{n} p_{ji}\, c_j \;; \; c_i(o) = 0, \quad i = 1, 2, \ldots, m \qquad (3.1)$$

$$\frac{dc_i}{dt} = 2\, a_i + \sum_{j=1}^{n} p_{ji}\, c_j - \mathop{S}_{j=m+1}^{n} \gamma_{ij}\, c_i\, c_j - 2\, \gamma_{ii}\, c_i^2 \;;$$
$$c_i(o) = 0, \quad i = m+1, m+2, \ldots, n \qquad (3.2)$$

where

$$p_{ii} = - \underset{j=1}{\overset{n}{S}} p_{ij} \tag{3.3}$$

$$S_j \gamma_{ij} = \sum_{j \neq i} \gamma_{ij} \tag{3.4}$$

or, in vector form

$$\frac{dx}{dt} = I_1 + P_{11} x + P_{12} y, \quad x(o) = 0 \tag{3.5}$$

$$\frac{dy}{dt} = I_2 + P_{21} x + P_{22} y + Q_2 (y), \quad y(o) = 0 \tag{3.6}$$

Non-determining radicals are more reactive than determining ones, so we can readily assume that

$$\frac{1}{\varepsilon} = \min_{\substack{i=1,m \\ j=1,n \\ j \neq i}} p_{ij} \gg \max_{\substack{i=m+1,n \\ j=1,n \\ j \neq i}} p_{ij} \tag{3.7}$$

It is now quite sensible to seek asymptotic expansions of x and y of singular perturbation type :

$$x(t) \sim \sum_{\ell=0}^{\infty} (x_\ell (t) + X_\ell (\tau)) \varepsilon^\ell \tag{3.8}$$

$$y(t) \sim \sum_{\ell=0}^{\infty} (y_\ell (t) + Y_\ell (\tau)) \varepsilon^\ell \tag{3.9}$$

where $\tau = t/\varepsilon$ is the stretched variable and x_ℓ, y_ℓ and X_ℓ, Y_ℓ the coefficients of the outer and inner solutions respectively.

Following the well known procedure, one obtains the composite solution :

$$x = (e^{P_{11} t} - U) P_{11}^{-1} I_1 - P_{11}^{-1} P_{12} y_o + \dots \tag{3.10}$$

$$y = y_o + \dots \tag{3.11}$$

4. PSEUDO-STATIONARY STATE APPROXIMATION (PSSA)

Let us now make the hypothesis $\tau >> 1$. It follows that $e^{P_{11}t}$ is negligible in (3.10) and then :

$$x \simeq - P_{11}^{-1} (I_1 + P_{12} y_o) \tag{4.1}$$

This PSSA allows us to calculate x as function of y_o even during the boundary layer (or induction period).

5. QUASI-STATIONARY STATE APPROXIMATION (QSSA)

From now on, we are assuming that the PSSA for x radicals is granted, i.e. (3.1) is replaced by :

$$0 = 2 a_i + \sum_{j=1}^{n} p_{ji} c_j, \qquad i = 1, 2, \ldots, m \tag{5.1}$$

Let us define three adimensional numbers :

$$\theta = t (a \gamma)^{1/2} \tag{5.2}$$

$$\pi = t p \tag{5.3}$$

$$\lambda = \pi/\theta = p/(a \gamma)^{1/2} \tag{5.4}$$

Where :

$$a = \sum_{i=1}^{n} a_i \tag{5.5}$$

$$p = \min_{\substack{i = m+1, n \\ j = 1, n \\ j \neq i}} p_{ij} \tag{5.6}$$

γ is defined by eq. (5.11).

We shall suppose that the chains are long, so that $\lambda >> 1$. We introduce the new variables λ and θ in (5.1) and (3.2) and we seek an expansion of c_i in inverse powers of λ :

$$c_i = c_{io} + c_{i1}/\lambda + c_{i2}/\lambda^2 + \ldots \tag{5.7}$$

We first get :

$$c_{io} = \xi_i c_o \tag{5.8}$$

The ξ_i s are only functions of rate constants and :

$$\sum_{i=1}^{n} \xi_i = 1 \tag{5.9}$$

$$c_o = \left(\frac{a}{\gamma}\right)^{1/2} \frac{1 - e^{-4\theta}}{1 + e^{-4\theta}} \tag{5.10}$$

with :

$$\gamma = \sum_{i=m+1}^{n} (\gamma_{ii}^{1/2} \xi_i)^2 \tag{5.11}$$

$$\gamma_{ij} = \gamma_{ji} = 2\,(\gamma_{ii}\,\gamma_{jj})^{1/2} \tag{5.12}$$

If $e^{-4\theta} \ll 1$, we can see that the QSSA is valid, i.e. the boundary layer or induction period t_i is over and :

$$c_o \simeq (a/\gamma)^{1/2} \tag{5.13}$$

For a six digit precision, one obtains the expression of t_i :

$$t_i = 3.6/(a\,\gamma)^{1/2} \tag{5.14}$$

6. LONG CHAIN APPROXIMATION (LCA)

The expressions of the first "normalized" perturbation terms, with θ being small, can be written :

$$\left\{\begin{array}{ll} \displaystyle\sum_{j=1}^{n} \frac{p_{ji}}{p}\,\frac{c_{j1}}{\lambda c_o} = -\frac{a_i}{a}\cdot\frac{1}{\pi} \;; & i = 1, m \\ \displaystyle\sum_{j=1}^{n} \frac{p_{ji}}{p}\,\frac{c_{j1}}{\lambda c_o} = (\xi_i - \frac{a_i}{a})\frac{1}{\pi} + \frac{\theta}{\lambda}\,\frac{\displaystyle S_{j=m+1}^{n}\gamma_{ij}\,\xi_i\,\xi_j + 2\gamma_{ii}\,\xi_i^2}{\gamma} \;; & \\ & i = m+1, n \end{array}\right. \tag{6.1}$$

Consequently, we can see that, if $\pi \gg 1$ and, of course, $\lambda \gg 1$, LCA can be used (eq. 5.8 to 5.12).

7. REACTANT STATIONARITY CRITERION (RCA)

In the beginning of this theory, we were supposing that $u \simeq u\,(o)$, and $v \simeq v\,(o)$. For material balance reasons $u \simeq u\,(o)$ implies $v \simeq v\,(o)$. It follows that the reactant stationarity criterion is written :

$$\rho = k\,(a/\gamma)^{1/2}\,t \qquad (7.1)$$

where k is the maximum determining propagation rate constant. So, if $\rho \ll 1$, it follows that $u \simeq u\,(o)$, and we can use the preceding approximations if their idiosyncratic conditions of use are met.

8. ALGORITHM

We can now build up a numerical algorithm :

$\rho \ll 1$? ⟶ No ⟶ No approximations

↓

Yes

↓

$\tau \ll 1$? ⟶ No ⟶ Apply RSA

↓

Yes

↓

$\pi, \lambda \gg 1$? ⟶ No ⟶ Apply RSA, PSSA

↓

Yes

↓

$\theta > 3{,}6$? ⟶ No ⟶ Apply RSA, PSSA, LCA

↓

Yes ⟶ Apply RSA, PSSA, LCA, QSSA

Each criterion is evaluated for the local condition (e.g. conditions (1.2) for the first step) and the local step size h (i.e. set t = h to calculate the criteria). Depending on the test results, equations (4.1), (5.10) or (5.13) are used. The complete system can now be solved by standard numerical algorithms, because the equations are no longer stiff. The procedure is then repeated on the following interval and so on. Let us notice that the method can be used equally in the inner region (in the inner region, the stiffness disappears because the step size is necessarily small) and outer one.

9. CONCLUSION

In the 1920's, Bodenstein formulated the QSSA, which has been, since then, the mathematical technique most commonly used in chemical kinetics. Numerical analysis of chemical kinetics equations has been performed by Snow (1966) using QSSA. But QSSA has been questionned lately by Farrow and Edelson (1974), who have asked whether QSSA is fact or fiction, and who recommend that stiff algorithms (such as GEAR's algorithm) be adopted in future works.

As a conclusion, we can say : a) A boundary layer type method has been used in chemical kinetics a long time ago, but without justification. b) A general purpose method does exist (Miranker, 1973), but some questions remain unanswered, as in Aiken and Lapidus (1975). c) This work gives the precise conditions of use of a similar method and an easy way to perform the calculationsfor chemical kinetics models.

10. REFERENCES

Aiken, R.C. and Lapidus, L. (1974). An effective numerical integration method for typical stiff systems. *A.I.Ch.E. Journal*, 20, 368-375.

Côme, G.M. (1977). Radical reaction mechanisms : mathematical theory. *J. Phys. Chem.*, 81, 2560-2563.

Farrow, L.A. and Edelson, D. (1974). The steady-state approximation : fact or fiction ? *Int. J. Chem. Kin.*, VI, 787-800.

Miranker, W.L. (1973). Numerical methods of boundary layer type for stiff systems of differential equations. *Computing*, 11, 221-234.

Snow, R.H. (1966). A chemical kinetics computer program for homogeneous and free-radical systems of reactions. *J. Phys. Chem.*, 70, 2780-2786.

A DIRECT THREE DIMENSIONAL APPROACH FOR LINEAR PLATE BY NAGHDI'S METHOD

Ph. Destuynder

Centre de Mathématiques Appliquées
Ecole Polytechnique
91128 Palaiseau Cédex, France

ABSTRACT

We deal here with the classical equations of linear elasticity for plates. Using the three dimensional approach we give a scheme of approximation concerning the dependance upon the coordinate in the thickness of the plate. We give also an error bound.

The difficulty lays in the singular behaviour of the three dimensional problem with respect to the thickness of the plate.

INTRODUCTION

We usually consider that the bending effect in linear plate is obtained by solving a biharmonic equation, meanwhile the membrane effect is governed by the classical equations of elasticity in two dimensions. From a numerical point of view, it is wellknown that the approximation of a biharmonic equation is both difficult and expensive. For these reasons, one has tried to treat numerically the complete system of elasticity considered as a three dimensional system. On the one hand the advantage is that we deal with a second order operator, on the other hand, the thickness of the plate still appears as a small parameter, and the approximate problem is a singular problem with respect to this small parameter. In the present approach, we study a particular scheme of approximation proposed by Naghdi.

We give an error estimate between the solution of the three dimensional problem and the approximate one.

As a rule Greek indices α,β take their values in the set $\{1,2\}$. While latin indices i,j take their values in the set $\{1,2,3\}$.

1 - NOTATIONS

The plate considered here will be supposed to occupy the open set Ω^ε of $\mathbb{R}^2$. We set : $\Omega^\varepsilon = \omega \times]-\varepsilon, \varepsilon[$, where ω denotes the middle surface of the plate, and 2ε the thickness. Moreover we set (Fig. 1) :

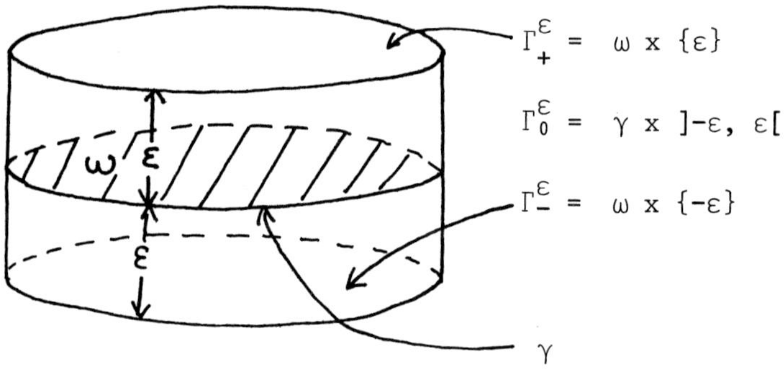

Fig. 1

We use the notation $\partial_\alpha f = \dfrac{\partial f}{\partial x_\alpha}$.

The displacement field, $u_i(x_1, x_2, x_3)$ and the stress field $\sigma_{ij}(x_1, x_2, x_3)$ are the unknowns.

The three-dimensional problem for plates can be formulated as follows : Let us set

$$\underset{\sim}{\Sigma}^{\varepsilon} = \{\tau = (\tau_{ij}) \in (L^2(\Omega^{\varepsilon}))^9 \;\; \tau_{ij} = \tau_{ij}\} \qquad (1)$$

$$\underset{\sim}{V}^{\varepsilon} = \{v = (v_i) \in (H^1(\Omega^{\varepsilon}))^3 \;;\; v = 0 \text{ sur } \Gamma_0^{\varepsilon}\} \qquad (2)$$

If E, ν denotes the mechanical coefficients of Young and Poisson (with $E > 0$, $0 < \nu < \frac{1}{2}$), then the problem is :

Find $(\sigma,u) \in \underset{\sim}{\Sigma}^{\varepsilon} \times \underset{\sim}{V}^{\varepsilon}$, such that

$$\forall\, \tau \in \underset{\sim}{\Sigma}^{\varepsilon}, \int_{\Omega^{\varepsilon}} \left(\frac{1+\nu}{E}\sigma_{ij} - \frac{\nu}{E}\sigma_{pp}\delta_{ij}\right)\tau_{ij} - \int_{\Omega^{\varepsilon}} \partial_i u_j \tau_{ij} = 0, \qquad (3)$$

$$\forall\, v \in \underset{\sim}{V}^{\varepsilon}, \int_{\Omega^{\varepsilon}} \sigma_{ij}\partial_i v_j = \int_{\Omega^{\varepsilon}} f_i v_i + \int_{\Gamma_+^{\varepsilon} \cup \Gamma_-^{\varepsilon}} g_i v_i \;. \qquad (4)$$

In the above equations, f_i (respectively g_i) denote the body forces (resp. surface forces) which act on the plate.

It can be easily shown that equations (3) and (4) have one and only one solution.

2 - THE APPROXIMATE THREE-DIMENSIONAL PROBLEM

In order to define the approximate problem let us first introduce the approximate spaces of functions.

$$\underset{\sim}{\Sigma}^1 = \{\tau = (\tau_{ij}) \in \underset{\sim}{\Sigma},\; \tau_{ij} = \tau^0_{ij} + x_3\, \tau^1_{ij}, \qquad (5)$$

with $\underline{\tau^1_{33} = 0}$, and $\tau^0_{ij},\; \tau^1_{ij} \in L^2(\omega)\}$

$$\underset{\sim}{V}^1 = \{v = (v_i) \in \underset{\sim}{V},\; v_i = v_i^0 + x_3\, v_i^1, \text{ with} \qquad (6)$$

$$v_i^0,\; v_i^1 \in H_0^1(\omega)\}.$$

These spaces are closed subspaces of $\underset{\sim}{\Sigma}$ and $\underset{\sim}{V}$, and they can be equipped with the norms of $\underset{\sim}{\Sigma}$ and $\underset{\sim}{V}$ which are clearly equivalent to some norms of product spaces of $L^2(\omega)$ type for $\underset{\sim}{\Sigma}$, and $H_0^1(\omega)$ for $\underset{\sim}{V}$.

The approximate problem (with self explanatory notations) is then defined by :

Find $(\sigma^{(1)}, u^{(1)}) \in \underset{\sim}{\Sigma}^1 \times \underset{\sim}{V}^1$ such that

$$\forall\, \tau^{(1)} \in \underset{\sim}{\Sigma}^1,\ a(\sigma^{(1)}, \tau^{(1)}) + b(\tau^{(1)}, u^{(1)}) = 0\ , \tag{7}$$

$$\forall\, v^{(1)} \in \underset{\sim}{V}^1,\ b(\sigma^{(1)}, v^{(1)}) = F(v^{(1)}). \tag{8}$$

If we set $(\sigma^{(1)}, u^{(1)}) = (\sigma^0, u^0) + x_3(\sigma^1, u^1)$, then we have an equivalence between the two following sets of assertions :

. $(\sigma^{(1)}, u^{(1)})$ is a solution of (7), (8), on one hand,

. and on the other hand :

1) - The element $(u^0{}_\alpha, u^1_3, \sigma^0{}_{\alpha\beta}\ \sigma^1{}_{\alpha 3}, \sigma^0{}_{33})$ which belongs to $(H^1_0(\omega))^3 \times (L^2(\omega))^7$ is a solution of

$\forall\, \tau\alpha\beta \in L^2(\omega), \tau_{12} = \tau_{21}$,

$$\int_\omega \left(\frac{1+\nu}{E}\sigma^0_{\alpha\beta} - \frac{\nu}{E}\sigma^0_{\mu\mu}\ \delta_{\alpha\beta}\right)\tau_{\alpha\beta} - \int_\omega \frac{\nu}{E}\sigma^0_{33}\tau_{\mu\mu} - \int_\omega \partial_\alpha\ u^0_\beta\ \tau_{\alpha\beta} = 0, \tag{9}$$

$$\forall\, \tau\alpha_3 \in L^2(\omega),\ 2\,\frac{\varepsilon^2}{3}\int_\omega \frac{1+\nu}{E}\sigma^1\alpha_3\tau_{\alpha_3} - \frac{\varepsilon^2}{3}\int_\omega \tau\alpha_3 . \partial\alpha u_3 = 0, \tag{10}$$

$$\forall\, \tau_{33} \in L^2(\omega),\ \int_\omega \frac{1}{E}\sigma^0_{33}\ \tau_{33} - \int_\omega \frac{\nu}{E}\sigma^0_{\alpha\alpha}\tau_{33} - \int_\omega u^1_3\tau_{33} = 0, \tag{11}$$

$$\forall\, v\ \alpha \in H^1_0(\omega),\ \int_\omega \sigma^0\alpha\beta\ \partial\alpha v\beta = \int_\omega F^0_\alpha\ v\alpha, \tag{12}$$

$$\forall\, v \in H^1_0(\omega),\ \int_\omega \sigma^0_{33}v_3 + \frac{\varepsilon^2}{3}\int_\omega \sigma^1_{\alpha_3}\ \partial\alpha v_3 = \int_\omega F^1_3 v_3 . \tag{13}$$

with
$$F^0\alpha = \frac{1}{2\varepsilon}\int_{-\varepsilon}^{+\varepsilon} f\alpha dx_3 + \frac{1}{2\varepsilon}(\overset{+}{g\alpha} + \overset{-}{g\alpha}),$$
$$F^1_3 = \frac{1}{2\varepsilon}\int_{-\varepsilon}^{+\varepsilon} x_3\ f_3\ dx_3 + \frac{\overset{+}{g_3} - \overset{-}{g_3}}{2}.$$

2) - The element $(u^1\alpha, u^0_3, \sigma^1\alpha\beta, \sigma^0\alpha_3) \in (H^1_0(\omega)^3 \times (L^2(\omega))^6$ is a solution of

$\forall\, \tau\alpha\beta \in L^2(\omega), \tau_{12} = \tau_{21}$,

$$\frac{\varepsilon^2}{3}\int_\omega \left(\frac{1+\nu}{E}\sigma^1\alpha\beta - \frac{\nu}{E}\sigma^1\mu\mu\delta\alpha\beta\right)\tau\alpha\beta - \frac{\varepsilon^2}{3}\int_\omega \partial\alpha u^1{}_\beta\ \tau\alpha\beta = 0, \tag{14}$$

$$\forall\ \tau_{\alpha 3} \in L^2(\omega),\ 2\int_\omega \frac{1+\nu}{E}\sigma^0{}_{\alpha 3}\tau_{\alpha 3} - \int_\omega (\partial_\alpha u_3^0 + u^1{}_\alpha)\ \tau_{\alpha 3} = 0, \quad (15)$$

$$\forall\ v_\alpha \in H_0^1(\omega), \frac{\varepsilon^2}{3}\int_\omega \sigma^1{}_{\alpha\beta}\ \partial_\alpha v_\beta + \int_\omega \sigma^0_{\alpha 3}\ v_\alpha = \int_\omega F^1_\alpha\ v_\alpha, \quad (16)$$

$$\forall\ v_3 \in H_0^1(\omega), \int_\omega \sigma^0{}_{\alpha 3}\partial_\alpha v_3 = \int_\omega F_3^0 v_3. \quad (17)$$

where
$$F_3^0 = \frac{1}{2\varepsilon}\int_{-\varepsilon}^{+\varepsilon} f_3\ dx_3 + \frac{\overset{+}{g}_3 - \overset{-}{g}_3}{2\varepsilon}$$

$$F^1{}_\alpha = \frac{1}{2\varepsilon}\int_{-\varepsilon}^{+\varepsilon} x_3\ f_\alpha\ dx_3 + \frac{\overset{+}{g}_\alpha - \overset{-}{g}_\alpha}{2}.$$

Remark 1 : From now on, we shall consider that the second member $(F_3^0, F_3^1, F^0_\alpha, F^1_\alpha)$, doesn't depend any more on ε. Of course this is an hypothesis, but which doesn't restrict the applicability of our method owing to the linearity of the problem studied here.

Remark 2 : The existence and unicity of a solution (6), (7) is equivalent to the existence and unicity of a solution of (9), (13) on the one hand, and (14), (17) on the other hand which is easy to obtain.

3 - ERROR ESTIMATE WITH THE THREE DIMENSIONAL PROBLEM

We shall apply the following result, usually called "A posteriori estimate", and which has been demonstrated first by Aubin [1], in a more general frame than the one we need here.

THEOREM 1

Let $(\sigma, u) \in \Sigma^\varepsilon \times V^\varepsilon$ *be the solution of (3), (4) and* $(\tilde{\sigma}^{(1)}, \tilde{u}^{(1)})$ the solution of (7), (8). Then for each stress field which equilibrates the exterior load σ^e (i.e. such that :

$$\forall\, v \in \underset{\sim}{V}^{\varepsilon} \int_{\Omega^{\varepsilon}} \sigma^{e}_{ij}\,\partial_i v_j = \int_{\Omega^{\varepsilon}} f_i v_i + \int_{\Gamma^{\varepsilon}_{+} \cup \Gamma^{\varepsilon}_{-}} q_i v_i)$$

the following inequalities hold :

$$\left\{ \begin{array}{l} \| \sigma - \underset{\sim}{\sigma}^{(1)} \|_{\underset{\sim}{\Sigma}^{\varepsilon}} \leqslant C \| \sigma^{e} - \underset{\sim}{\sigma}^{(1)} \|_{\Sigma^{\varepsilon}}{}' \\ \| \gamma(u) - \gamma(\underset{\sim}{u}^{(1)}) \|_{\underset{\sim}{\Sigma}^{\varepsilon}} \leqslant C \| \sigma^{e} - \underset{\sim}{\sigma}^{(1)} \|_{\underset{\sim}{\Sigma}^{\varepsilon}} \end{array} \right. \tag{18}$$

where C *is a constant which depends only on the coefficients* E *and* ν, *and where we have set :*

$$\begin{aligned} \underset{\sim}{\sigma}^{(1)}_{ij} &= \frac{E}{1+\nu} \{\gamma_{ij}(u^{(1)} + \frac{\nu}{1-2\nu} \gamma_{pp} (u^{(1)})\gamma_{ij}\}, \\ \gamma_{ij}(u^{(1)}) &= \frac{1}{2} (\partial_i u_j^{(1)} + \partial_j u_i^{(1)}). \end{aligned} \tag{19}$$

We point out here that $\underset{\sim}{\sigma}^{(1)}_{ij} \neq \sigma^{(1)}_{ij}$. ∎

The proof is a standard manipulation of the inequalities which characterize the solution of equations (3) and (4).

Theorem 1 can be used to obtain an error estimate with the three-dimensional problem in the following way :

1) – We set $\sigma^{e}\alpha\beta = \sigma^{(1)}\alpha\beta = \sigma^{0}\alpha\beta + x_3\, \sigma^{1}\alpha\beta$, which is the solution of the approximate problem (7), (8).

2) – Considering now the equation :

$$\forall\, v\,\alpha \in H^1(\Omega^{\varepsilon}),\ v_{\alpha} = 0 \text{ on } \Gamma^{\varepsilon}_{0}\ ;\ \int_{\Omega^{\varepsilon}} \sigma^{e}\alpha_3 \partial_3 v_{\alpha} = \int_{\Omega^{\varepsilon}} f_{\alpha} v_{\alpha} + $$

$$+ \int_{\Gamma^{\varepsilon}_{+} \cup \Gamma^{\varepsilon}_{-}} g_{\alpha} v_{\alpha} - \int_{\Omega^{\varepsilon}} \sigma^{(1)}_{\alpha\beta}\, \partial_{\alpha}\, v_{\beta},$$

where $\sigma^{e}\alpha_3$ is the unknown, since this equation is satisfied for $v\,\alpha \in H^1_0(\omega)$, there exists one unique solution. We have :

$$\begin{aligned} \sigma^{e}_{\alpha_3} &= \frac{\overset{+}{g\alpha} - \overset{-}{g\alpha}}{2} + \frac{1}{2} \int_{-\varepsilon}^{+\varepsilon} f\alpha - \int_{-\varepsilon}^{x_3} f\alpha + \\ &+ x_3 F^{0}_{\alpha} + 3\frac{(x_3^2 - \varepsilon^2)}{2\varepsilon^2} (F^{1}_{\alpha} - \sigma^{0}\alpha_3). \end{aligned} \tag{20}$$

3) - Finally from

$\forall\ v_3 \in H^1(\Omega^\varepsilon),\ v_3 = 0$ on Γ_0^ε

$$\int_{\Omega^\varepsilon} \sigma^e{}_{33}\partial_3 v_3 = \int_{\Omega^\varepsilon} f_3 v_3 + \int_{\Gamma_+^\varepsilon \cup \Gamma_-^\varepsilon} g_3 v_3 - \int_{\Omega^\varepsilon} \sigma^e{}_{\alpha 3}\partial_\alpha v_3 ,$$

we deduce the existence of $\sigma^e{}_{33}$

$$\sigma^e{}_{33} = \frac{1}{2}\int_{-\varepsilon}^{+\varepsilon} f_3 - \int_{-\varepsilon}^{x_3} f_3 + \frac{\overset{+}{g_3} - \overset{-}{g_3}}{2} + \frac{1}{2}\int_{-\varepsilon}^{+\varepsilon} \partial\alpha\sigma^e{}_{\alpha 3} \quad \int_{-\varepsilon}^{x_3} \partial\alpha\sigma^e{}_{\alpha 3} . \tag{21}$$

With this expression of σ^e, we obtain by theorem 1

$$\|\gamma(u)-\gamma(u^{(1)})\|_{\underset{\sim}{\Sigma}^\varepsilon} + \|\sigma - \underset{\sim}{\sigma}^{(1)}\|_{\underset{\sim}{\Sigma}^\varepsilon} \leqslant C \sum_{\alpha=1,2} \|\sigma^e_{\alpha 3} \quad \sigma^{(1)}_{\alpha 3}\|_{L^2(\Omega^\varepsilon)} + \|\sigma^e_{33} - \sigma^{(1)}_{33}\|_{L^2(\Omega^\varepsilon)} \tag{22}$$

Using the explicit expression of $\sigma^e{}_{\alpha 3}$ and $\sigma^e{}_{33}$, and equations (9), (17), we have as soon as $F_3^0 \in L^2(\omega)$, $F_3^1 \in L^2(\omega)$, $F_\alpha^0 \in H^1(\omega)$, $F_\alpha^1 \in H^1(\omega)$:

$$\|\gamma(u) - \gamma(u^{(1)})\|_{\underset{\sim}{\Sigma}^\varepsilon} + \|\sigma - \underset{\sim}{\sigma}^{(1)}\|_{\underset{\sim}{\Sigma}^\varepsilon} \leqslant C\sqrt{\varepsilon} . \tag{23}$$

Our final result will be stated as follows :

THEOREM 2

Setting $n_{\alpha\beta} = \int_{-\varepsilon}^{+\varepsilon} \sigma_{\alpha\beta}\, dx_3$, *which are the resulting stresses in the thickness of the plate, and* $m_{\alpha\beta} = \int_{-\varepsilon}^{+\varepsilon} x_3\, \sigma_{\alpha\beta}\, dx_3$, *for the bending moments then we have :*

$$\left\{\begin{aligned} &\frac{\|n_{\alpha\beta} - \underset{\sim}{n}^{(1)}_{\alpha\beta}\|_{L^2(\omega)}}{\|n_{\alpha\beta}\|_{L^2(\omega)}} \leqslant C\,\varepsilon \\ &\frac{\|m_{\alpha\beta} - \underset{\sim}{m}^{(1)}_{\alpha\beta}\|_{L^2(\omega)}}{\|m_{\alpha\beta}\|_{L^2(\omega)}} \leqslant C\,\varepsilon . \quad \blacksquare \end{aligned}\right. \tag{24}$$

4 - CONCLUSION

We have proved here that the direct three-dimensional approach for plates can be efficient if we choose a good scheme of approximation with respect to the coordinate in the thickness of the plate.

A more complete study [3], of the equations (9 - 13) and (14 - 17) gave us the behaviour of the solutions when ε tends to zero. Furthermore we obtained the expression of the boundary layer which appears near the lateral boundary of the plate when it is clamped [3]. This allows us to give a rule of mesh refinement near the boundary to obtain a better accuracy in the solution of equations (9 - 13) - (14 - 17). Finally we mentioned numerical esperiments concerning equations (14 -17) performed by Bercovier 1978, [2] and Hugues, Taylor, Kanahnuhulchai 1977, [4].

REFERENCES

[1] J.P. Aubin, *Approximation of boundary elliptic value problem*, J. Wiley, (1972) p. 244.

[2] M. Bercovier, to appear.

[3] Ph. Destuynder, *Direct three dimensional approach for linear plate by Naghdi's method*, Rapport N° 36 du Centre de Mathématiques Appliquées, Mai 1978, Ecole Polytechnique.

[4] T.J.R. Hugues, R.L. Taylor, W. Kanahnuhulchai, *A simple and efficient finite element for plate bending*, Int. J. Numer. Method in Eng., Vol. 11, (1977), p. 1529-1543.

[6] R.C. Koeller, *On a formulation of the bending of elastic plates*, Int. J. Solids Structures (1973), Vol. 9, p. 1053-1074.

[7] P.M. Naghdi, *Handburch der physik*, Springer Verlag (1972) Band v/a/2, p. 425-640.

SOLUTIONS TO THE DIFFERENCE EQUATIONS OF SINGULAR PERTURBATION PROBLEMS

G. C. Hsiao and K. E. Jordan
Department of Mathematics
University of Delaware
Newark, Delaware, U.S.A.

ABSTRACT.

This paper discusses two numerical schemes for singularly perturbed boundary value problems. Both schemes are motivated by the asymptotic behavior of singular perturbation problems. One is based on the method of matched asymptotic expansions for treating singular perturbations for difference equations and the other one involves a modification of the boundary layer problem so that one can treat it as a boundary value problem in a finite domain by any suitable method such as the Galerkin's method. Both schemes have been tested on examples. Numerical results show that the accuracy predicated can always be achieved with very little computational effort.

1. INTRODUCTION.

The numerical treatment for singular perturbation problems has always been far from trivial, because of the boundary layer behavior of the solutions. The conventional finite difference approach is, of course, to reduce the mesh size in the neighborhood of the boundary layer, so that typical features of the boundary layer will not be lost. This procedure has been successfully applied to several problems by Pearson (1968) and many others. However, the disadvantage of this approach is that it requires considerable computational effort. In fact, often because of the limitation of the computer system, the discrete problems involved may become ill-posed numerically when mesh sizes are getting too small.

In this note, we propose two schemes. Neither one of them needs very fine mesh size. Both schemes are motivated by the asymptotic behavior of singular perturbation problems. In essence, our schemes use singular perturbation theory to construct the leading terms in formal asymptotic expansions of the solution. Thus they are more accurate as the small parameter in the equations become smaller.

Throughout the paper, for convenience, we shall term our schemes the *discrete (matched asymptotic) expansions scheme* (DES) and the *boundary layer scheme* (BLS). For simplicity, we shall confine ourself to the two-point boundary value problem for the second-order linear equation. To be specific, we consider the problem:

$$L_\varepsilon u := -\varepsilon u''+p(x)u'+q(x)u = f(x), \quad 0 < x < 1, \tag{E}$$

$$u(0) = \alpha, \qquad u(1) = \beta. \tag{B}$$

Here $0 < \varepsilon < 1$ is the small parameter, p, q, and f are given sufficiently smooth functions; α and β are given constants which may or may not be equal. Furthermore, we assume that $p(x) \geq p_0 > 0$ throughout the interval $[0,1]$. This assumption merely implies that the boundary layer will be in the neighborhood of $x = 1$. Then from the standard singular perturbation theory (see e.g. O'Malley, Jr. (1974)), it follows that the solution of the problem (E), (B) admits the representation

$$u = U(x)+V(\tilde{x})+O(\varepsilon) \quad \text{as} \quad \varepsilon \to 0^+ \tag{*}$$

uniformly in $[0,1]$. Here $\tilde{x} = \frac{1-x}{\varepsilon}$ is the stretched variable, and U and V are respectively the solutions of the *reduced problem* and the *boundary layer problem*. That is,

$$L_0 U := p(x)U' + q(x)U = f(x), \quad x > 0, \quad U(0) = \alpha \tag{I}$$

and

$$\tilde{L}V := -V'' - p(1)V' = 0, \quad \tilde{x} > 0, \tag{II}$$

$$V(0) = \beta - U(1), \quad \lim_{\tilde{x} \to \infty} V(\tilde{x}) = 0.$$

We note that (II) is an exterior boundary value problem in terms of the stretched variable $\tilde{x}$. Hence the usual numerical schemes cannot be directly applied to (II) without some modification. As will be seen, our BLS is one such modification. It should be emphasized that although here both (I) and (II) are linear and hence the solutions can be constructed explicitly, our aim is really for those nonlinear problems where U and V cannot be constructed explicitly.

2. DISCRETE EXPANSIONS SCHEME.

Comstock and Hsiao (1976) obtained a representation for the solution of the boundary value problem for the difference equation of the form:

$$\begin{aligned} &\varepsilon Y_{k+2} + a_k Y_{k+1} + b_k Y_k = 0 \quad (k = 0,1,2,\ldots,N-2) \\ &Y_0 = \alpha, \quad Y_N = \beta, \end{aligned} \tag{2.1}$$

where a_k and b_k are non-zero discrete functions which are assumed to be bounded. In analogue to (*), it was shown by Comstock et al (1976) that

$$Y_k = U_k + V_k + O(\varepsilon) \quad \text{as} \quad \varepsilon \to 0^+ \tag{2.2}$$

uniformly for $k = 0,1,2,\ldots,N$, where U_k is the solution of the *discrete reduced problem* defined by

$$a_k U_{k+1} + b_k U_k = 0 \quad (k = 0,1,\ldots,N-1); \quad U_0 = \alpha \tag{2.3}$$

and V_k is the solution of the *discrete boundary layer*

problem defined by

$$\tilde{V}_{k+2}+a_k\tilde{V}_{k+1} = 0 \quad (k = N-2,N-3,\ldots,0); \quad \tilde{V}_N = (\beta-U_N). \quad (2.4)$$

Here the proper stretched variable $\tilde{V}_k$ is related to V_k via $\tilde{V}_k = \varepsilon^{k-N}V_k$ and the boundary layer is always at the right end point, regardless of the sign of the coefficients a_k in (2.1).

The essence of our DES is to construct the corresponding U_k and V_k for the problem (E) and (B) so that the solution $u_k = u(x_k)$ can be approximated by Y_k in (2.2). Hence let us begin by considering the discretization of (E). As pointed out by Il'in (1969), for $p(x) > 0$, one should use centered difference approximation for the second derivative u'' and the backward difference approximation for the first derivative u'. This leads to the difference equation:

$$-\varepsilon u_{k+1}+(2\varepsilon+p_k h+h^2 q_k)u_k+(-\varepsilon-p_k h)u_{k-1} = h^2 f_k+h^2\tau_k,$$

$$k = 1,2,\ldots,N-1$$

with the local trunction error τ_k. Here as usual $h = 1/N$ denotes the mesh size, $x_k = kh$, $f_k = f(x_k)$, and similarly for p_k and q_k.

By introducing the new parameter $\bar{\varepsilon} = \varepsilon/h$, and neglecting τ_k, this can then be rewritten in the form of (2.1):

$$\begin{aligned} &\bar{\varepsilon}Y_{k+2}+a_k(\bar{\varepsilon})Y_{k+1}+b_k(\bar{\varepsilon})Y_k = F_k \quad (k = 0,1,2,\ldots,N-2) \\ &Y_0 = \alpha \quad \text{and} \quad Y_N = \beta \end{aligned} \quad (2.5)$$

where

$$a_k(\bar{\varepsilon}) := -(p_{k+1}+hq_{k+1}+2\bar{\varepsilon}), \quad b_k(\bar{\varepsilon}) := p_{k+1}+\bar{\varepsilon},$$

$$F_k := -hf_{k+1}.$$

From (2.5) we can define the proper U_k and V_k according to (2.3) (with the nonhomogeneous term F_k) and (2.4) with $\bar{\varepsilon}$ where a_k and b_k are replaced by $a_k(0)$ and $b_k(0)$ respectively. Then the solution of (E) and (B) at the node points x_k is approximated by $u_k^h = U_k+V_k$.

The basic error estimate can be stated as

Theorem 1. For sufficiently small $\bar{\varepsilon} = \varepsilon/h$, the approximation u_k^h, derived by the DES, satisfies

$$|u_k-u_k^h| \leq c(h^{-1}\bar{\varepsilon}^2+Mh+\bar{\varepsilon}), \quad k = 0,1,\ldots,N,$$

with $M := \sup_{x \varepsilon [0,1]} |U''(x)|$, where U is the solution of the reduced problem (2.3). The proof follows easily from the stability inequality given by Comstock et al (1976) (see also Reinhart (1977)). We shall only comment that the first two terms on the RHS are due to the trunction error τ_k. Our result here is comparable to the one given by Il'in (1969). However, it should be pointed out that one may obtain the accuracy up to $O(\bar{\varepsilon})$ when $M = 0$, provided $\bar{\varepsilon} \leq h < 1$. A typical example of this is the one used by many authors where $p = 1$, $q = 0$, and $f = 0$. In this case, numerical results show that accuracy can be achieved particularly with a moderate step size h.

3. BOUNDARY LAYER SCHEME.

The idea of the boundary layer scheme is simple. We attempt to construct U and V in (I) and (II) numerically by any suitable numerical scheme with a moderate step size h (and $\tilde{h}$) so that the numerical solution

$$u^h := U^h+V^{\tilde{h}} \tag{3.1}$$

can be used to approximate u of (E) and (B), and that $u-u^h = O(\varepsilon)$ uniformly for all x throughout the interval $[0,1]$. This certainly is possible for the reduced problem (I) since no small parameter ε is involved. On the other hand, however, for the boundary layer equation (II), although ε does not appear explicitly in the equation, the semi-infinite domain causes some difficulty. Hence our first step is to modify (II) by a similar problem ($\tilde{\text{II}}$) but in a finite interval, so that it can be handled easily by any conventional numerical scheme and that the solution $\tilde{V}^h$ serves the same purpose as V^h in (3.1). Since from the theory of singular perturbations, we know that V is insignificant outside "the boundary layer", we can simply consider the following modification:

$$\tilde{L}\tilde{V} = 0, \quad m > \tilde{x} > 0; \quad \tilde{V}(0) = \beta - U(1), \quad \tilde{V}(m) = 0, \qquad (\tilde{\text{II}})$$

where m is a constant to be determined. We choose m so that $m << \varepsilon^{-1}$ and $\tilde{V}-V = O(\varepsilon)$.

For the present situation, an easy computation from the explicit solutions of (II) and ($\tilde{\text{II}}$) shows that

$$m \geq \frac{1}{p(1)} \, (\ell n\ 2 - \ell n\ \varepsilon) \qquad (3.2)$$

will be sufficient for our purpose. In general, including some nonlinear problems, one may determine m without the help of the explicit solutions by consider simply an inequality such as $e^{-p(1)m} \leq \varepsilon$, and obtain the crude estimate

$$m \geq \frac{\gamma\ \ell n\ 10}{p(1)} \simeq \frac{3\gamma}{p(1)}, \quad \text{for} \quad \varepsilon = 10^{-\gamma}. \qquad (3.2)'$$

Now let us consider the accuracy of the BLS consisting of any numerical schemes for solving (I) and ($\tilde{\text{II}}$) with m given by (3.2) or (3.2)'. First we note that the accuracy of U^h is independent on the choice of m. Hence we believe that the error analysis here is simpler than the one given by Lorenz (1978). It is easy to see that the error bound $|u-u^h|$ is in general dominated by the sum $|U-U^h|+|\tilde{V}-\tilde{V}^h|+$ $O(\varepsilon)$. In practice, however, one does not know a priori the value of $U(1)$ in ($\tilde{\text{II}}$), and hence one is actually seeking the approximation $\tilde{V}_*^h$ for the solution $\tilde{V}_*$ of ($\tilde{\text{II}}$) with $\tilde{V}_*(0) = \beta - U^h(1)$. This, of course, introduces additional error for $u_*^h = U^h + \tilde{V}_*^h$, although $u_*^h(1) = u^h(1) = u(1)$. In general, if a particular numerical scheme is employed in the BLS so that it achieves $O(\tilde{h}^t)$ approximation for $\tilde{V}_*^{\tilde{h}}$ and $O(h^r)$ or $O(h^s)$ for U^h depending on whether $x \leq 1-\varepsilon m$ or $x > 1-\varepsilon m$, then the following estimates can easily be obtained:

Theorem 2. Let $u_*^h = U^h + \tilde{V}_*^{\tilde{h}}$ be this approximation. Then we have

$$|u-u_*^h| = O(h^r)+O(\varepsilon) \quad \text{for} \quad 0 < x \leq 1-\varepsilon m,$$

and for $1-\varepsilon m < x < 1$,

$$|u-u_*^h| \leq C\{|U(1)-U^h(1)|+ \frac{\varepsilon m}{p_0} h^{s-1}\}+O(\tilde{h}^t)+O(\varepsilon).$$

We have applied the BLS successfully to several examples. The particular numerical scheme we used for solving both (I) and ($\tilde{II}$) is the Galerkin method with linear finite elements as trial functions. We were motivated by the work of Hulme (1972) to apply the Galerkin method for the problem (I). We note that (I) is an initial-value problem. In general it is more suitable to use approximate schemes which are primarily designed for the IVP's, e.g. Runge-Kutta method and to use different schemes such as the Galerkin method for the boundary value problem ($\tilde{II}$). For the special case where the coefficients p = const and $q = 0$, a simple computation shows that $|U_k-U_k^h| = O(h^4)$, $k = 0,1,\ldots,k = N-1$ while $|U_N-U_N^h| = O(h^2)$. Here we have used the notations: $Nh = 1$ and $x_k = kh$. As for the solution $\tilde{V}^{\tilde{h}}$ (with * being suppressed), we have the estimate

$$|\tilde{V}_k-\tilde{V}_k^{\tilde{h}}| = O(\tilde{h}^2), \quad k = 1,2,\ldots,N-1$$

if the mesh size $\tilde{h}$ satisfies the restriction $\frac{\tilde{h}}{2} p(1) \leq 1$. Here we recall that $\tilde{N}\tilde{h} = m$ and $\tilde{x}_k = k\tilde{h}$ which corresponds to $x = 1-\varepsilon\tilde{x}_k$. In this example from Theorem 2, h and $\tilde{h}$ can be chosen optimally by the relations $h^2 = \tilde{h}^2 = \varepsilon$, provided $-\ell n\ \varepsilon \leq m \leq \varepsilon^{-1/2}$, and the asymptotic error of the BLS will be $O(\varepsilon)$.

For the numerical experiments, we refer the readers to our earlier report, Hsiao and Jordan (1978).

4. ACKNOWLEDGEMENTS.

This research was supported by the Air Force Office of Scientific Research through AFOSR Grant No. 76-28-79

REFERENCES.

Comstock, C., and Hsiao, G. C. (1976). Singular perturbations for difference equations. *Rocky Mountain J. of Math.* 6, 561-567.

Hsiao, G. C., and Jordan, K. E. (1978). Solutions to the difference equations of singular perturbation problems. *University of Delaware IMS Technical Report 36.*

Hulme, B. L. (1972). One-step piecewise polynomial Galerkin methods for initial value problem. *Math Computation.* 26, 415-426.

Il'in, A. M. (1969). Differencing scheme for a differential equation with a small parameter affecting the highest derivative. *Math Notes.* 6, 596-602.

Lorenz, J. (1978). Numerical treatment of singular perturbations by combinations of initial and boundary value methods. Invited lecture. *Conf. on Numer. Anal. of Sing. Pertur. Prob.*

O'Malley, R. E., Jr. (1974). *Introduction to Singular Perturbations.* Academic Press, New York.

Pearson, C. E. (1968). On a differential equation of the boundary layer type. *J. Math. and Phys.* 47, 134-154.

Reinhart, H. J. (1978). Singular perturbations of difference methods for linear ordinary differential equations. Contributed paper. *Conf. on Numer. Anal. of Sing. Pertur. Prob.*

ITERATIVE METHODS FOR A CLASS OF VARIATIONAL INEQUALITIES

K. Inayat Noor and M. Aslam Noor
Mathematics Department, Kerman University, Iran

ABSTRACT

The main aim of this paper is to prove that the solution of variational inequalities can be found by an iterative scheme, which is obtained as a projection of a real Hilbert space into a convex subset. It is shown that the approximate solution obtained by the iterative scheme does converge strongly to the exact solution. We also show that this method can be extended to study a class of singularly perturbed boundary value problems, which are required to satisfy some extra conditions.

INTRODUCTION

It is well known that many physical situations are described by singularly perturbed differential or integral equations in which the highest derivative is multiplied by a small parameter, see Cole (1968), Cohen (1971) and O'Malley (1968). Sattinger (1972), using the concept of upper and lower solutions, has established the validity of the singular perturbation argument for nonlinear boundary value problems in chemical reactor theory.

Recently much attention has been given to study the linear and nonlinear boundary value problems arising in fluid dynamics, chemical reactor theory, plasticity and elasticity, when their solutions are required to satisfy some extra auxiliary constraint conditions. It has been shown by Lions (1971) and Fichera (1972) that these problems can be considered by the variational inequalities approach. The problem of finding the approximate solutions of variational inequalities is not an easy one. Due to the constraint, the solution of variational inequalities cannot be obtained as a projection. Indeed very few methods exist to compute the solutions of variational inequalities, see Sibony (1970). He has shown that the solution of variational inequalities can be obtained by the iterative scheme provided there exists a linear coercive continuous operator with its inverse.

In section 2, we show that the solution of a variational inequality can be obtained by an iterative scheme without assuming the existence of the linear operator. The convergence of the approximate solution to the exact solution is also proved. Furthermore, it is shown that by this method a class of singularly perturbed boundary value problems satisfying the constraint conditions can be studied.

ITERATIVE METHODS

First of all, we introduce some notions and definitions. Let H be a real Hilbert space with its dual H', whose inner product and norm are denoted by (.,.) and $\|\cdot\|$ respectively. The pairing between $f \in H'$ and $u \in H$ is denoted by $\langle f,u\rangle$. Let M be a closed convex subset of H.

For a given element $f \in H'$, we consider the problem of finding $u \in M$ such that

$$\langle Tu, v-u\rangle \geq \langle f, v-u\rangle, \text{ for all } v \in M, \tag{1}$$

where T is a nonlinear operator. We note that if T is a strongly monotone and Lipschitz continuous operator, then one can show the existence of a unique solution $u \in M$ of (1), see Browder (1965), Noor (1975a) and Sibony (1970).

Definitions

A nonlinear operator $T : H \to H'$ is said to be

(i) Strongly monotone, if there exists a constant $\alpha > 0$ such that

$$\langle Tu-Tv,u-v\rangle \geq \alpha \,\| u-v \|^2 \quad \text{for all } u,v \in H.$$

(ii) Lipschitz continuous, if there exists a constant $\mu > 0$ such that

$$\| Tu-Tv \| \leq \mu \,\|u-v \| \quad \text{for all } u,\ v \in H.$$

Note that if T is a linear operator, then (i) and (ii) are replaced by

(i)' Coercive, if there exists a constant $\alpha>0$ such that

$$\langle Tv,v\rangle \geq \alpha \|v \|^2 \quad \text{for all } v \in H$$

(ii)' Continuous, if there exists a constant $\mu > 0$ such that

$$\langle Tu,v\rangle \leq \mu \,\| u \| \,\| v \| \quad \text{for all } u,\ v \in H.$$

In particular, it follows that $\alpha \leq \mu$. Furthermore, let Λ be the canonical isomorphism from H' onto H defined by the relation

$$\langle f,v\rangle = (\Lambda f,v), \quad \text{for all } f \in H' \text{ and } v \in H. \tag{2}$$

Then

$$\|\Lambda\|_{H'} = \|\Lambda^{-1}\|_{H} = 1.$$

We need the following result, see Noor (1975b).

Lemma 1. Let M be a closed convex subset of H. Then for a given $z \in H$, $u \in M$ satisfies

$$(u-z, v-u) \geq 0 \text{ for all } v \in M$$

if and only if

$$u = P_M z,$$

where P_M is a projection of H into M.

We now state and prove the main result of this section.

Theorem 1. $u \in M$ is a solution of the variational inequality (1) if and only if $u \in M$ satisfies the relation

$$u = P_M(u-\rho\Lambda(Tu-f)), \tag{3}$$

where $\rho > 0$ is a constant.

Proof: Suppose that $u \in M$ satisfies (1), i.e.,

$$\langle Tu, v-u\rangle \geq \langle f, v-u\rangle, \text{ for all } v \in M.$$

By using (2), the above relation is equivalent to finding $u \in M$ such that

$$(\Lambda(Tu-f), v-u) \geq 0.$$

This relation can be written as

$$(\rho\Lambda(Tu-f), v-u) \geq 0, \tag{4}$$

where $\rho > 0$ is constant. Hence by lemma 1, (4) is equivalent to finding $u \in M$ such that

$$u = P_M(u-\rho\Lambda(Tu-F)),$$

where P_M is a projection of H into M.

From theorem 1, it follows that the solution $u \in M$ satisfying (1) can be computed by the iteration scheme

$$u_{n+1} = P_M(u_n - \rho\Lambda(Tu_n - f). \tag{5}$$

In the next theorem, we show the convergence of the solution u_{n+1} defined by (5) to the exact solution u of (1).

Theorem 2. Let the operator T be strongly monotone and Lipschitz continuous. If u and u_{n+1} are the solutions of (1) and (5) respectively, then

$$u_{n+1} \to u \quad \text{strongly in } H,$$

provided $0 < \rho < 2\alpha / \mu^2$.

Proof: Suppose that u and u_{n+1} are solutions of (1) and (5), then we have

$$u_{n+1} - u = P_M(u_n-\rho\Lambda(Tu_n-f)) - P_M(u-\rho\Lambda(Tu-f)).$$

Since P_M is non-expansive, see Mosco (1973), we obtain

$$\| u_{n+1} - u \|^2 \leq \| u_n - u - \rho\Lambda(Tu_n - Tu) \|^2$$

$$= \| u_n - u \|^2 - 2\rho <Tu_n - Tu, u_n - u> + \rho^2 \| Tu_n - Tu \|^2$$

$$\leq (1 - 2\rho\alpha + \rho^2\mu^2) \| u_n - u \|^2,$$

by using strongly monotonicity and Lipschitz continuity of T. Thus we have

$$\| u_{n+1} - u \|^2 = \theta \| u_n - u \|^2$$

where $\theta = 1-2\rho\alpha + \rho^2\mu^2 < 1$ for $0 < \rho < 2\alpha/\mu^2$. Thus it follows from the above relation that u_{n+1} does converge strongly to u, the solution of (1), from the fixed point theorem.

From theorem 2, we get a natural algorithm to compute the approximate solution as follows:

(1) $u_o \in M$, is given.

(2) $u_{n+1} = P_M(u_n - \rho\Lambda(Tu_n - f))$,

where ρ is required to satisfy the condition,

$$0 < \rho < 2\alpha / \mu^2.$$

We note that the scope of this algorithm in applications is limited due to the reason that it may be very difficult to find the projection of the solution u.

APPLICATIONS

Many physical situations in biology, elasticity, fluid

dynamics and chemical reactor theory can be described by singularly perturbed equations. For simplicity, we consider the problem of the following type.

$$\left.\begin{aligned} -\varepsilon\Delta u &= f, & x &\in \Omega \\ u(x) &= 0, & x &\in \partial\Omega \\ u(x) &> \psi(x) \text{ on } \Omega \end{aligned}\right\} \qquad (6)$$

where $\psi(x)$ is a given function on the simply connected domain and ε is a small parameter. It is assumed that the boundary $\partial\Omega$ and the function f satisfy some smoothness conditions to ensure the existence of the unique solution of (6).

The problem (6) is studied via the variational inequality in the Sobolev space $H_0^1(\Omega)$, which is a Hilbert space. The Sobolev space $H^1(\Omega)$ is the space of functions which together with their generalized derivatives of order one are in $L_2(\Omega)$. The space of functions from H^1 which in a generalized sense satisfy the homogeneous boundary conditions on $\partial\Omega$ is $H_0^1(\Omega)$. The set $M \equiv \{u | u \in H_0^1(\Omega) : u > \psi \text{ on } \Omega\}$ is a closed convex subset of $H_0^1(\Omega)$, see Lions (1971). Thus the weak solution of (6) can be characterized by the variational inequality of the following type

$$\varepsilon a(u,v-u) > \langle f,v-u\rangle, \quad \text{for all } v \in M. \qquad (7)$$

In this case $\varepsilon a(v,v) = \varepsilon \|v\|^2$ and $\varepsilon a(u,v) \le \varepsilon \|u\| \|v\|$, so that $\alpha = \varepsilon$ and $\mu = \varepsilon$. Thus theorem 2 can be applied for $0 < \rho < 2/\varepsilon$ to find the approximate of (6). Also it follows that opt.$\rho = 1/\varepsilon$.

We also like to remark that this technique can be applied

to study the wider class of nonlinear singularly perturbed boundary value problems arising in plasma physics, and other branches of continuum mechanics.

REFERENCES

Browder, F.E. (1965). Nonlinear monotone operators and convex sets in Banach spaces, *Bull.Amer.Math.Soc.*, 71, 780-785.

Cohen, D.S. (1971). Multiple stable solutions of nonlinear boundary value problems in chemical reactor theory, *Siam. J.Appl.Math.*, 20.

Cole, J.D. (1968). *Perturbation methods in applied mathematics*, Blaisdell, Waltham, Mass.

Fichera, G. (1972). Boundary value problems of elasticity with unilateral constraints in *Handbuch der physik*, Bd/2, Springer-Verlag, Berlin, New York.

Lions, J.L. (1971). *Optimal control systems governed by partial differential equations*, Springer-Verlag, Berlin, New York.

Lions, J.L. and Stampacchia, G. (1967). Variational inequalities. *Comm.Pure Appl.Math.*, 20, 493-519.

Mosco, U. (1973). An introduction to the approximate solution of variational inequalities in *Constructive aspects of functional analysis*, Edizione Cremonese, Roma.

Noor, M.A. (1975a). *On variational inequalities*, Ph.D. Thesis, Brunel University, U.K.

Noor, M.A. (1975b). Variational inequalities and approximation, *Math.J.* 8, 24-40.

O'Malley, R. (1968). Topics in singular perturbations, *Advances in Mathematics*, 2, 365-470.

Sattinger, D.H. (1972). Monotone methods in nonlinear elliptic parabolic bound value problems, *Indiana Univ.Math.J.*, 21, 979-1000.

Sibony, M. (1970). Methods iteratives pour les equations et inequations aux derivees partielles nonlineares de type monotone, *Calcolo*, 7, 65-183.

NUMERICAL ESTIMATES IN ASYMPTOTIC SERIES BY H^o-STRICTLY SINGULAR PERTURBATIONS

K. Ingólfsson
Science Institute, Reykjavik, Iceland

ABSTRACT

Two strongly continuous one-parameter semigroups of contraction operators, $\{Z(t);t\geq 0\}$ and $\{Z_o(t);t\geq 0\}$, determine related evolutions in the Hilbert spaces H and H_o respectively. We consider the mappings under $Z(t)$ and $Z_o(t)$, when the infinitesimal generators G and G_o belong to a product class, properly defined with respect to an H^o-norm. The inversed transform of the identity $R(\lambda,G)=R(\lambda,-iH^o)-iR(\lambda,-iH^o)VR(\lambda,G)$ in the resolvent of G converges on Ψ iff $\Psi\varepsilon\mathcal{D}(G)$. By iteration an asymptotic series emerges when $t\to 0$. Numerical estimates of the approximation in the first order explain the deviation from the exponential decay in the integral transform, which corresponds to a certain H_o in H_o.

1. INTRODUCTION

Perturbation methods have long been an important aid in solving quantum mechanical problems. The occurrence of analytic expansions, however, has been greatly missed in many cases, and the search for these still prevails wide areas of investigation. Recently physicists have begun to accept singular perturbations as a property of nature (Ezawa, Klauder and Shepp, 1975). The physical formulations can in these cases sometimes stimulate a mathematical interest of its own (Symanzik, 1975). In the following we will sketch a fundamental problem in quantum physics, relevant to singular perturbations.

The consistent theory does not imply a pure exponential form in the decay of a radiating particle, if the wave equations are governed by a self-adjoint energy operator. This is even true, if the problem is looked at in a more general context: Let H be a direct sum of the orthogonal Hilbert spaces H_1 and H_2, where the latter refers to the excited eigenvalue of some unperturbed energy operator. If $\{Z(t); t \geq 0\}$ is a strongly continuous one-parameter semigroup of contraction operators on H with a skew-symmetric infinitesimal generator, G, the evolution

$$\Psi_t = Z(t)\Psi_2 \text{ with } \Psi_2 \varepsilon D(G) \cap H_2 \tag{1}$$

and the usual exponential decay mode

$$\Psi_t = \exp(-\gamma t/2)\Psi_2 + \sigma(t)U(t)\Psi_1 \text{ with } \Psi_1 = \overline{W}\Psi_2 \tag{2}$$

should be understood as the same state. The operator U(t) is unitary in H_1 for finite values of t, where $\overline{W}$ is an isometry from H_2 to H_1. Similar to a theorem of Stone, which characterizes the generators of groups of unitary operators, we will prove that a closed, densly defined G generates an isometry iff $G^{\dagger} \supset -G$. From (1) and (2) follows that $|\sigma(t)|^2$ has the form $1-\exp(-\gamma t)$ and $\sigma(t)/t$ diverges when $t \to 0$. Therefore $\Psi_2 \varepsilon D(G)$ is a contradiction.

It has been emphasized in radiation theory from the beginning that the exponential decay is an extremely well established fact (Heitler, 1954). It is also a principle in quantum theory that symmetric operators should correspond to observable quantities (Dirac, 1947). We will therefore have to modify some of our assumptions, if we intend to achieve a consistent theory of the decay. In the sequel we will introduce two semigroups, one being an isometry the other produces the exponential decay. The evolutions under both are asymptotically related for small values of t.

For the formal work we need a relation between the resolvents of the infinitesimal generators. In theoretical physics this is a well known practice:

Geometric series connecting the resolvents of unperturbed and perturbed self-adjoint energy operators have been developed in the theory of line breadth phenomena (Messiah, 1962) and in scattering theory (Yakubovski, 1967). The inversed transform of the series, however, offers considerable difficulties. A solution method was proposed by the author (Ingólfsson, 1967) and occasionally carried out (Ingólfsson, 1973 and 1976). The problem may be looked at generally on the basis of the analytic theory of semigroups over the B-space (Yosida, 1965).

In this paper we will discuss the convergence of a series, actually fabricated by an iteration of the following integral equation, when $\Psi \varepsilon \mathcal{D}(G)$:

$$Z(t)\Psi = U^{o}(t)\Psi + U^{o}(t)\,{}_{o}\!\int^{t} ds\, U^{o}(s)^{\dagger}(-iV)Z(s)\Psi \quad . \qquad (3)$$

$U^{o}(t)$ is unitary in H and V is derived from the difference of the infinitesimal generators. If we consider V proportional to some scalar, ε, the solution will not be analytic in ε for all t. By imposing the condition (2) on the evolution $Z(t)\Psi_2$ we can determine the set of the infinitesimal generators. This will be at the cost of the size of the system involved, because $|\sigma(t)|<(1-\exp(-\gamma t))^{1/2}$ for t positive and small enough. As a matter of fact $||Z(t)\Psi_2||= 1-1/2t\gamma+O(t^2)$, a well known behavior in singular perturbation problems (O'Malley, 1974).

Finally we will consider a time evolution in terms of an integral transform, which has been used in order to to explain the deviation from the exponential decay (Ingólfsson, 1968). We get numerical estimates of the asymptotic convergence through comparison of the solution of theses integrals with $Z(t)\Psi_2$ and the lowest remainderterm of the series.

2. THE ASYMPTOTIC SERIES

As starting point we consider the two strongly continuous one-parameter semigroups of contraction operators on the Hilbert spaces H and H_o, $\{Z(t);t\geq 0\}$ and $\{Z_o(t);\geq 0\}$, with the infinitesimal generators G and G_o respectively. For any H_o, which is a proper subspace of H, we define the class $S=\{G x G_o\}$ of generator-pairs by the following assumptions:

a) G_o *is skew-symmetric in* H_o.
b) Let W *be an isometry from* H *onto* H_o, *which maps* $D(G)$ *onto* $D(G_o)$. *For any such* W *exists* H^o, *a self-adjoint extension of* $iG^o = iW^{\dagger}G_oW$, *and* V, *an extension of* $i(G-G^o)$ *on* $D(V) \supseteq D(G)$, *so that* V *is* H^o-*strictly singular.*

The following theorem characterizes the generators of semigroups of isometric operators in a Hilbert space:

Theorem 1: A closed, linear operator G_o *with a dense domain generates a one-parameter strongly continuous semigroup of isometric operators iff* $G_o^{\dagger} \supset -G_o$.

Proof: When $\{Z_o(t); t \geqq 0\}$ is a strongly continuous semigroup of isometric operators in H_o with the infinitesimal generator G_o, we may carry out

$$\lim_{\Delta t \to +0} \frac{Z_o^{\dagger}(\Delta t)-1}{\Delta t}\Psi = \lim_{\Delta t \to +0} -Z_o^{\dagger}(\Delta t)\frac{Z_o(\Delta t)-1}{\Delta t}\Psi$$

on both sides, if and only if $\Psi \varepsilon D(G_o)$. The generator of the adjoint semigroup is the adjoint of the generator of the semigroup. Therefore is $G_o^{\dagger} \supset -G_o$. Conversely, when G_o is skew-symmetric and λ real and positive, is $||(\lambda - G_o)\Psi|| \geqq \lambda ||\Psi||$ for $\Psi \varepsilon D(G_o)$. As $\lambda - G_o$ is also closed and densely defined, it is one-to-one and onto with $||(\lambda - G_o)^{-1}|| \leqq \lambda^{-1}$. According to a theorem of Hille and Yosida (Yosida, 1965) G_o generates a semigroup, $\{Z_o(t); t \geqq 0\}$. We find that the derivative of $||Z_o(t)\Psi||^2$ vanishes for $t \geqq 0$ and $\Psi \varepsilon D(G_o)$. Therefore is $||Z_o(t)\Psi|| = ||\Psi||$ for any positive t and Ψ in H_o.

We consider again the class S. Then follows

Lemma 1: If the real part of z *is* λ *and* $\{G, G_o\}$ *a pair in* S, *then is the relation*

$$R(z,G) = R(z,-iH^o) - iR(z,-iH^o)VR(z,G) \quad , \lambda > 0 \qquad (4)$$

an identity in the resolvent of G *on the entire* H.

The proof is based on the fact that

$$R(z,G)\Psi = \int_o^{\infty} \exp(-zt)Z(t)\Psi \, dt \text{ with } \lambda > 0,$$

where the integral on the right converges for any Ψ. The inversed transform, however, i.e.

$$Z(t)\Psi = (2\pi i)^{-1} \int_{\lambda - i\infty}^{\lambda + i\infty} \exp(zt)R(z,G)\Psi dz, \lambda > 0 \quad (5)$$

converges as a differentiable $Z(t)\Psi$ if and only if $\Psi\varepsilon D(G)$ (Ingolfsson, 1967). Therefore we get
<u>*Corollary 1:*</u> *The inversed transform of* (4) *on* Ψ *is the relation* (3) *for the semigroup iff* $\Psi\varepsilon D(G)$.

Through the iteration of $R(z,G)$ in (4) we can find geometric series of a finite number of terms with a remainderterm, which is again an identity on the Hilbert space because $D(V)\supset D(H^o)$. When it exists, the inversed transform of the series is

$$Z(t)\Psi=\sum_{o}^{n-1} A_\nu(t)\Psi+R_n(t)\Psi \quad , \tag{6}$$

where the terms $A_\nu(t)\Psi$ are determined by the recurrance relations

$$A_\nu(t)\Psi=-iU^o(t)\,{}_o\!\int^t ds\; U^o(s)^\dagger V\, A_{\nu-1}(s)\Psi,\nu\geqq 1$$

$$A_o(t)\Psi=U^o(t)\Psi \quad ,$$

and the remainderterm $R_n(t)\Psi$ obeys for $n\geqq 1$ a similar relation with $R_o(t)$ being $Z(t)$. Therefore is

$$R_n(t)\Psi=A_n(t)\Psi+R_{n+1}(t)\Psi \tag{7}$$

and for n=0 this is just the equation (3). Formally we have therefore got (6) through the iteration of $Z(t)\Psi$ in (3). Concerning the existence we have
<u>*Lemma 2:*</u> *Let* $\{G,G_o\}$ *be a pair in* S. *The equation* (6) *gives an expression for a strongly differentiable* $Z(t)\Psi$ *for all positive* t *and any natural number* n *iff* $\Psi\varepsilon D(G)$.
Proof: Let $A_{\nu-1}(t)\Psi\varepsilon D(H^o)$ and $\frac{d}{dt} A_{\nu-1}(t)\Psi\varepsilon D(V)$ be correct, when $\Psi\varepsilon D(G)$. By partial integration we find that $A_\nu(t)\Psi\varepsilon D(H^o)$. Therefore the derivative of $A_\nu(t)\Psi$ converges and equals $-iH^oA_\nu(t)\Psi-iVA_{\nu-1}(t)\Psi$. The difference quotient $\Delta t^{-1}(A_\nu(t+\Delta t)-A_\nu(t))\Psi$ is in $D(H^o)$ and the limit of its mapping under the closed operator V therefore also converges, when $\Delta t\to 0$, because V is H^o-compact (Goldberg, 1966). V is therefore defined on the derivative of $A_\nu(t)\Psi$. It is easy to prove that $A_o(t)\Psi\varepsilon D(H^o)$ and that the derivative of $A_o(t)\Psi$ is in $D(V)$ by using the same properties as before of the H^o-strictly singular V. $\Psi\varepsilon D(G)$ will contradict the differentiability. The induction is then completed. To prove the existence

of the remainderterm we may use similar arguments.

The asymptotic mode can now be explained by

Theorem 2: *Let* $\{G,G_o\}$ *be a pair in* S, $\Psi\varepsilon D(G)$ *and* $Z(t)\Psi$ *expressed by the series* (6) *with a remainderterm. Then exists for any* n *a real number* M *such that the following estimate is correct for* $t>0$*:*

$$||R_n(t)\Psi|| \leqq t\ M \sup_{s<t}(||R_{n-1}(s)\Psi||+||H^oR_{n-1}(s)\Psi||)$$

For the proof we only need to remember that V is H^o-bounded. Together with the relation (7) this theorem shows that the series in (6) is an asymptotic expression for $Z(t)\Psi$, when t is small enough.

We will now impose the condition (2) on $Z(t)\Psi_2$. By using the abbreviation $\Psi_1(t)$ for $\sigma(t)U(t)\overline{W}\Psi_2$ this means for any positive t_1 and t_2 that

$$Z(t_1)\Psi_1(t_2)=\Psi_1(t_1+t_2)-\exp(-1/2\cdot\gamma t_2)\Psi_1(t_1).$$

If the limit of $\sigma(t)/t$ is kalled k, when t tends to zero, and the infinitesimal generator of U is -iH, we find the following relations:

$$\begin{aligned} G\Psi_2 &= (-1/2\cdot\gamma+k\overline{W})\Psi_2 \\ G\Psi_1(t) &= \left(\frac{\sigma'}{\sigma}-iH\right)\Psi_1(t)-\exp(-1/2\cdot\gamma t)k\overline{W}\Psi_2 \end{aligned} \qquad (8)$$

Any point in the Hilbert space H_1 belongs to only one path, $(U\sigma)(t)\overline{W}\Psi_2$ for $t>0$. The function $|\sigma(t)|$ is increasing in a neighbourhood of 0 and the norm of the vector therefore determines the value of t. The relations (8) describe in this sense infinitesimal generators for all positive values of k.

3. COMPARISON WITH A SOLVABLE MODEL

If $U_o(t)$ is the transform under W of the unitary $U^o(t)$, its generator, $-iH_o$, is a skew self-adjoint extension of G_o. We may therefore transform the asymptotic series (6) into H_o and the behaviour of the remainderterm, given by theorem 2, will be preserved by the isometry. Especially the equation (3), which explains the first approximation by the series, may be written in H_o in the form

$$U_o(t)\Psi_2 = WZ(t)W^{\dagger}\Psi_2 - R(t)\Psi_2 \quad , \qquad (9)$$

where the first term on the right represents the exponential decay. The author proposed earlier a model for the decay, $U_o(t)\Psi_2$ written under the inversed transform (5) and solvable in a closed form (Ingólfsson, 1968). It was based on

$$\Psi_2(t) = (2\pi i)^{-1}\exp(-i\omega_2 t) \int_{\xi-i\infty}^{\xi+i\infty} dz \frac{\exp(zt)}{z+i\omega_o+\Gamma(z)}$$

$$\Psi_1(k,r,t) = (2\pi i)^{-1}\exp(-i\omega_2 t) \int_{\xi-i\infty}^{\xi+i\infty} dz \frac{\exp(zt)g(r,k)}{(z+i\omega_o+\Gamma(z))(z+ik)}$$

$$\Gamma(z) = \int \sum_{r=1}^{3} \frac{d^3k}{2k} \frac{|\phi(r,k)|^2}{z+ik} \quad \text{with } \xi \text{ real and positive.}$$

The function $\phi(r,k)$ is essentially the amplitude of the current density under the Fourier transform for an electron bound in a central field. By assuming $\phi(r,x)\varepsilon L^1 \cap L^2$ it was possible to describe some important analytic properties of $\Gamma(z)$ and evaluate the integrals. The physically relevant constants

$$\gamma/2 = \mathrm{Re} \lim_{\xi\to+o} \Gamma(\xi - i\omega_o)$$

$$\Delta\omega_o = \mathrm{Im} \lim_{\xi\to+o} \Gamma(\xi - i\omega_o)$$

represent the natural line breadth and the Lamb shift respectively. The solution is an exponential decay mode of the type (2) with a deviation from the exponential form computed as an integral around the branchpoint of $\Gamma(z)$. The deviation is

$$-f^2/4\pi^2\omega_o^2 \exp(-i\omega_2 t)\overline{(t^2-\rho^2)^{-1}} \quad , \quad \rho = |x-x'| \qquad (10)$$

where the bar refers to a mean by the distribution $f_1^{-2}\Sigma\phi(r,x)\phi^{\dagger}(r,x')$ in $\{x,x'\}$-space and f is the L^1-norm of ϕ.

We can prove that the energy operator occuring under the above integrals is symmetric. Its self-adjoint extension may be carried out under the Fouriertransform. Let this symmetric operator be iG_o. If we claim the existence of G, which could fit to this G_o as a pair in S, we must essentially only adapt $U_o(t)$ to this situation, i.e. we have to secure the H^o-compactness. We will, however, be free to choose any real value for k, because the norm of $R(t)\Psi_2$ in the estimate (10) is independent of k.

REFERENCES

Dirac, P.A.M. (1947). *The Principles of Quantum Mechanics*. Oxford University Press. Chap.II, §8.
Ezawa, H., Klauder, J.R. and Shepp, L.A. (1975). Vestigal effects of singular potentials in diffusion theory and quantum mechanics. *Journal of Mathematical Physics*. 16, 783-799.
Goldberg, S. (1966). *Unbounded Linear Operators*. McGraw-Hill, New York. Chap.V, §2.
Heitler, W. (1954). *The Quantum Theory of Radiation*. Oxford University Press. Chap.IV, §16.
Ingólfsson, K. (1967). Zur Formulierung der mathematischen Theorie der natürlichen Linienbreite. *Helvetica Physica Acta*. 40, 237-263.
Ingólfsson, K. (1968). Die Abweichung vom Exponentialzerfall angeregter Zustände. *Communications in Mathematical Physics*. 11, 168-180.
Ingólfsson, K. (1973). On the mathematical structure of a model converging in a space of semidefinite metric. *Notices of the American Mathematical Society*. 20, A 681. (1974). Preprint Sci.Inst.B4.
Ingólfsson, K. (1976). Notes on the classical and the nonrelativistic limits in quantum mechanics. *Letters in Mathematical Physics*. 1, 351-359.
Messiah, A. (1962). *Quantum Mechanics*. North-Holland Publishing Company, Amsterdam. Chap.XXI, §13.
O'Malley, R.E. (1974). *Introduction to Singular Perturbations*. Academic Press, New York.
Symanzik, K. (1975). Renormalization problems in nonrenormalizable massless Φ^4 theory. *Communications in Mathematical Physics*. 45, 79-98.
Yakubovski, O.A. (1967). *Soviet Journal of Nuclear Physics*. 5, 937.
Yosida, K. (1965). *Functional Analysis*. Springer, Berlin. Chap. IX.

ON APPROXIMATING SMOOTH SOLUTIONS OF LINEAR SINGULARLY PERTURBED ODE

R.M.M. Mattheij

Mathematisch Instituut, Katholieke Universiteit, Toernooiveld, Nijmegen, The Netherlands

ABSTRACT

A method is given for approximating smooth solutions of the Cauchy problem for linear singularly perturbed (or stiff) ODE, by using explicit multisteps, which have fairly small stability regions. The method is based on decoupling "regular" and "singular" components and computing them separately. In this way the instability is ruled out so that the global error is essentially determined by the local discretization error, if only the step size is not too small.

1. STATEMENT OF THE PROBLEM

Let $x^1(t)$ be an m dimensional and $x^2(t)$ be an (n-m) dimensional vector for all t. Consider the following linear ODE

$$\begin{pmatrix} \varepsilon\, \dot{x}^1 \\ \dot{x}^2 \end{pmatrix} = A(t,\varepsilon)x + f(t,\varepsilon) \tag{1.1}$$

where $A(t,\varepsilon)$ will be partitioned correspondingly. Define

$$\hat{A}(t,\varepsilon) = \begin{pmatrix} \varepsilon^{-1} & 0 \\ 0 & 1 \end{pmatrix} \begin{pmatrix} A^{11}(t,\varepsilon) & A^{12}(t,\varepsilon) \\ A^{21}(t,\varepsilon) & A^{22}(t,\varepsilon) \end{pmatrix} \tag{1.2}$$

Assume that $A(t,\varepsilon)$ is uniformly bounded and $A^{12}(t,\varepsilon) = O(\varepsilon)$. Let ε be small enough, $\varepsilon \leq \varepsilon_0$ say, such that $\hat{A}(t,\varepsilon)$ has m eigenvalues with "large" negative real part, $\lambda_1(t,\varepsilon),\ldots,\lambda_m(t,\varepsilon)$, of order $\frac{1}{\varepsilon}$ (so this also holds for those of $\frac{1}{\varepsilon}A^{11}(t,\varepsilon)$). Finally assume that the remaining eigenvalues of $\hat{A}(t,\varepsilon)$, viz. $\lambda_{m+1}(t,\varepsilon),\ldots,\lambda_n(t,\varepsilon)$ have a "small" negative real part and a small modulus as well (for all interesting values of t and ε). Then the homogeneous part of (1.1) will have rapidly decaying solutions and fairly slowly decaying solutions, to be referred to as singular and regular solutions respectively.

We are interested in approximating the smooth solution x of (1.1) on $[t_0,T]$, say, for which an initial value $x(t_0) = x_0$ is given. A consistent discretization method will then produce moderate local errors for fairly large stepsizes h. For certain methods, however, the stiffness of the problem (as characterized by $\lambda_1(t,\varepsilon),\ldots,\lambda_m(t,\varepsilon)$) may cause unstable error growth for such values of h.

To find out which values of h can be expected to produce growth factors that are bounded by unity one can take recourse to pictures of the stability region S (see e.g. Gear (1971)); fig 1 gives the region for the two step Adams-Bashford. For our type of problems

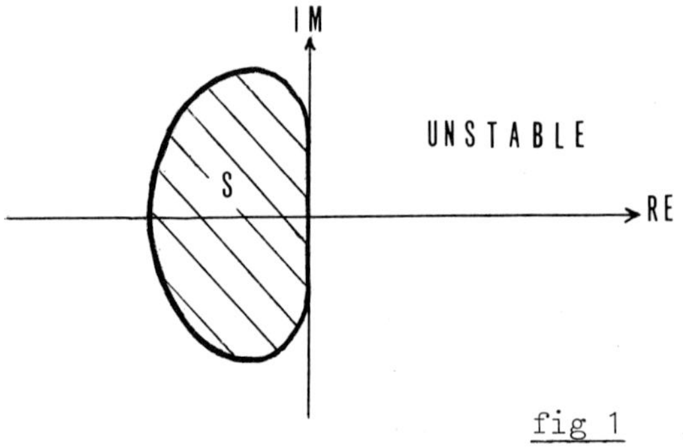

fig 1

a reasonable minimum requirement is that the negative real axis should be contained in the region; it turns out that such methods are implicit. However, this means that one is faced with inversion of possibly ill conditioned matrices. On account of the latter problem we direct our attention to explicit methods and try to "remove" the singular components first, rather than approximating the solution straightforward. Other examples of methods based on this idea can be found in Dahlquist (1968), Alfeld and Lambert (1977), O'Malley and Anderson (1979).

2. REFORMULATION OF A DIFFERENCE EQUATION

Discretize (1.1) using a multistep (of which the coefficients $\{\alpha_j\}$ and $\{\beta_j\}$ will be investigated lateron):

$$\sum_{j=0}^{k} \alpha_j \begin{pmatrix} \varepsilon\, x^1_{i-j} \\ x^2_{i-j} \end{pmatrix} = h \sum_{j=1}^{k} \beta_j \{A_{i-j} x_{i-j} + f_{i-j}\}, \qquad (2.1)$$

where $A_{i-j} = A(t_0+(i-j)h,\varepsilon)$, $f_{i-j} = f(t_0+(i-j)h,\varepsilon)$ and x_{i-j} is the approximant of $x(t_0+(i-j)h)$. Let $\alpha_0 = 1$. Rewrite (2.1) as

$$\begin{pmatrix} \varepsilon\, x^1_i \\ x^2_i \end{pmatrix} = \sum_{j=1}^{k} \tilde{A}_{i,j} x_{i-j} + \tilde{f}_{i-j} \qquad (2.2)$$

In order to show how the singular part can be isolated it will be convenient to write (2.2) as an nk-th order matrix vector recursion (by which we do not mean that one should employ such a form for actual computation). We find

formula (2.3)

(see over the page)

$$\left[\begin{array}{c} \varepsilon x^1_i \\ \hline x^1_{i-1} \\ \vdots \\ x^1_{i-k+1} \\ \hline x^2_i \\ \vdots \\ x^2_{i-k+1} \end{array}\right] = \left[\begin{array}{c|ccc:ccc} \tilde{A}^{11}_{i,1} & \tilde{A}^{11}_{i,2} & \cdots & \tilde{A}^{11}_{i,k} & \tilde{A}^{12}_{i,1} & \cdots & \tilde{A}^{12}_{i,k} \\ \hline I & 0 & & & & & \\ & \ddots & & & & & \\ & & I & 0 & & & \\ \hdashline \tilde{A}^{21}_{i,1} & & \cdots & \tilde{A}^{21}_{i,k} & \tilde{A}^{22}_{i,1} & \cdots & \tilde{A}^{22}_{i,k} \\ & & & & I & & \\ & & & & & \ddots & \\ & & & & & I & 0 \end{array}\right] \left[\begin{array}{c} x^1_{i-1} \\ \hline \vdots \\ x^1_{i-k} \\ \hdashline x^2_{i-1} \\ \vdots \\ x^2_{i-k} \end{array}\right] + \left[\begin{array}{c} \tilde{f}^1_i \\ \hline \vdots \\ \tilde{f}^1_{i-k+1} \\ \hdashline \tilde{f}^2_i \\ \vdots \\ \tilde{f}_{i-k+1} \end{array}\right]$$

With the partitioning as indicated in (2.3) and introducing new variables we can write for (2.3)

$$\begin{pmatrix} \varepsilon y^1_1 \\ y^2_i \end{pmatrix} = \begin{pmatrix} B_i & C_i \\ D_i & E_i \end{pmatrix} \begin{pmatrix} y^1_{i-1} \\ y^2_{i-1} \end{pmatrix} + \begin{pmatrix} g^1_i \\ g^2_i \end{pmatrix} \updownarrow m \tag{2.4}$$

So $y^1_i = x^1_i$ is an m dimensional and y^2_i an (nk-m) dimensional vector.

Now let $\{T_i\}_{i\geq 0}$ be a sequence of nonsingular matrices, such that the sequence of first m columns of the T_i, denoted by $\{T^1_i\}_{i\geq 0}$ satisfies the homogeneous part of (2.4) i.e.,

$$T^1_i = \begin{pmatrix} \frac{1}{\varepsilon}B_i & \frac{1}{\varepsilon}C_i \\ D_i & E_i \end{pmatrix} T^1_{i-1}. \tag{2.5}$$

then for each i the transformed matrix

$$T^{-1}_i \begin{pmatrix} \frac{1}{\varepsilon}B_i & \frac{1}{\varepsilon}C_i \\ D_i & E_i \end{pmatrix} T_{i-1} \tag{2.6}$$

has a zero lower left (nk-m)×m block. A similar transformation result also holds if a sequence $\{T^1_i L_i\}_{i\geq 0}$ satisfies the homogeneous part of (2.4), where for each i L_i may be any nonsingular m th order matrix, cf Mattheij (1978a). Presently, we are particularly interested in transformations, where

T_i has the form

$$T_i = \begin{pmatrix} I & \emptyset \\ M_i & I \end{pmatrix} \updownarrow m \tag{2.7}$$

(so M_i is an $(nk-m)\times m$ matrix). If such T_i exist, then we can derive a transformed decoupled recursion: indeed define

$$\begin{aligned} z_i^1 &= y_i^1 \\ z_i^2 &= -M_i y_i^1 + y_i^2 \\ h_i &= T_i^{-1} g_i \end{aligned} \tag{2.8}$$

Then

$$\begin{pmatrix} \varepsilon z_i^1 \\ z_i^2 \end{pmatrix} = \begin{pmatrix} B_i + C_i M_{i-1} & C_i \\ \emptyset & E_i - \frac{1}{\varepsilon} M_i C_i \end{pmatrix} \begin{pmatrix} z_{i-1}^1 \\ z_{i-1}^2 \end{pmatrix} + \begin{pmatrix} h_i^1 \\ h_i^2 \end{pmatrix} \tag{2.9}$$

The sequence $\{\begin{pmatrix} I \\ M_i \end{pmatrix}\}$ satisfies (2.5), apart from right multiplication by m th order nonsingular matrices, iff

$$M_i\{B_i + C_i M_{i-1}\} - \varepsilon\{D_i + E_i M_{i-1}\} = 0. \tag{2.10}$$

The recursion (2.10) is the well known Ricatti-formula. The existence of a set $\{M_i\}$ is given by the following

Theorem 1. Let $\forall_i B_i$ be nonsingular. Choose $M_0 = 0$. Then there exists an $\varepsilon_1 > 0$, such that for all $0 < \varepsilon \leq \varepsilon_1$ the matrices $B_i - C_i M_{i-1}$ are nonsingular. Moreover, $||M_i|| = O(\varepsilon)$. □

Using power method like arguments (cf QR algorithm) it can be shown that the columns of $\begin{pmatrix} I \\ M_i \end{pmatrix}$ (with M_i as in Th.1) have the directions of respective dominant solutions of the homogeneous part of (2.4); hence their computation is expected to be stable. Under suitable (weak) conditions it can now be shown that the discretized singular components correspond to solutions of the recursion $\varepsilon z_i^1 = \{B_i + C_i M_{i-1}\} z_{i-1}^1$, whereas the discretized regular components essentially correspond to

$z_i^2 = \{E_i - \frac{1}{\varepsilon}M_iC_i\}z_{i-1}^2$. The next theorem gives a more precise formulation for the case that $A(t,\varepsilon)$ is constant (which is customary in stability theory).

Theorem 2. Let $\forall_t\ A(t,\varepsilon) \underset{D}{=} A(\varepsilon)$, $\forall_t, \forall_s\ \lambda_s(t,\varepsilon) \underset{D}{=} \lambda_s(\varepsilon)$. Assume (i) h is such that $\forall_{\varepsilon\leq\varepsilon_0}\ \forall_{s,m+1\leq s\leq n}\ h\lambda_s(\varepsilon) \in S$ (S is the stability region) and (ii) there exists an $\varepsilon_1 \leq \varepsilon_0$ such that for $\varepsilon \leq \varepsilon_1$ and $1 \leq s \leq m$, the characteristic equations $\rho^k + \sum_{j=1}^{k} \{\alpha_j - h\lambda_s(\varepsilon)\beta_j\}\rho^{k-j} = 0$ have precisely one single root outside the unit disk which is of order $\frac{1}{\varepsilon}$. Then there exists an $\varepsilon_2 \leq \varepsilon_1$ such that for all $\varepsilon \leq \varepsilon_2$ the eigenvalues of the matrix $\frac{1}{\varepsilon}\{B_i+C_iM_{i-1}\}$ are all outside the unit disk, whereas those of $\{E_i - \frac{1}{\varepsilon}M_iC_i\}$ are within the unit disk. □

Corollary 1. For all $\varepsilon \leq \varepsilon_2$, the columns of $\binom{I}{M_i}$ span approximately the same space as the i-th iterand of the discretized singular solutions. □

The assumption (i) in Th.2 is quite reasonable in this context. The assumption (ii) has to be evaluated further; we shall see, however, in §3 that it can easily be satisfied.

The above mentioned decoupling of the recursion now leads to the following algorithm (2.11), cf. Mattheij (1978b):

(i) Choose an integer $N > (T-t_0)/h$

(ii) Compute $\{z_1^2,\ldots,z_N^2\}$ from $z_i^2 = \{E_i - \frac{1}{\varepsilon}M_iC_i\}z_{i-1}^2 + h_i^2$ (i.e. with increasing index)

(iii) Define $z_N^1(N) = 0$ and compute a sequence $\{z_{N-1}^1(N),\ldots,z_0^1(N)$ from $z_i^1(N) = \{B_i+C_iM_{i-1}\}^{-1}\{\varepsilon z_i^1(N) - C_iz_i^2(N) - h_i^1\}$ (i.e. with decreasing index).

The stability of (2.11) can be shown using a theorem like Th.2 (the solutions of the homogeneous part will not outgrow the particular solution in the indicated direction). Since the solutions of the homogeneous part of the recursion in

(2.11) (iii) decay very rapidly it appears that $z_i^1(N)$ approximates z_i^1 to any desired accuracy if N is chosen large enough (if this is allowed of course).
For the constant case we have (cf Th.2).

Theorem 3. Assume (i) and (ii) of Th.2. Then for any i we have $\lim_{N\to\infty} z_i^1(N) = z_i^1$. □

Theorem 4. Assume (i) and (ii) of Th.2. Denote the ε-dependance of $z_i^1(N)$ and z_i^1 by $z_i^1(N,\varepsilon)$ and $z_i^1(\varepsilon)$ respectively. Then for any $N > (T-t_0)/h$ and any i we have $\lim_{\varepsilon\to 0} z_i^1(N,\varepsilon) = z_i^1(0)$. □

3. SELECTING MULTISTEPS

A detailed analysis of multisteps satisfying requirements as in Th.1 and Th.2 has not been carried out yet. However, it is not difficult to find examples, some of which are given below

Theorem 5. Let $\forall_t\ A(t,\varepsilon) = A(\varepsilon)$. Consider the 2-step Adams-Bashford formula, i.e. $\alpha_0 = 1$, $\alpha_1 = -1$, $\alpha_2 = 0$, $\beta_1 = \frac{3}{2}$ and $\beta_2 = -\frac{1}{2}$. If for $\varepsilon \leq \varepsilon_0$ and $s = 1,\dots,m$ $h\lambda_s(\varepsilon)$ is outside the stability region S (see fig 1), then (i) $\alpha_1 - h\beta_1\lambda_s(\varepsilon) \neq 0$, i.e. $\forall_i B_i$ is nonsingular, and (ii) assumption (ii) of Th.2 is satisfied with $\varepsilon_1 = \varepsilon_0$. □

Theorem 6. Let the multistep be such that (i) $\beta_i \neq 0$ and (ii) the polynomial $\sum_{j=1}^{k} \beta_j \rho^{k-j-1}$ has no roots outside the unit disk. Then for each fixed h there exists an $\varepsilon_1(h) \leq \varepsilon_0$ such that for all $\varepsilon \leq \varepsilon_1(h)$ (i) $\forall_i B_i$ is nonsingular and (ii) assumption (ii) of Th.2 is satisfied with $\varepsilon_1 = \varepsilon_1(h)$. □

The proof of Th 5 follows from inspection of the stability region, whereas Th 6 can be seen by letting $\varepsilon \to 0$.

4. FURTHER COMMENTS

The main issue in this talk is that we are interested in a reasonable accuracy for a fairly large stepsize rather than in a method giving convergence as $h \to 0$. Some of the prerequisites in §1 may be too restrictive. However, it can directly be seen that a lot can be generalized. Thus the stepsize needs not necessarily be fixed. Moreover, we can allow positive eigenvalues as well, even with large real part. ODE with such eigenvalues occur in particular types of boundary value problems, see O'Malley (1979); naturally our approach involves the computation of solutions of boundary value problems with boundary conditions of a certain type.

If the parameter ε is not explicitly present, i.e. if we have a stiff ODE $\dot{x} = A(t)x + f(t)$, we may transform the discretized version (cf. (2.1)), $y_i = A_i y_{i-1} + f_i$ say, by recursively determined matrices T_i (cf. §2) of which the first m columns are directionally close to directions of solutions corresponding to "singular" solutions; the recursion $z_i \underset{D}{=} T_i y_i = T_i A_i T_{i-1}^{-1} z_{i-1} + T_i f_i$ is then expected to have similar properties as (2.1) (cf. Mattheij (1978b), §4).

REFERENCES

Alfeld, P. and Lambert, J.D. (1977). Correction in the dominant space: A numerical technique for a certain class of stiff initial value problems. *Math.Comp.* 31, 922-938.

Dahlquist, G. (1969). A numerical method for some ordinary differential equations with large Lipschitz constants. In: *Information Processing 68*, pp 183-186. North Holland, Amsterdam.

Gear, C.W. (1971). *Numerical initial value problems in ordinary differential equations.* Prentice Hall, Englewood Cliffs, N.J.

Mattheij, R.M.M. (1978a). *Report 7802*. Mathematisch Instituut, Katholieke Universiteit, Nijmegen, The Netherlands.

Mattheij, R.M.M. (1978b). *Report 7803*. Mathematical Katholieke Universiteit, Nijmegen, The Netherlands.

ON THE CONVERGENCE, UNIFORMLY IN ε, OF DIFFERENCE SCHEMES FOR A TWO POINT BOUNDARY SINGULAR PERTURBATION PROBLEM

John J.H. Miller

School of Mathematics, Trinity College, Dublin, Ireland

ABSTRACT. We analyze the convergence, uniformly in ε, of a standard and of an exponentially fitted finite difference scheme for a two point boundary singular perturbation problem without a first derivative term. Sufficient conditions for the convergence, uniformly in ε on the whole domain, are given for a general three point finite difference scheme.

Let $\Omega = (a, b)$ and consider the two point boundary singular perturbation problem

$$(P)\quad \begin{cases} Lu \equiv -\varepsilon u'' + a_o u = f & \text{on } \Omega \\ u(a),\ u(b) \ \text{given} \end{cases}$$

where the functions a_o and f are smooth, and the parameter ε is positive. We assume that

$a_o(x) = \alpha^2(x)$ for all $x \in \Omega$

and that

$\alpha(x) \geq \alpha > 0$ for all $x \in \Omega$.

To solve (P) numerically we introduce the uniform mesh $\Omega_h = \{x_i\}_1^{N-1}$, $x_i = a + ih$, $Nh = b - a$, $0 < h \leq h_o$ and the general three point finite difference scheme

$$(P^h)\quad \begin{cases} L^h u_i \equiv b_i u_i - a_i(u_{i+1} + u_{i-1}) = f_i \quad \text{on } \Omega_h \\ u_o,\ u_N \ \text{ given} \end{cases}$$

where $u_i = u^h(x_i)$, $f_i = f^h(x_i)$ are regarded as approximations to $u(x_i)$, $f(x_i)$ respectively. (P^h) may be written in the equivalent form

$$(P^h)\quad \begin{cases} L^h u_i \equiv -\varepsilon_i h^{-2}\delta^2 u_i + \alpha_i^2 u_i = f_i \quad \text{on } \Omega_h \\ u_o,\ u_N \ \text{ given.} \end{cases}$$

$\varepsilon_i = \varepsilon^h(x_i)$, $\alpha_i = \alpha^h(x_i)$ are regarded as approximations to ε, $\alpha(x_i)$ respectively and δ is the central difference operator defined by

$$\delta g(x) = g(x + h/2) - g(x - h/2).$$

We shall also need the averaging operator μ, where

$$2\mu g(x) = g(x + h/2) + g(x - h/2).$$

The equivalence of the two forms is a consequence of the following relations between the coefficients

$$\varepsilon_i = h^2 a_i$$

$$\alpha_i = b_i - 2a_i.$$

Two particular cases of (P^h) are

Scheme A.

$$\begin{cases} -\varepsilon h^{-2}\ \delta^2 u_i + a_o(x_i)u_i = f(x_i) \\ u_o = u(a), \quad u_N = u(b) \end{cases}$$

the standard central difference scheme

and

Scheme B.

$$\begin{cases} -\varepsilon h^{-2}\left((\alpha(x_i)\rho/2)/\sinh(\alpha(x_i)\rho/2)\right)^2\ \delta^2 u_i + \\ \qquad + a_o(x_i)u_i = f(x_i) \\ u_o = u(a), \ u_N = u(b) \end{cases}$$

where $\rho = h/\sqrt{\varepsilon}$. This is an exponentially fitted scheme.

Definition 1. Let u, u^h be the solutions of (P), (P^h) respectively. Then u^h converges to u, uniformly in ε, if $(u^h - u)(x_i) \to 0$ as $h \to 0$ for all $x_i \in \overline{\Omega}_h$, all $0 < h \leq h_o$ and all $\varepsilon > 0$. Furthermore, u^h converges to u, uniformly in ε, with order h^p if, for some constant C independent of h and ε, $|(u^h - u)(x_i)| \leq Ch^p$ for all $x_i \in \overline{\Omega}_h$, all $0 < h \leq h_o$ and all $\varepsilon > 0$.

Remark 1. Let u be the solution of the equation $-\varepsilon u'' + u = 1$ with homogeneous boundary conditions and u^h the solution of Scheme A applied to this problem. Then a simple calculation shows that $(u^h - u)(h) \to e^{-\rho} - t^{-1}$ as $h \to 0$ with ρ fixed; here $t = 1 + \frac{1}{2}\rho^2 + \rho(1 + \rho^2/4)^{\frac{1}{2}}$. Thus Scheme A does not converge uniformly in ε.

Remark 2. Let u be the solution of the equation $-\varepsilon u'' + u = x^2$ with homogeneous boundary conditions and u^h the solution of Scheme B applied to this problem. Then it is not hard to show that for all $0 < h \leq h_o$ and all $\varepsilon > 0$ we have
$(u^h - u)(\frac{1}{2}) = 2\varepsilon\left(1 - \left((\rho/2)\sinh\rho/2\right)^2\right)\left(1 - \operatorname{sech} 1/2\sqrt{\varepsilon}\right) \geq C(\rho)h^2$. This shows that the convergence of Scheme B, uniformly in ε, is of order h^2 at best.

For convenience we introduce the notation

$$A_i = a_i/b_i \qquad\qquad F_i = f_i/b_i.$$

For a continuous function g on $\overline{\Omega}$ we use the norm

$$||g|| = \max_{x\in\overline{\Omega}} |g(x)|$$

and for a mesh function $w^h = \{w_i\}_0^N$ on Ω_h we use the semi-norm

$$|w^h| = \max_{1\leq i\leq N-1} |w_i|.$$

The local truncation error τ^h in approximating L by L^h is defined by

$$\tau^h = L^h - L.$$

Definition 2. (P^h) is consistent with (P) if

$$\lim_{h\to 0} \left(|\tau^h u| + |f^h - f| + |u_o - u(a)| + |u_N - u(b)|\right) = 0.$$

Remark 3. A sufficient condition for (P^h) to be consistent with (P) is

(i) $\lim_{h\to 0} \left(|\varepsilon^h - \varepsilon| + |\alpha^h - \alpha| + |f^h - f| + |u_o - u(a)| + \right.$
$\left. + |u_N - u(b)|\right) = 0.$

Definition 3. L^h is positive if, for $1 \leq i \leq N-1$, we have $b_i \geq 2a_i \geq 0$ with at least one inequality strict for each i.

Remark 4. L^h is positive iff the following two conditions hold

(ii) $b_i > 0$ for $1 \leq i \leq N - 1$

(iii) $0 \leq A_i \leq \frac{1}{2}$ with at least one inequality strict for each i.

Definition 4. The maximum principle holds for L^h if, for all mesh functions, the inequalities

$$L^h w_i \leq 0 \quad \text{for} \quad 1 \leq i \leq N - 1$$

imply that

$$w_i \leq \max \{0, w_o, w_N\} \text{ for } \quad 0 \leq i \leq N.$$

Lemma 1. The maximum principle holds for L^h if L^h is positive.

Definition 5. L^h is stable, with stability constant C, if for all mesh functions w^h and all $0 < h \leq h_o$ we have

$$|w_i| \leq \max \{|w_o|, |w_N|\} + C\,|L^h w^h| \quad \text{for} \quad 0 \leq i \leq N.$$

where C is independent of h.

Remark 5. If the maximum principle holds for L^h and the following condition holds

(iv) $b_i(1 - 2A_i) \geq C^{-1} > 0$ for $1 \leq i \leq N - 1$

then L^h is stable, with stability constant C. Furthermore, if u, u^h are the solutions of (P), (P^h) respectively, then for $0 \leq i \leq N$ we have

$$|u_i| \leq \max \{|u_o|, |u_N|\} + C|f^h|$$

and

$$|u_i - u(x_i)| \leq \max \{|u_o - u(a)|, |u_N - u(b)|\} + + C(|\tau^h u| + |f^h - f|).$$

Theorem 1. Assume that (P^h) is consistent with (P) and that L^h is stable, then u^h converges to u.

Corollary. Conditions (i) - (iv) guarantee that u^h converges to u.

Remark 6. Let u be the solution of (P) and u^h the solution of either Scheme A or Scheme B. Then, for each fixed $\varepsilon > 0$, u^h converges to u with order h^2.

All the results so far are classical and the proofs are simple and well known. We now obtain sufficient conditions for the convergence of u^h to u, uniformly in ε. The results may be proved by methods similar to those in |4|.

We begin with a general principle.

Convergence Principle. Let F^h be any quantity depending on h, p a positive number and C_1, C_2 constants independent of h. Then, for all $0 < h \leq h_o$, we have

$$|F^h - F| \leq C_1 h^p$$

iff (a) $\lim_{h\to 0} |F^h - F| = 0$ and (b) $|F^h - F^{h/2}| \leq C_2 h^P$.

Moreover, either $C_2 = 2C_1$ or $C_1 = 2C_2$.

An immediate consequence of this is the following theorem, which gives necessary and sufficient conditions for convergence, uniformly in ε.

Theorem 2. Let u, u^h be the solutions of (P), (P^h) respectively, and let C_1, C_2 be as above. Then, for all $x_i \in \overline{\Omega}_h$, all $0 < h \leq h_o$, and all $\varepsilon > 0$

(1) $|(u^h - u)(x_i)| \leq C_1 h^p$

iff

(2)(a) $\lim_{h\to 0} |(u^h - u)(x_i)| = 0$ and (b) $|(u^h - u^{h/2})(x_i)| \leq C_2 h^P$

Furthermore, C_1 is independent of ε iff C_2 is.

From the Corollary to Theorem 1 we know that (2a) in Theorem 2 holds if (P^h) fulfills conditions (i) - (iv). We now determine what additional conditions will imply (2b) in Theorem 2.

The following lemma is immediate.

Lemma 2. For all $0 < h \leq h_o$ and all $\varepsilon > 0$, assume that L^h is stable with stability constant C and that

$$\max\{|(u^h - u^{h/2})(a)|, |(u^h - u^{h/2})(b)|\} + C|L^h(u^h - u^{h/2})| \leq C_2 h^P.$$

Then (2b) in Theorem 2 holds.

We now give sufficient conditions on (P^h) to guarantee that (2b) in Theorem 2 holds with p = 1 in general and p = 2 in a special case. In the following lemma δ and μ operate on functions defined on the mesh $\Omega_{h/2}$ and they denote respectively the central difference and averaging operators corresponding to this mesh.

Lemma 3. For all $x_i \in \Omega_h$, $0 < h \leq h_o$, $\varepsilon > 0$, assume that (ii) - (iv) hold and that for some constant C_3 independent of h and ε

(v) $|f_i^{h/2}| + |h^{-1}\ \delta f_{i+\frac{1}{2}}^{h/2}| + |h^{-2}\ \delta^2 f_i^{h/2}| \leq C||f||$

(vi) $|f_i^h - f_i^{h/2}| \leq Ch^2\ ||f||$

(vii) $|(b_i^{h/2})^{-1}| + |h^{-1}A_i^{h/2}\ \delta(b_{i\pm\frac{1}{2}}^{h/2})^{-1}| + |h^{-2}\ \delta^2(b_i^{h/2})^{-1}| \leq$
$$\leq C(b_i^h A_i^{h/2})^{-1}$$

(viii) $|h^{-1}\mu\delta A_i^{h/2}| + |h^{-2}\delta^2 A_i^{h/2}| \leq C(b_i^h A_i^{h/2})^{-1}$

(ix) $|(b_i^h)^{-1}(1 - 2(A_i^{h/2})^2) - (b_i^{h/2})^{-1}(1 + 2A_i^{h/2})| +$
$$+ |A_i^h(1 - 2(A_i^{h/2})^2) - (A_i^{h/2})^2| \leq Ch^2(b_i^h)^{-1}.$$

Then there is a constant C_4, independent of h and ε, such that

$$|L^h\ (u^h - u^{h/2})| \leq C_4\ h.$$

If, in addition, a_i and b_i are independent of i, then

$$|L^h\ (u^h - u^{h/2})| \leq C_4\ h^2.$$

Remark 7. The condition that a_i and b_i are independent of i is fulfilled in the case where a_o is a constant, and normally only in that case.

We then deduce immediately

Theorem 3. Let u, u^h be the solutions of (P), (P^h) respectively. For all $x_i \in \Omega_h$, $0 < h \leq h_o$, $\varepsilon > 0$, assume that conditions (i) - (ix) hold. Then u^h converges to u, uniformly in ε, with order h. If, in addition, the coefficients a_i, b_i in (P^h) are independent of i, then u^h converges to u, uniformly in ε, with order h^2.

A simple calculation shows that Scheme B fulfills conditions (i) - (ix). This scheme is the analogue of the scheme considered by Il'in |2| for a differential equation containing also a first order derivative term. It may be obtained as a special case of a difference scheme studied by Hemker |1| and also of another scheme in Shishkin and Titov |5|.

The analysis of convergence, uniformly in ε, given above is similar to that in |4|. The latter is a generalization of the convergence results of Il'in |2| for a differential equation having also a first derivative term; this problem has also been considered independently by Kellogg and Tsan |3|. For the problem studied in the present paper Shishkin and Titov |5| have shown that Scheme B converges, uniformly in ε, with order h^r in Ω and order h^s outside the boundary layers for all $r < 0.2$ and all $s < 2$.

This paper was written while the author was the Professor of Numerical Analysis at the Mathematical Institute, Catholic University, Nijmegen, The Netherlands.

REFERENCES

|1| P. W. Hemker "A numerical study of stiff two-point boundary problems" *Thesis*, Math. Cent. Amsterdam (1977)

|2| A. M. Il'in "Differencing scheme for a differential equation with a small parameter affecting the highest derivative" *Math. Notes Acad. Sci. USSR* 6 (1969) 596 - 602.

|3| R. Bruce Kellogg and Alice Tsan "Analysis of some difference approximations for a singular perturbation problem without turning points" *Preprint*.

|4| John J. H. Miller "Sufficient conditions for the convergence, uniformly in ε, of a three point difference scheme for a singular perturbation problem" in Proc. Conf. Oberwolfach 11-17 Dec. 1977. *Lecture Notes in Maths*, Springer Verlag (to appear)

|5| G. I. Shishkin and V. A. Titov "A difference scheme for a differential equation with two small parameters affecting the derivatives" *Num. Meth. Mechs. Cont. Media* 7, 2 (1976) 145 - 155 (in Russian).

A SINGULARLY PERTURBED WEAKLY-SINGULAR INTEGRO-DIFFERENTIAL PROBLEM FROM ANALYTICAL CHEMISTRY

S.L. Paveri-Fontana and Rossella Rigacci
Istituto Matematico "U.Dini",
Viale Morgagni 67a, 50134 Florence, Italy

ABSTRACT

A singularly-perturbed integro-differential weakly-singular problem is treated. Making use of two different equivalent sets of equations, external solutions and boundary layer corrections are obtained formally.

INTRODUCTION

As discussed by Paveri-Fontana, Tessari, Torsi (1978), the release and redeposition of a given substance between an electro-thermal rod atomizer and the surrounding gas can be described according to the following equations:

$$(\partial/\partial t)c(z,t) = -(\partial/\partial z)J(z,t) \text{ for } z > 0,\ t > 0;$$

$$(d/dt)\phi(t) = -J(0^+,t), \text{ for } t > 0;$$

$$(d/dt)\phi(t) = -J^r(t) + J^c(t), \text{ for } t > 0;$$

to be studied with appropriate initial and boundary conditions, subject to the quasi-equilibrium assumption (namely, $J^r \simeq J^c$). Here, $\phi(t)$ is the concentration of the substance on the surface of the atomizer; $c(z,t)$ is the (volume) concentration of the substance in the gas; $J^+(J^-)$ is the upward (downward) current in the gas; $J = J^+ - J^-$ is the net upward current; J^r is the release current at the surface; J^c is the redeposition current on the surface. It is assumed that: i) $J^r(t) = K^r(t)\phi(t)$, where K^r is the temperature controlled release parameter; ii) $J^c(t) = K^*(t)f(\phi(t))\, J^-(t)$, where $K^*(t)\in(0,1]$ is the probability that a returning molecule will have a velocity which permits reattachment, $f(\phi) = 1 - \phi/\phi_M$ is the probability for a returning molecule to find an empty location on the surface; iii) $J^{+,-} = (-,+)D(\partial/\partial z)c + wc/2$, where the diffusion coefficient D and the mean square thermal speed $w = (2KT/m\pi)^{1/2}$ are time independent.

After some changes of variables the problem to be studied can be written as follows:

$$(\partial/\partial t - \partial^2/\partial x^2)u(x,t) = 0,\ t > 0,\ x > 0, \tag{1}$$

$$(d/dt)\psi(t) = (\partial u/\partial x)(0^+,t),\ t > 0, \tag{2}$$

$$\varepsilon \frac{d}{dt}\psi(t) = \frac{-\lambda_r(t)\psi(t) + \lambda_c(t)u(0^+,t)(1-\theta\psi(t))}{1-q\lambda_c(t)(1-\theta\phi(t))},\ t > 0 \tag{3}$$

where $\lambda_r(t) > 0$ for $t > 0$; $0 \le \theta < 1$; $0 \le q\lambda_c(t) \le \frac{1}{2}$ for $t > 0$; $\lambda_c(t) > 0$ for $t > 0$; $\lambda_r(0) = \lambda_c(0) = 1$; subject to

$$\psi(0^+) = 1. \tag{4}$$

$$u(x,0^+) = 0,\ x > 0, \tag{5}$$

$$u(+\infty,t) = 0,\ t > 0. \tag{6}$$

Now it is well known that if u(x,t) is a sufficiently

regular solution of eqs. (1), (4) then

$$\int_0^t \frac{\partial u}{\partial x}(0^+,t')\,dt' = -\int_0^t \frac{u(0^+,t')}{\sqrt{\pi(t-t')}}\,dt'.$$

This permits to cast system (1) through (6) into the form

$$\psi(t) = 1 - \int_0^t \frac{u(t')}{\sqrt{\pi(t-t')}}\,dt', \quad t > 0, \tag{7a}$$

$$\varepsilon \frac{d}{dt}\psi(t) = K(t,\psi(t),u(t)), \quad t > 0, \tag{7b}$$

to be studied subject to initial conditions

$$\left.\begin{aligned} \psi(0^+) &= 1, \\ u(0^+) &= 0. \end{aligned}\right\} \tag{8}$$

Here $u(t) \equiv u(0^+,t)$ is the (volume) concentration in the immediate neighbourhood of the atomizer; also,

$$k(t,\psi,u) \equiv \frac{-\lambda_r(t)\,\psi + \lambda_c(t)(1-\theta\psi)u}{1 - q\lambda_c(1-\theta\psi)}\,. \tag{9}$$

In eqs. (3), (7b), (9), $\varepsilon \equiv K^r(0)D/(K^c(0))$; $\lambda_r(0) = \lambda_c(0) = 1$. As analyzed by Paveri-Fontana *et al.* (1978), the above-mentioned assumption of quasi-equilibrium corresponds to the assumption that the non-dimensional parameter ε is small. For $\varepsilon \to 0^+$ problem (7), (8) is singularly perturbed. From the point of view of numerical analysis, it is "stiff". The remaining parts of this paper are devoted to a *formal* asymptotic study of the singularly-perturbed weakly singular integro-differential problem (7), (8) under the assumption that the assigned functions λ_r, λ_c are sufficiently smooth.

It can be shown - by means of a procedure which is common in the treatment of Abel-like integral equations - that eq. (7a) can be written in the equivalent form

$$u(t) = -\frac{1}{\varepsilon}\int_0^t \frac{K(t',\psi(t'),u(t'))}{\sqrt{\pi(t-t')}}\,dt', \quad t > 0. \tag{10}$$

Eqs (7) will be employed for the "external" solution. On the other hand, for the "boundary layer" corrections we shall make use of eqs (7b) and (10).

To a large extent we shall follow the notation and the "philosophy" of Chapter 4 of O'Malley's monograph (1974).

The formal treatment outlined here will be completed in a forthcoming paper containing a more precise mathematical study. The development of numerical schemes for the problem discussed here is under way.

3. THE ASYMPTOTIC EXPANSIONS

It is clear that the reduced problem is

$$\psi_o(t) = 1 - \int_0^t \frac{(\lambda_r(t')/\lambda_c(t'))\psi_o(t')}{\sqrt{\pi(t-t')}\,(1-\theta\psi_o(t'))}\,dt'\,. \tag{11}$$

We proceed now to the construction of an asymptotic representation of ψ and u in terms of a generalized asymptotic expansion. Let

$$\psi(t) = \overline{\psi}(t,\varepsilon) + \varepsilon m(\tau,\varepsilon),$$
$$u(t) = \overline{u}(t,\varepsilon) + n(\tau,\varepsilon). \tag{12}$$

where the "stretched" time variable is given by

$$\tau \equiv t/\varepsilon^2, \tag{13}$$

and seek, for $\varepsilon \to 0^+$,

$$\overline{\psi}(t,\varepsilon) \sim \sum_{k=0}^{\infty} \psi_k(t)\varepsilon^k , \tag{14a}$$

$$\overline{u}(t,\varepsilon) \sim \sum_{k=0}^{\infty} u_k(t)\varepsilon^k , \tag{14b}$$

$$m(\tau,\varepsilon) \sim \sum_{k=0}^{\infty} m_k(\tau)\varepsilon^k , \tag{14c}$$

$$n(\tau,\varepsilon) \sim \sum_{k=0}^{\infty} n_k(\tau)\varepsilon^k . \tag{14d}$$

The requirements

$$\psi_0(t) \to 1 \text{ for } t \to 0^+ , \tag{15a}$$

$$\psi_k(t) + m_{k-1}(t/\varepsilon^2) \to 0 \text{ for } t \to 0^+ \ (k=1,2,\ldots), \tag{15b}$$

$$u_k(t) + n_k(t/\varepsilon^2) \to 0 \text{ for } t \to 0^+ \ (k=0,1,\ldots), \tag{15c}$$

$$m_k(\tau) \to 0 \text{ for } \tau \to +\infty \ (k=0,1,2,\ldots..), \tag{15d}$$

$$n_k(\tau) \to 0 \text{ for } \tau \to +\infty \ (k=0,1,2,\ldots..), \tag{15e}$$

must be satisfied. Asymptotic representations (14) can now be substituted into eqs (7) and (10).

THE LINEAR CASE ($\theta = 0$)

Inserting ansatz (14a), (14b) into eqs (7) we obtain for $k \geq 1$

$$\psi_k(t) = \alpha_k - \int_0^t \frac{u_k(t')}{\sqrt{\pi(t-t')}}\, dt' ,$$

$$u_k(t) = \frac{\mu_r(t)}{\mu_c(t)} \psi_k(t) + \frac{1}{\mu_c(t)} \frac{d}{dt} \psi_{k-1}(t) , \qquad (16)$$

whereas for k=0 we obtain the reduced problem, eq. (11). Here $\mu_r(t) = \lambda_r(t)(1-q\lambda_c(t))^{-1}$; $\mu_c(t) = \lambda_c(t)(1-q\lambda_c(t))^{-1}$. The parameters α_k will be evaluated employing requirements (15). Equations (16) can be solved recursively. One can show that parameters γ and c_k, d_k (for k = 1,2,...) exist such that

$$\psi_k(t) = \alpha_k - c_k + d_k t^{\frac{1}{2}} + \theta(t), \text{ for } t \to 0^+ \quad \text{(with } k \geq 1) \qquad (17a)$$

$$u_k(t) = c_k(\pi t)^{-\frac{1}{2}} + \gamma d_k + \theta(t^{\frac{1}{2}}), \text{ for } t \to 0^+ \quad \text{(with } k \geq 1) \qquad (17b)$$

Now we turn to the "boundary layer corrections", m and n. Proceeding in a formal way we find that they should obey the equations

$$\frac{d}{d\tau} m(\tau,\varepsilon) = - \mu_r(\varepsilon^2\tau)m(\tau,\varepsilon) + \mu_c(\varepsilon^2\tau)n(\tau,\varepsilon) , \tag{18}$$

$$n(\tau,\varepsilon) = - (\pi\tau)^{-\frac{1}{2}}(c_1 + \varepsilon c_2 + \varepsilon^2 c_3 + \ldots.) + + \int_0^\tau \frac{\mu_r(\varepsilon^2\tau')m(\tau') - \mu_c(\varepsilon^2\tau')n(\tau')}{\sqrt{\pi(\tau-\tau')}} d\tau' , \tag{19}$$

as well as requirements (15). Here eq. (19) is obtained from eq. (10). Now, by inserting ansatz (14c) and (14d) into eq. (19) and by equating the coefficients of equal powers of ε one can obtain equations for the coefficients $m_k(\tau)$, $n_k(\tau)$ provided one assumes that $\mu_r(t)$, $\mu_c(t)$ admit the asymptotic representations

$$\mu_r(t) \sim \mu_{ro} + \mu_{r1}t^{\frac{1}{2}} + \mu_{r2}t + \ldots. \quad \text{for } t \to 0^+,$$

$$\mu_c(t) \sim \mu_{co} + \mu_{c1}t^{\frac{1}{2}} + \mu_{c2}t + \ldots. \quad \text{for } t \to 0^+.$$

The equations must be supplemented with initial conditions (15b), (15c). Finally, one can show that "matching" requirements (15d), (15e) are satisfied if one lets $\alpha_k=0$ for all values of k.

As an example consider the case of constant coefficients (that is, assume μ_r, μ_c to be time-independent.) Here one obtains

$$\frac{d}{d\tau} m_k (\tau) = -\mu_r m_{k-1}(\tau) + \mu_c n_k(\tau), \text{ for } k=0,1,2,\ldots. ,$$

$$n_k(\tau) = - (\pi\tau)^{-\frac{1}{2}}c_{k+1} + \int_0^\tau \frac{\mu_r m_{k-1}(\tau')-\mu_c n_k(\tau')}{\sqrt{\pi(\tau-\tau')}} d\tau' \quad \text{for } k=0,1,2,\ldots \quad , \tag{20}$$

with $m_{-1} \equiv 0$, to be solved subject to

$$m_k(0^+) = \delta_{ko} - \psi_k(0^+),$$

with all $\alpha_k = 0$. Observe however the following: if an asymptotic approximation up to terms of order N is sought, - that is, if representations (12a), (14b), (14d) are truncated after the N-th term and representation (14c) is truncated after the (N-1)th term - then the N-th term n_N in the asymptotic representation for n is not the solution $\tilde{n}_N$ of problem (20) for the case k = N. On the contrary, one has to set

$$n_N(\tau) = \tilde{n}_N(\tau) + (\pi\tau)^{-\frac{1}{2}} c_{N+1}. \tag{21}$$

THE LEADING TERMS FOR THE NON LINEAR PROBLEM

For the complete non-linear problem - that is, for $\theta \neq 0$ in eq. (9) - the equations for the successive terms of the asymptotic expansion (14) are quite involved. Here we restrict our attention to the leading-term representation:

$$\begin{aligned} \psi(t) &= \psi_o(t) + O(\varepsilon) \text{ for } \varepsilon \to 0^+ , \\ u(t) &= u_o(t) + n_o(t\varepsilon^{-2}) + O(\varepsilon) \text{ for } \varepsilon \to 0^+ . \end{aligned} \tag{22}$$

We find that if ψ_o obeys the 'reduced problem', eq. (11); if $K(t,\psi_o(t),u_o(t)) = 0$ (that is, if

$$u_o(t) = \frac{\lambda_r(t)}{\lambda_c(t)} \frac{\psi_o(t)}{1-\theta\psi_o(t)} \quad) ;$$

finally, if n_o obeys the integral equation

$$n_o(\tau) = \frac{-1}{1-\theta} - \frac{1-\theta}{1-q(1-\theta)} \int_o^{\tau} \frac{n_o(\tau')}{\sqrt{\pi(\tau-\tau')}} d\tau' \quad , \qquad (23)$$

then indeed result (22) holds uniformly with respect to t for t belonging to some interval [0,T].

CONCLUSIONS

The heuristic approach followed here leads to results which can be shown to be correct from the point of view of asymptotic analysis. However, they have unsatisfactory features. For instance, singularities appear in the "external" solution - eq. (17b) - which are "artificial" since they cancel out with corresponding singularities in the "boundary layer" corrections (see eqs (15c)). In addition, correction (21) turns out to be confusing when actual calculations are performed. Actually, it can be shown that the whole problem can be reconsidered and that the above-mentioned shortcomings can be weeded out. In this case, however, the interpretation of the new set of equations for the terms of the asymptotic expansion (14) is not straightforward. We shall report on these questions in a forthcoming paper devoted to a more precise mathematical study of the problem treated here.

ACKNOWLEDGEMENTS

S.L. Paveri-Fontana has participated in this research under the auspices of the National Group for Mathematical Physics of CNR, Italy.

REFERENCES

O'Malley, R.E., Jr. (1974). Introduction to Singular

Perturbations, Academic Press, New York and London.
Paveri-Fontana, S.L., Tessari, G. and Torsi, G. (1978). Equilibrium Release of Metal Atoms at the Electrothermal Rod Atomizer: A Theoretical Model (*to appear*).

STABILITY OF SINGULARLY PERTURBED LINEAR DIFFERENCE EQUATIONS

H.-J. Reinhardt

Fachbereich Mathematik, Goethe-Universität,
Frankfurt am Main, Federal Republic of Germany

INTRODUCTION

By this paper we want to contribute to the treatment of singular perturbation problems by an approach using methods of functional analysis. For linear problems the underlying principles are rather simple. However, before using them, it is an important task to prove a stability inequality of the form (6) for suitable norms. In this paper we shall prove such a stability inequality (uniformly with respect to ε) for singularly perturbed difference-boundary value problems of the form

$$u_o = \alpha_\varepsilon,\ u_n = \beta_\varepsilon,\ B_{k,\varepsilon} u_{k+1} + A_{k,\varepsilon} u_k + \varepsilon u_{k-1} = f_{k,\varepsilon},\ k=1,..,n-1. \quad (1)$$

They include difference approximations of the following class of boundary value problems for second order, ordinary differential equations

$$v(o) = \alpha_\varepsilon,\ v(T) = \beta_\varepsilon,\ \varepsilon v'' + a(t)v' + b(t)v = w_\varepsilon(t),\ t \in [o,T]. \quad (2)$$

Stability inequalities are used to answer the fundamental question for approximations of the solution of singular perturbation problems. The concept of stability together with the concepts of formal approximation, matching, and corrector seem to be the appropriate ones in order to construct and prove asymptotic approximations. This is carried out for the above problems in another paper which also includes some numerical examples (see Reinhardt (1978)). Let us finally announce that nonlinear singular perturbation problems can be analogously treated where deeper results from functional analysis are needed.

1. A STABILITY INEQUALITY

In this first chapter we shall prove a uniform estimate for the solutions of (1) by means of the right-hand sides (cf. Theorem 1) which implies the stability inequality (6). Such inequalities with respect to other norms are proved by Dorr(197o), Miller (1977), and others by virtue of maximum principles or by Abrahamsson, Keller and Kreiss (1974), Kreiss (1975) with respect to L^2-norms. The idea for the stability inequality proved here has arisen from an unproved estimate in Comstock and Hsiao (1976) which is only correct in a modified version. Contrary to Dorr (197o), Miller (1977) the difference operators considered here are not necessarily of positive type.

For given numbers $A_{k,\varepsilon}$, $B_{k,\varepsilon} \neq 0$, $k=1,\ldots,n-1$, $(n \geq 2)$, and right-hand sides $f_{k,\varepsilon}$, $k=1,\ldots,n-1$, α_ε, β_ε, $0<\varepsilon\leq\varepsilon_0$, let us consider difference - boundary value problems of the form (1). Their solutions are denoted by $u_\varepsilon = (u_{0,\varepsilon},\ldots,u_{n,\varepsilon})$. For an estimate of u_ε by means of the right-hand sides we essentially need the following lemma, whose proof together with those of the other lemmas will be carried out in chapter 3.

<u>LEMMA 1</u>. *Defining* $B_{0,\varepsilon}=A_{1,\varepsilon}$, $\gamma_0=0$, $\gamma_1=\max(|A_{n-1,\varepsilon}|^{-1},|B_{n-2,\varepsilon}|^{-1})$

and $\gamma_k = \dfrac{|B_{n-k,\varepsilon}|\gamma_{k-1}}{\min(|A_{n-k,\varepsilon}|,|B_{n-k-1,\varepsilon}|) - \varepsilon|B_{n-k,\varepsilon}|\gamma_{k-1}}$, $k=2,\ldots,n-1$,

then for all $k=1,\ldots,n-1$ *and all* ε *in*

$$o < \varepsilon < \min(|A_{k,\varepsilon}|, |B_{k-1,\varepsilon}|)/(|B_{k,\varepsilon}|\gamma_{n-k-1}) ,$$

$|B_{k-1,\varepsilon}|\gamma_{n-k} \geqslant 1$ *holds and the solution of* (1) *satisfies*

$$|u_{k,\varepsilon}| \leqslant \gamma_{n-k} (|B_{n-1,\varepsilon}||u_{n,\varepsilon}| + \varepsilon |u_{k-1,\varepsilon}| + \sum_{m\geqslant k} |f_{m,\varepsilon}|) . \quad (3)$$

The reciprocals $\rho_k = \gamma_k^{-1}$ satisfy the recursion formula

$$\rho_{k+1} = \min(\mu_k^{-1}, \nu_k^{-1})\rho_k - \varepsilon , \quad k=1,\ldots,n-1,$$

which gives the explicit representation

$$\gamma_k = \prod_{m=0}^{k} \max(\mu_m, \nu_m) \{ |B_{n-1,\varepsilon}| - \varepsilon \sum_{m=1}^{k-1} \prod_{j=0}^{m} \max(\mu_j, \nu_j) \}^{-1} \quad (4)$$

where $\mu_m = |B_{n-m-1,\varepsilon}|/|A_{n-m-1,\varepsilon}|$, $\nu_m = |B_{n-m-1,\varepsilon}|/|B_{n-m-2,\varepsilon}|$. As a last step before proving a stability inequality we need a uniform bound (with respect to ε) for these constants. This can be obtained for sufficiently small ε.

LEMMA 2. *Let* $b_0, b_1 > o$, $c \geqslant o$, $h > o$ *be numbers independent of* ε *such that* $b_0 \leqslant |B_{n-1,\varepsilon}| \leqslant b_1$, *and* $\max(\mu_k, \nu_k) \leqslant 1 + ch$, $k=0,\ldots,n-2$, $o<\varepsilon\leqslant\varepsilon_0$. *Then for all* ε *in* $o<\varepsilon\leqslant\varepsilon_1$ *and all* $k=1,\ldots,n-1$ *the estimates* $\gamma_k \leqslant \gamma$ *hold where* $\gamma = 2\exp(cnh)/b_0$, $\varepsilon_1 = b_0\exp(-cnh)/(2n)$.

We are now in the position to estimate the solutions of (1) by the right-hand sides uniformly with respect to ε.

THEOREM 1. *Under the assumptions of Lemma 2 and* $h \leqslant (2c)^{-1}$ *the solutions* u_ε *of* (1) *can be estimated in the following form*

$$|u_{k,\varepsilon}| \leqslant 2\gamma(\varepsilon^k\gamma^{k-1}|\alpha_\varepsilon| + b_1|\beta_\varepsilon| + \sum_{m=1}^{k-1} (\varepsilon\gamma)^{k-m}|f_{m,\varepsilon}| + \sum_{m=k}^{n-1} |f_{m,\varepsilon}|) \quad (5)$$

for $k=1,\ldots,n-1$ *and all* ε *in* $o<\varepsilon\leqslant\varepsilon_1$.

PROOF. By the assumptions we obtain $\varepsilon\gamma_k \leqslant \varepsilon\gamma \leqslant 1/n \leqslant 1/2 < (1+ch)^{-1} \leqslant \min(\mu_{m-1}^{-1}, \nu_{m-1}^{-1})$ for all $k,m=1,\ldots,m-1$ and $\varepsilon \leqslant \varepsilon_1$. Thus the restriction for ε in Lemma 1 is fulfilled and (3) is valid. By induction (3) implies (the subscript ε is omitted)

$$|u_k| \leqslant \gamma(\varepsilon^k\gamma^{k-1}|u_o| + \sum_{j=o}^{k-1}(\varepsilon\gamma)^j|B_{n-1}||u_n| +$$
$$+ \sum_{m=1}^{k-1}(\varepsilon\gamma)^{k-m}\sum_{j=o}^{m-1}(\varepsilon\gamma)^j|f_m| + \sum_{j=o}^{k-1}(\varepsilon\gamma)^j\sum_{m=k}^{n-1}|f_m|) ,$$

for all $k=1,\dots,n$, $o<\varepsilon\leqslant\varepsilon_1$. Noting $\varepsilon_1\gamma\leqslant 1/2$ and estimating the sums over powers of $\varepsilon\gamma$ by 2, inequality (5) is proved. □

Taking the maximum over all net points, inequality (5) yields the following *stability inequality*

$$\max_{o\leqslant k\leqslant n}|u_{k,\varepsilon}| \leqslant 2\gamma(|\alpha_\varepsilon| + |\beta_\varepsilon| + \sum_{m=1}^{n-1}|f_{m,\varepsilon}|) , \quad o<\varepsilon\leqslant\varepsilon_1. \tag{6}$$

The factor ε before the left boundary value in (5) must obviously be cancelled if one considers the maximum over all $o\leqslant k\leqslant n$. The constant γ in (6) coincides with that in (5) provided $b_1\leqslant 1$; otherwise the latter is multiplied by b_1.

2. STABILITY OF DIFFERENCE APPROXIMATIONS FOR SECOND ORDER, ORDINARY DIFFERENTIAL EQUATIONS

We shall now verify the assumptions of Theorem 1 for a special difference approximation of the boundary value problem (2). Then Theorem 1 yields stability inequalities of the form (5) and (6), where the constants can be explicitly determined by bounds of the coefficients of the differential equation.

For a fixed net width $h = T/n$ independent of ε, equidistant net points are denoted by $t_k=kh$, $k=o,\dots,n$. Using the central difference quotient for the second derivative and the forward difference quotient for the first derivative, this leads to difference equations of the form (1) where ε is replaced by ε/h, the coefficients are determined by

$$A_{k,\varepsilon} = -(a(t_k) - h\,b(t_k) + 2\frac{\varepsilon}{h}), \quad B_{k,\varepsilon} = a(t_k) + \frac{\varepsilon}{h} , \tag{7}$$

and the right-hand sides are given by $f_{k,\varepsilon}=h\,w_\varepsilon(t_k)$, $k=1,\dots,n-1$.

The following Lemma ensures the assumptions of Lemma 2 for the coefficients $A_{k,\varepsilon}$, $B_{k,\varepsilon}$ requiring the usual conditions for the coefficients of the differential equation (cf. Erdélyi (1968), O'Malley (1974)).

LEMMA 3. *Let us assume that* (a) $o < m_o \leqslant a(t) \leqslant m_1$, (b) $|a(t) - a(t')| \leqslant m_2 |t - t'|$, *and* (c) $|b(t)| \leqslant m_3$, $t, t' \in [o, T]$, *with constants* m_i, $i = o,1,2,3$, *independent of* ε. *Then the coefficients of* (7) *do not vanish and with* $b_o = m_o$, $b_1 = m_1 + \varepsilon/h$, $c = \max(m_2, 2m_3)/m_o$ *the assumptions of Lemma 2 hold.*

Let us remark that for $a(t) \geqslant m_o > o$ and $b(t) \leqslant o$ the associated difference operators are of positive type which satisfy a maximum principle (see Dorr (197o), 2.). In this case the proof of Lemma 3 shows that c can even be chosen as zero. For not necessarily positive typed operators, Lemma 3 together with Theorem 1 yields the following Stability Theorem for the difference approximations defined by (7).

THEOREM 2. *Under the assumptions of Lemma 3 and* $h \leqslant m_o \min(m_2^{-1}, (2m_3)^{-1})/2$, *for all* ε *in* $\varepsilon/h^2 \leqslant m_o \exp(-cT)/(2T)$ *the solutions of* (1) *satisfy*

$$\max_{1 \leqslant k \leqslant n} |u_{k,\varepsilon}| \leqslant \gamma_1 \left(\frac{\varepsilon}{h}|\alpha_\varepsilon| + (m_1 + \frac{\varepsilon}{h})|\beta_\varepsilon| + h\sum_{m=1}^{n-1} |w_\varepsilon(t_m)|\right) \tag{8}$$

$$\max_{o \leqslant k \leqslant n} |u_{k,\varepsilon}| \leqslant \gamma_2 \left(|\alpha_\varepsilon| + |\beta_\varepsilon| + h\sum_{m=1}^{n-1} |w_\varepsilon(t_m)|\right) \tag{9}$$

where $\gamma_1 = 4\exp(cT)/m_o$, $\gamma_2 = \gamma_1 \max(1, m_1 + m_o \frac{h}{2T})$.

The proof immediately follows by (5) if one replaces ε by ε/h and notes that $\varepsilon/h \leqslant h\, m_o \exp(-cT)/(2T) \leqslant h\, m_o/(2T)$.

Finally let us emphasize that an analogous stability estimate for the differential equation can be proved with the maximum norm on the left-hand side and the integral norm on the right-hand side. In a corresponding pointwise estimate a factor $\exp(-t/\varepsilon)$ occurs at the left boundary value which can be viewed as a replacement of ε/h in (8) (see Erdélyi (1968)).

3. PROOFS OF THE LEMMAS

(a) *PROOF of Lemma 1.* In the proof we omit the subscript ε. With $k = n-1$ in (1) one obtains

$$|A_{n-1}|\,|u_{n-1}| \leqslant |f_{n-1}| + |B_{n-1}|\,|u_n| + \varepsilon|u_{n-2}| .$$

If $|A_{n-1}| \leqslant |B_{n-2}|$, inequality (3) follows with $\gamma_1 = 1/|A_{n-1}|$. Moreover $|B_{n-2}|\gamma_1 = |B_{n-2}|/|A_{n-1}| \geqslant 1$. If $|A_{n-1}| \geqslant |B_{n-2}|$ we obtain

$$|B_{n-2}|\,|u_{n-1}| \leqslant |A_{n-1}|\,|u_{n-1}| \leqslant |f_{n-1}| + |B_{n-1}|\,|u_n| + \varepsilon|u_{n-2}|$$

and hence (3) is obtained with $\gamma_1 = 1/|B_{n-2}|$, i.e. $|B_{n-2}|\gamma_1 = 1$. Let us assume that the assertion is proved up to $k = n-j$. Then by (1) we obtain $|B_{k-1}|\gamma_{n-k} \geqslant 1$ and

$$\begin{aligned}|A_{k-1}|\,|u_{k-1}| &\leqslant |f_{k-1}| + \varepsilon|u_{k-2}| + \\ &\quad + |B_{k-1}|\gamma_{n-k}\Big(|B_{n-1}|\,|u_n| + \varepsilon|u_{k-1}| + \sum_{m\geqslant k}|f_m|\Big).\end{aligned} \tag{10}$$

For $|A_{k-1}| \leqslant |B_{k-2}|$ and $\varepsilon < |A_{k-1}|/(|B_{k-1}|\gamma_{n-k})$, this implies

$$\begin{aligned}(|A_{k-1}| - \varepsilon|B_{k-1}|\gamma_{n-k})\,|u_{k-1}| &\leqslant \\ &\leqslant |B_{k-1}|\gamma_{n-k}\Big(|B_{n-1}|\,|u_n| + \varepsilon|u_{k-2}| + \sum_{m\geqslant k-1}|f_m|\Big).\end{aligned}$$

According to the definition of $\gamma_{j+1} = \gamma_{n-k+1}$, this yields (3) for $k-1$ and

$$|B_{k-2}|\gamma_{n-k+1} = |B_{k-2}|\{|A_{k-1}|/(|B_{k-1}|\gamma_{n-k}) - \varepsilon\}^{-1} \geqslant |B_{k-2}/A_{k-1}| > 1.$$

In the alternative case $|A_{k-1}| \geqslant |B_{k-2}|$, inequality (10) and $|B_{k-2}|\,|u_{k-1}| \leqslant |A_{k-1}|\,|u_{k-1}|$ imply

$$(|B_{k-2}| - \varepsilon|B_{k-1}|\gamma_{n-k})|u_{k-1}| \leqslant$$

$$\leqslant |B_{k-1}|\gamma_{n-k}(|B_{n-1}||u_n| + \varepsilon|u_{k-2}| + \sum_{m\geqslant k-1}|f_m|)$$

provided $\varepsilon < |B_{k-2}|/(|B_{k-1}|\gamma_{n-k})$. Thus (3) is proved and finally

$$|B_{k-2}|\gamma_{n-k+1} = |B_{k-2}|\{|B_{k-2}|/(|B_{k-1}|\gamma_{n-k}) - \varepsilon\}^{-1} \geqslant |B_{k-2}/B_{k-2}| = 1. \square$$

(b) *PROOF of Lemma 2.* By assumption the denominator of γ_k can be estimated by

$$|B_{n-1,\varepsilon}| - \varepsilon\sum_{m=1}^{k-1}\prod_{j=0}^{m}\max(\mu_j,\nu_j) \geqslant b_0 - \varepsilon\sum_{m=1}^{k-1}\prod_{j=0}^{m}(1 + ch) \geqslant$$

$$\geqslant b_0 - \varepsilon n \exp(cnh) \geqslant b_0/2 \;, \quad k=1,\dots,n-1,$$

provided $\varepsilon \leqslant \varepsilon_1 = b_0 \exp(-cnh)/(2n)$. Therefore γ_k satisfies the inequalities

$$\gamma_k \leqslant \frac{2}{b_0}\prod_{m=0}^{k-1}(1 + ch) \leqslant \frac{2}{b_0}\exp(cnh) \;, \quad k=1,\dots,n-1. \square$$

(c) *PROOF of Lemma 3.* We will again omit the subscript ε and define $a_k = a(t_k)$, $b_k = b(t_k)$. By definition and assumption (a) the inequalities for B_k obviously hold.

(i). It will be now proved that $A_k \neq o$, $k=1,\dots,n-1$. If $b_k \leqslant o$, then (a) yields $A_k = hb_k - (a_k + 2\varepsilon/h) < -m_0 < o$. If $b_k > o$, then by means of (a) and $h \leqslant m_0/(2m_3)$ one obtains $b_k = |b_k| \leqslant m_3 < < m_0/h \leqslant a_k/h$ and $A_k = h(b_k - a_k/h - 2\varepsilon/h^2) < o$.

(ii). In this part the quotients B_k/A_k will be estimated. Under the assumptions (a), (c), and $h \leqslant m_0/(2m_3)$ we conclude that

$$\frac{|B_k|}{|A_k|} = \frac{B_k}{-A_k} = \frac{a_k + \varepsilon/h}{a_k + 2\varepsilon/h - h\,b_k} \;, \quad k=1,\dots,n-1.$$

If $b_k \leqslant o$, one obtains $|B_k/A_k| \leqslant 1$. If $b_k > o$, we write $|B_k/A_k| = = \{1 - (hb_k - \varepsilon/h)/B_k\}^{-1}$ and estimate $(hb_k - \varepsilon/h)/B_k \leqslant$

$\leqslant (hm_3-\varepsilon/h)/(m_o+\varepsilon/h) \leqslant hm_3/m_o$. The assumption $hm_3/m_o \leqslant 1/2$ yield $(1-hm_3/m_o)^{-1} \leqslant 1+2hm_3/m_o$. Altogether in the case $b_k>o$ one obtains

$$|B_k/A_k| \leqslant (1 - hm_3/m_o)^{-1} \leqslant 1 + 2hm_3/m_o .$$

(iii). Finally $|B_k/B_{k-1}|$ will be estimated. For k=1, we have $B_k/B_{k-1} = B_1/A_1$ which is estimated in (ii). By definition $|B_k/B_{k-1}| = (a_k + \varepsilon/h)/(a_{k-1} + \varepsilon/h) = 1 + (a_k - a_{k-1})/(a_{k-1} + \varepsilon/h)$, $k=2,\ldots,n-1$. Assumptions (a), (b) yield the desired estimates

$$|B_k/B_{k-1}| \leqslant 1 + hm_2/(a_{k-1} + \varepsilon/h) \leqslant 1 + hm_2/m_o, \; k=2,\ldots,n-1. \square$$

REFERENCES

Abrahamsson, L.R., Keller, H.B. and Kreiss, H.-O. (1974). Difference approximations for singular perturbations of systems of ordinary differential equations. *Numer. Math.* 22, 367-391.

Comstock, C. and Hsiao, G.C. (1976). Singular perturbations for difference equations. *Rocky Mountain J. Math.* 6, 561-567.

Dorr, F.W. (197o). The numerical solution of singular perturbations of boundary value problems. *SIAM J. Num. Anal.* 7, 281-313.

Erdélyi, A. (1968). Approximate solutions of a nonlinear boundary value problem. *Arch. Rational Mech. Anal.* 29, 1-17.

Hsiao, G.C. and Jordan, K.E. (1978). Solution to difference equations of singular perturbation problems. *IMS Technical Report* 36, University of Delaware, Newark, Delaware.

Kreiss, H.-O. (1975). Difference approximations for singular perturbation problems. In: *Numerical solution of boundary value problems for ordinary differential equations* (Proc. Sympos. Univ. Maryland, Baltimore, 1974), pp. 199-211.

Miller, J.J.H. (1977). Sufficient conditions for the convergence, uniformly in ε, of a three point difference scheme for a singular perturbation problem. To appear in: *Praktische Behandlung von Differentialgleichungen in Anwendungsgebieten* (Proc. Conf. Oberwolfach, 1977). Lecture Notes in Math., Springer, Berlin-Heidelberg-New York.

O'Malley, R.E., Jr. (1974). *Introduction to singular perturbations.* Academic Press, New York.

Reinhardt, H.-J. (1978). Singular perturbations of difference methods for linear ordinary differential equations. Submitted to *Applicable Anal.*

VANISHING VISCOSITY METHOD FOR A QUASI LINEAR FIRST ORDER EQUATION WITH BOUNDARY CONDITIONS

A.Y. LE ROUX
Laboratoire d'Analyse Numérique, I.N.S.A., B.P. 14 A
35031 RENNES CEDEX FRANCE

ABSTRACT. We consider the first order equation $\partial u/\partial t + \text{div } f(u) = 0$ on $\Omega \times]0,T[$, with prescribed boundary and initial conditions, and propose a definition which characterizes the solution obtained by the vanishing viscosity method. Existence and uniqueness are proved, and the convergence of some numerical schemes are established.

Let Ω be a bounded open set of $\mathbb{R}^p$, with a piecewise regular boundary Γ. The outside normal unit vector is denoted by n. Let $f \in (C^2(\mathbb{R}))^p$ and $T > 0$; for $\varepsilon > 0$ destined to vanish, the function $u_\varepsilon \in C^2(\bar{\Omega} \times]0,T[)$ is the unique solution of the quasilinear parabolic equation

$$\frac{\partial u_\varepsilon}{\partial t} + \text{div } f(u_\varepsilon) = \varepsilon \, \Delta u_\varepsilon \quad , \tag{1}$$

on $\Omega \times]0,T[$, with a given initial condition

$$u_\varepsilon(.,0) = u_o \in C^2(\bar{\Omega}) \quad , \tag{2}$$

and a boundary condition on $\Gamma \times]0,T[$

$$u_\varepsilon \Big|_{\Gamma \times]0,T[} = a \in C^2(\Gamma \times [0,T]) \tag{3}$$

$BV(\Omega \times]0,T[)$ is the space of bounded variation functions on $\Omega \times]0,T[$ in the sense of Tonelli Cesari. For such a function u, we can define a trace γu on $\Gamma \times]0,T[$, which is given by

$$\iint_{\Gamma \times]0,T[} \gamma u \; n.\phi \; ds dt = \iint_{\Omega \times]0,T[} u \; \text{div}\,\phi \; dxdt + \iint_{\Omega \times]0,T[} \phi . du(x,t) \tag{4}$$

for all $\phi \in \{C^1(\bar{\Omega} \times [0,T])\}^p$, where $du(x,t)$ is the Radon measure associated with the distribution (grad u, $\partial u/\partial t$), and the scalar product is denoted by a dot. To prove it, we use the fact that over an interval, a bounded variation function has at each point a left and a right limit.

We study here the passage to the limit on the family $\{u_\varepsilon\}_{\varepsilon>0}$, when ε tends towards zero, allowed by the following compacity result.

<u>Theorem 1</u>. The family $\{u_\varepsilon\}_{\varepsilon>0}$ is sequentially compact in $L^1(\Omega \times]0,T[)$ and a function u which is the L^1-limit of a converging sequence $\{u_{\varepsilon_m}\}$ belongs to $BV(\Omega \times]0,T[)$ and its trace for t=0 satisfies (2).

The proof of Theorem 1 uses as a main argument that u_ε is bounded in $W^{1,1}(\Omega \times]0,T[)$, uniformly with respect to $\varepsilon > 0$, and then is sequentially compact in $L^1(\Omega \times]0,T[)$ by Riesz - Tamarkin Theorem. See [1].

The function u we have just obtained as the limit of $\{u_{\varepsilon_m}\}$ is obviously a weak solution (i.e. a solution in distribution sense which is measurable) on $\Omega \times]0,T[$ of the quasi-linear first order equation

$$\frac{\partial u}{\partial t} + \operatorname{div} f(u) = 0 \quad , \tag{5}$$

which verifies the same initial condition (2) and some boundary condition sometimes quite different from (3). Moreover u is the only weak solution that satisfies a so called entropy condition. Such a condition ensures uniqueness for the Cauchy problem on $\mathbb{R}^p$ (see [3] or [7]) and we prove that it also yields uniqueness for the problem on Ω with a correctly prescribed boundary condition, which we shall now study on two examples.

As a first example, the right weak solution of the Burgers equation, i.e. for $f(u) = u^2/2$ on $\Omega =]0,1[$ with the conditions

$$u(x,o) = 1 \; ; \; u(o,t) = -1 \; ; \; u(1,t) = t \; ; \tag{6}$$

is the following for $t < T = 1$

$$u(x,t) = \begin{cases} x/t & \text{if} \quad 0 < x < t \quad , \\ 1 & \text{if} \quad t < x < 1 \quad . \end{cases} \tag{7}$$

The boundary condition at x=0 is not satisfied, and this comes from the propagation of the data, the speed of which is given by the characteristics equation

$$\frac{dx}{dt} = f'(u) \quad . \tag{8}$$

For $x = t = 0$, the data present a jump $[-1,1]$ and from (8) the $[-1,0]$-part of that jump is going to the left ($x<o$), and leaves Ω while the $[0,1]$-part enters Ω as a rarefaction wave.

For the linear problem with $f(u) = \lambda u$, on $]0,1[$, the second member of (8) is a constant and then the boundary condition at $x=0$ is operative for $\lambda > 0$, but not for $\lambda \leqslant 0$. We find the reverse at $x=1$. In the general case, the right boundary condition is to be written almost everywhere on $\Gamma \times]0,T[$

$$\underset{k \in I(\gamma u,a)}{\text{Min}} \left\{ sg(\gamma u-a) \ (f(\gamma u)-f(k)).n \right\} = 0 \ , \tag{9}$$

where $I(\gamma u,a)$ denotes the closed interval $[\text{Min}(\gamma u,a),\text{Max}(\gamma u,a)]$, and sg is the sign function, defined on $\mathbb{R}$ by

$$sg(v) = \begin{cases} 1 & \text{if} \quad v > 0 \quad , \\ 0 & \text{if} \quad v = 0 \quad , \\ -1 & \text{if} \quad v < 0 \quad . \end{cases} \tag{10}$$

We have now a well posed problem given by (5), (2) and (9), and the following definition characterizes its right weak solution. We shall afterwards prove that it is the limit u obtained at Theorem 1.

Definition 1. A function $u \in BV(\Omega \times]0,T[)$ is a solution of Problem (5), (2), (9) when it satisfies for all $k \in \mathbb{R}$ and for all non negative test function $\phi \in C^2(\bar{\Omega} \times]0,T[)$ with a compact support contained in $\bar{\Omega} \times]0,T[$

$$\iint_{\Omega\times]0,T[} |u-k| \frac{\partial\phi}{\partial t} + sg(u-k) \ (f(u)-f(k)).\text{grad}\ \phi \ dxdt \geqslant \iint_{\Gamma\times]0,T[} sg(a-k) \ (f(\gamma u)-f(k)).n \ \phi \ \ dsdt \ , \tag{11}$$

and when γu satisfies (2) almost everywhere on Ω.

Theorem 2. Problem (5), (2), (9) has a unique solution, which is the L^1-limit of the solutions of (1), (2), (3) when ε vanishes.

To prove existence, we introduce a regular approximation sg_η of sg, multiply (1) by $sg_\eta(u_\varepsilon-k)\phi$, and make ε to tend to zero, by using

$$\lim_{\varepsilon\to o} \iint_{\Gamma\times]0,T[} \varepsilon \frac{\partial u_\varepsilon}{\partial n} sg_\eta(a-k)\phi \, ds\, dt = \iint_{\Gamma\times]0,T[} n.(f(a)-f(u))sg_\eta(a-k)\phi \; dsdt \tag{12}$$

and then take the limit for η, to obtain (11).

Uniqueness follows from the fact that two solutions u and v satisfy

$$t_o < t_1 \Rightarrow |u(t_1)-v(t_1)|_{L^1(\Omega)} \leqslant |u(t_o)-v(t_o)|_{L^1(\Omega)} \tag{13}$$

which means that the semigroup associated with (9) and (5) is a contraction on $L^1(\Omega)$. This proof is similar to [3] and is detailed in [1].

Formula (12) allows us to pass to the limit on the boundary layers of u_ε, which appear on these parts of $\Gamma \times]0,T[$ where the boundary conditions for u_ε (i.e. (3)) are no longer satisfied by u.

Let us now study two finite difference schemes (Lax scheme and Godunov scheme) which give a convergent approximation of the solution u for the one dimension problem on $\Omega =]0,1[$. We take $I \in \mathbb{N}$, $h = 1/I$ (space mehsize) and $q > 0$, and write for $o \leqslant i \leqslant I$, $o \leqslant n \leqslant N = [1 + T/qh]$

$$I_i = [(i-1/2)h,\ (i+1/2)h[\ ;\ J_n = [(n-1/2)qh,(n+1/2)qh[\ ,$$

where qh is the time meshsize. The approximating solution is a step function u_h, equal to the constant u_i^n on $I_i \times J_n$.

Initial and boundary conditions are introduced through

$$u_i^o = \frac{1}{h}\int_{I_i} u_o(x)dx\ ,\quad u_o^n = \frac{1}{qh}\int_{J_n} a(o,t)dt\ ,\quad u_I^n = \frac{1}{qh}\int_{J_n} (1,t)dt \tag{14}$$

The Lax scheme is written as follows

$$u_i^{n+1} = \frac{1}{2}(u_{i+1}^n + u_{i-1}^n) - \frac{q}{2}(f(u_{i+1}^n)-f(u_{i-1}^n))\ , \tag{15}$$

and the Godunov scheme is to be given by

$$\left\{\begin{array}{l} u^n_{i+1/2} \in I(u^n_i, u^n_{i+1}) \text{ realizes } \underset{k \in I(u^n_i,u^n_{i+1})}{\text{Min}} \left\{ sg(u^n_{i+1}-u^n_i) f(k) \right\}, \qquad (16) \\ u^{n+1}_i = u^n_i - q(f(u^n_{i+1/2}) - f(u^n_{i-1/2})). \qquad (17) \end{array}\right.$$

The first step (16) is given by the solution of the Riemann problem on $I_i \cup I_{i+1}$ and the second step (17) follows from the integration of (5) on $I_i \times J_n$.
For $M_o = \text{Max}(|u_o|_{L^\infty(0,1)}, |a(o,.)|_{L^\infty(0,T)}, |a(1,.)|_{L^\infty(0,T)}$,
we have the following results :

Theorem 3. If the stability condition of Courant Friedrichs Lewy

$$q \sup_{|k|<M_o} |f'(k)| \leqslant 1 \qquad (18)$$

is satisfied, then u_h constructed by the Lax Scheme (15) or by the Godunov scheme (16) (17) converges towards the solution u of Problem (5) (2) (9) in $L^1(\Omega \times]0,T[)$ as h tends to zero.

The proof of Theorem 3 uses a priori estimates in $L^\infty(\Omega \times]0,T[)$ and $BV(\Omega \times]0,T[)$ to get convergence and (2). We also prove a discrete analogue of (11), which gives (11) at the limit. See [2,4,5,6].

We now give two numerical examples, where the Lax scheme, the Godunov scheme and the one-step Lax Wendroff scheme given by

$$u^{n+1}_i = u^n_i - \frac{q}{2}(f(u^n_{i+1})-f(u^n_{i-1})) + \frac{q^2}{2}|f'(\xi^n_{i+1/2})|^2(u^n_{i+1} - u^n_i)$$
$$- \frac{q^2}{2} |f'(\xi^n_{i-1/2})|^2 (u^n_i - u^n_{i-1}) \qquad (19)$$

with $\xi_{i+1/2} \in I(u^n_i, u^n_{i+1})$ given by

$$f(u^n_{i+1}) - f(u^n_i) = f'(\xi^n_{i+1/2}) (u^n_{i+1} - u^n_i) \qquad (20)$$

have been compared, usign the Burgers equation.

At first, for $u_o = a(1,.) = 0$, $a(o,.) = 1$, $q = .25$, $h = .02$ we get figure 1, where the exact solution presents a discontinuity. The second example approaches a boundary layer; we have taken $u_o = a(o,.) = 1$, $a(1,.) = 0$, $q = .o1$, $h = .o2$ and results are given by figure 2.

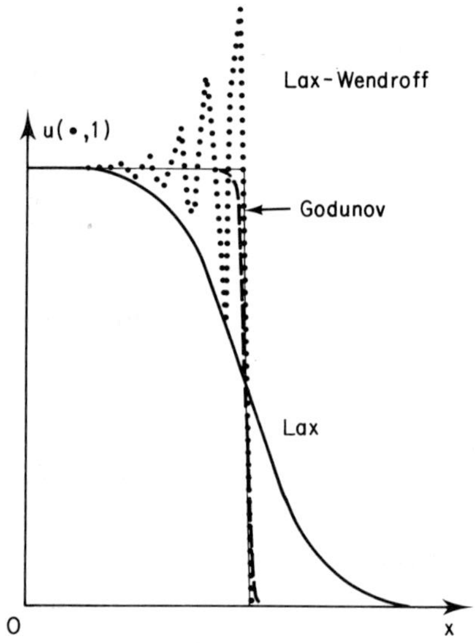

Figure 1

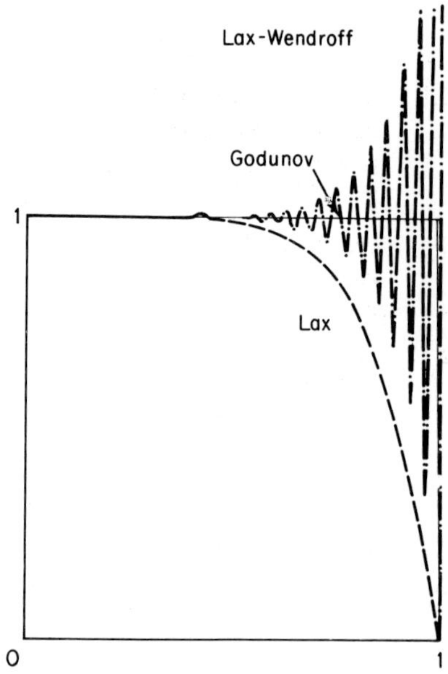

Figure 2

These three schemes have a numerical viscosity whose coefficient is of the form $h\,|qf'|^m/q$ with m=0 for the Lax scheme (order 1), m=1 for the Godunov scheme (order 1), m=2 for the Lax Wendroff scheme (order 2), and we easily see that a small value of the coefficient q makes the numerical viscosity too large for the Lax scheme and the discontinuities are spread out, and too small for the Lax Wendroff scheme, which becomes instable near the discontinuities. It is well known that the Lax Wendroff scheme may give an approximating solution that does not converge towards u satisfying (11) ; for example we get the stationary solution for the Burgers equation with $u_i^o = -1$ for $i < 0$, $u_i^o = 1$ for $i \geq 0$, $u_o^n = -1$ for $n > 0$, $u_I^n = 1$ for $n > 0$.

For the Lax scheme these results may be generalized for p-dimensional problems since the estimates for convergence are given in [2] and a numerical analogue of (11) is true, for a harder stability condition than (18).

REFERENCES

[1] Bardos C., Le Roux A.Y., Nedelec J.C. To appear (1978).

[2] Conway E., Smoller A. (1966). Global solutions of the Cauchy problem for quasi linear first order equations in several space variables. *Comm. Pure Appl. Math.* **19**, *pp. 95-105.*

[3] Kruzkov S.N. (1970). First order quasi linear equations with several independent variables. *Mat. Sb.* **81** *(123), pp. 228-255 = Math USSR Sbornik (10), pp. 217-243.*

[4] Le Roux A.Y. (1977). A numerical conception of entropy for quasi linear equations. *Math. Comp.* **31** *(10), pp. 848-872.*

[5] Le Roux A.Y. (1976). Convergence of the Godunov scheme for first order quasi linear equations. *Proc. Japan Acad.* **52** *(9), pp. 488-491.*

[6] Le Roux A.Y. (1977). Etude du problème mixte pour une équation quasi linéaire du premier ordre. *C.R.A.S.* **285**, *pp. 351-354.*

[7] Oleinik O.A. (1963). Uniqueness and stability of the generalized solution of the Cauchy problem for a quasi linear equation. *AMS Transl. Ser. 2,* **33**, *pp. 285-290.*